Acoustic Properties

FLASHPOINTS

The FlashPoints series is devoted to books that consider literature beyond strictly national and disciplinary frameworks, and that are distinguished both by their historical grounding and by their theoretical and conceptual strength. Our books engage theory without losing touch with history and work historically without falling into uncritical positivism. FlashPoints aims for a broad audience within the humanities and the social sciences concerned with moments of cultural emergence and transformation. In a Benjaminian mode, FlashPoints is interested in how literature contributes to forming new constellations of culture and history and in how such formations function critically and politically in the present. Series titles are available online at http://escholarship.org/uc/flashpoints.

A complete list of titles begins on page 314.

Acoustic Properties

Radio, Narrative, and the New Neighborhood of the Americas

Tom McEnaney

NORTHWESTERN UNIVERSITY PRESS | EVANSTON, ILLINOIS

THIS BOOK IS MADE POSSIBLE BY A COLLABORATIVE GRANT
FROM THE ANDREW W. MELLON FOUNDATION.

Northwestern University Press
www.nupress.northwestern.edu

Printed in the United States of America

10 9 8 7 6 5 4 3 2 1

Library of Congress Cataloging-in-Publication Data

Names: McEnaney, Tom, author.
Title: Acoustic properties : radio, narrative, and the new neighborhood of the Americas / Tom McEnaney.
Other titles: FlashPoints (Evanston, Ill.)
Description: Evanston, Illinois : Northwestern University Press, 2017. | Series: Flashpoints | Includes bibliographical references and index.
Identifiers: LCCN 2016038867 | ISBN 9780810135383 (pbk. : alk. paper) | ISBN 9780810135390 (cloth : alk. paper) | ISBN 9780810135406 (e-book)
Subjects: LCSH: Radio broadcasting—Political aspects—United States—History—20th century. | Radio broadcasting—Political aspects—Cuba—History—20th century. | Radio broadcasting—Political aspects—Argentina—History—20th century. | American literature—20th century—History and criticism. | Cuban literature—20th century—History and criticism. | Argentine literature—20th century—History and criticism. | Radio and literature. | Radio authorship.
Classification: LCC PN1991.8.L5 M37 2017 | DDC 302.23445 23
LC record available at https://lccn.loc.gov/2016038867

∞ The paper used in this publication meets the minimum requirements of the American National Standard for Information Sciences—Permanence of Paper for Printed Library Materials, ANSI Z39.4811992.

Contents

Preface: Wireless Cultures

This book is a prehistory of our wireless culture. It examines the coevolution of radio and the novel in the Americas from the early 1930s to the late 1960s, and the various populist political climates in which a new medium—radio—became the chosen means to produce the voice of the people. As I follow politicians, activists, businessmen, and, above all, writers across these turbulent decades, I also examine how they turned to radio—a technology that helped people transform the air into a commodity—to reconsider the meaning of property and, in the majority of cases, to agitate for more equitable access to, fairer distribution of, and improved living conditions in housing.

As with all studies of the past, this book is fundamentally concerned with understanding its present. When I began writing these pages in 2008, voters in the United States, amid the worst financial recession and housing crisis since the Great Depression, had just elected the country's first African American president on a platform that invoked the country's most ambitious social programs. While Barack Obama promised a "new, New Deal" in the United States, that same year the political movement known as "Kirchnerismo" in Argentina announced a new Peronist politics under that country's first directly elected female president, Cristina Fernández de Kirchner.[1] And in Cuba, where Fidel Castro had controlled the government since 1959, Raúl Castro officially assumed the office of the presidency, declaring both a new stage but also a familial continuity in Cuba's revolutionary history. These three dramatic political events carried with them a return to the twentieth century's most influential movements in populist politics: the New Deal, Peronism, and the Cuban Revolution. In this book I focus on those movements and how their different uses of radio sought to channel and control popular power.

While the 2008 presidential transitions dominated political headlines, a new revolution in technological communications and popular wireless computing altered how people connected with each other and shared their voices. Building on the initial release of the iPhone in the summer of 2007, Apple released its first 3G (Third Generation mobile communications) model in 2008, linking the most popular mobile computing device in history to international wireless standards. In the same year, 4G devices appeared: wireless speed and connectivity increased once again, and the new efficiency in connectivity was felt in politics, art, and theory.

Although the 2008 presidential transitions and the rapid increase in wireless efficiency that same year are not often discussed together, the connections between wireless computing and popular politics have become a contemporary cliché. There is still much to be said about the Arab Spring (2010), Occupy Wall Street (2011), Black Lives Matter (2012), and other social movements that have demonstrated how social media can enable powerful grassroots political uprisings that focus on people and issues previously invisible or unheard in print, radio, or television news. On the other hand, digital utopians who proclaim the arrival of "free culture" or celebrate the slogan "information wants to be free" ignore the material politics and economics involved in what one radio scholar has called "the *problem* of information" that radio history makes clear.[2]

In this book I do not directly take on these contemporary movements, but rather I turn back to investigate how previous generations seized on radio—the wireless technology of their day—as they helped create and connect with a popular voice that would call for a more just understanding and use of property. I will detail these claims in the following pages with the hope that what I am calling the prehistory of wireless culture will help us see the contemporary from another angle, or hear it through what Ralph Ellison famously called the "lower frequencies."[3] In bringing together literary theory, media archaeology, hemispheric studies, and political history, I aim to show that this historical context and my sustained analysis of novels, *radionovelas,* poems, and broadcasts that struggled to define their own moment can provide an alternative access point to make sense of our legislative fights over net neutrality, our experience of networked real-time interactivity, and our encounters with the loosely defined populist movements that seem to only increase in power and influence as we move through this new wireless age.[4] Against protocol, radio's new relevance, and its unexpected importance in the history of the novel and our contemporary culture, might remind us to question apparent obsolescence and to continue seeking out new resources in sustained actions against control.

Acknowledgments

Although listeners tend to hear the sound of a radio broadcast as if it came from nowhere, it requires a coalition of diligent workers, from the sound engineers, to the DJs, to the designers of the receiver in your car stereo, phone, or computer to assure that the signal reaches you. A book like this is no different. Without the support of multiple friends, colleagues, and institutions, my thoughts about sound would have never made it to the page.

Amid all the noise when I first began this book in another form, three people helped me tune in to the most important strains of literature and politics that have guided me through this process. Colleen Lye shared her time and her vast knowledge of socialism and U.S. literature to assure that I understood the power of activism and collectivity, but also the limits of the New Deal imagination. Michael Lucey's patient thoughtfulness and keen ear introduced me to entirely new ways to consider how literature could work in the world. Francine Masiello has been a professor, mentor, and friend whose critical enthusiasm and generous advice buoyed this project at key moments over many years. I will be forever thankful for her model of intellectual life as a practice of humor, inquisitiveness, and unending dedication to the public work of education.

I am incredibly grateful to Michelle Clayton, my editor at Flash-Points, who supported and encouraged my work at a very early stage, and whose boundless energy inspired and challenged me throughout the process of revising this book. I want to thank everyone at Northwestern Unviersity Press who helped bring this book into production, but especially Michelle and Gianna Mosser, my editor at Northwestern University Press, who both helped locate two extraordinary readers for the manuscript. Rachel Price's incisive suggestions and encouraging words alerted me to stylistic alterations and a key structural point that helped me better grasp my own sense of the "neighborhood" of the Americas. Jonathan Sterne's keen and bountiful suggestions ran the gamut from finely detailed readings to conceptual and structural arguments that forced me to think more seriously about literature's own importance as a media system.

In addition to the Press's readers, I was also fortunate to benefit from Mark Goble's and Graciela Montaldo's thorough reading of the manuscript. Their expert suggestions helped me return to my texts with fresh eyes and, I hope, something of their suggestive insights. In a less formal, but no less helpful mode, Brían Hanrahan, Paul Fleming, and Ted Martin offered valuable revisions of the introduction. Chris Ehrick helped me think more carefully about gender on the radio in Argentina. Jenny Stoever helped me hear the sonic color-line in Richard Wright. Raúl Fernandez lent his expert ear and deep knowledge of Cuban music and culture to help me broaden and nuance my arguments about Cuban radio. Roger Moseley helped assure me I wasn't crazy about Kittler, and Damien Keane, an early reader, offered key insights and a sympathetic ear to my search for a vocabulary to write about sound in print.

I was lucky to meet Brían, Damien, Jenny, and Roger during a year at Cornell's Society for the Humanities dedicated to the theme of "Sound." I am grateful to Tim Murray for including me in the group, and to Eliot Bates, Marcus Boon, Duane Corpis, Miloje Despic, Sarah Ensor, Nina Sun Eidsheim, Ziad Fahmy, Michael Jonik, Jeanette Jouli, Nick Knouf, Eric Lott, Jamie Nisbit, Jonathan Skinner, and Trevor Pinch for their excellent conversation, criticism, and collegial support. My colleagues in the Department of Comparative Literature at Cornell, especially Paul Fleming, Natalie Melas, Tracy McNulty, and Karen Pinkus, as well as Andrea Bachner, Anindita Bannerjee, Cathy Caruth, Debra Castillo, Cynthia Chase, Jonathan Culler, Brett de Bary, Naminata Diabate, Laurent Dubreuil, Bill Kennedy, Barry Maxwell, Jonathan Monroe, Nancy Pollak, and Neil Saccamano provided a supportive and welcoming atmosphere from the moment I came to Cornell.

Beyond my department, many colleagues and friends at Cornell and in Ithaca have sustained me throughout the writing of this book. Jeremy Braddock and Rayna Kalas made me feel at home in Ithaca as soon as I arrived. Merike Andre-Barrett, Theo Black, Julia Chang, Elisha Cohn, Tao DuFour, Eli Friedman, Kate Howe, Louis Hyman, Patty Keller, Oneka LaBennett, Annie Lewandowski, Alex Livingston, Philip Lorenz, Shawn McDaniel, Alejandro Madrid, Jenny Mann, Mukoma Wa Ngugi, Guy Ortolano, and Ben Piekut have made work more pleasurable and the winters a little warmer. Judith Peraino has been an inspiration in sound.

Books don't get written on hiking trails, or in corridors, seminar rooms, or bars, but I worked out many of these ideas there or on the streets of San Francisco, Berkeley, and Oakland in conversation and friendship with Mike Allan, Munia Bhaumik, Mayra Bottaro, Kathryn Crim, Katrina Dodson, Jonathan Dotan, Harris Feinsod, Andrea Gadberry, Zach Goldman, Lily Gurton-Wachter, Paul Haacke, Michael Holt, Javier Jiménez, Seth Kimmel, Ted Martin, Annie McClanahan, Blanca Misse, Karina Palau, Allison Schachter, David C. Simon, Ben Tran, Toby Warner, Tristram Wolff, Selby Wynn Schwartz, Avi Shapiro, and Matt Takata. Steve Goldsmith, Kevis Goodman, and Rob Kaufman were tremendous intellectual, emotional, and professional resources, and made me feel like an honorary Romanticist. In between Ithaca and the Bay Area, Philip Barnard, Lauren Berlant, Anke Birkenmaier, Alejandra Bronfman, Patrick Greaney, Nick Harkness, Emily Maguire, Mara Mills, and Neil Verma shared helpful advice and kind encouragement. Paul Brennan, Jenn Naegele, Mike O'Hare, Mike Oznowicz, Dave Steinberg, Phil Strnad, Katy Zaugg, and Mike Zaugg are great friends who made sure music was never far from my thoughts. I'm grateful to Marty Kirchner for years of intense debate and friendship, to Petro Petreas for his Odyssean spirit, and to Aaron Fischer for his skeptical questions but never-doubted support. Antoine Traisnel and Anna Watkins-Fisher provided brilliance and hilarity in many contexts. I'm thankful for Natalia Brizuela's friendship and advice, which opened doors all over and guided me to the best bookstores in Barcelona and Buenos Aires, and I'm grateful to Martina and Tony for all of their help over the years in La Habana. Special thanks to Martina Broner and Sam Carter for eleventh-hour proofreading

Despite my focus on notoriously ephemeral radio sound, archives across the world have been an invaluable contribution to my research for this book. I want to thank the library staffs at the Argentores archives and the Biblioteca Nacional, as well as Carlos Puig at the Casa Puig in Buenos Aires, the Biblioteca Nacional José Martí in Havana, the Bibliothèque Nationale in Paris, the Bancroft Library at the University of California, Berkeley, the Beinecke Library at Yale University, the Nettie Lee Benson Latin American Collection at the University of Texas at Austin, the Robert F. Williams archive at the Bentley Historical Library at the University of Michigan in Ann Arbor, and the Hoover Institute at Stanford University.

Audiences at Binghamton University, Brown University, Harvard University, the Massachusetts Institute of Technology, the University of Chicago, and the University of California, Los Angeles, as well as the American Comparative Literature Association, the American Studies Association, the Latin American Studies Association, and the Modernist Studies Association, the Society for Novel Studies, and the UC–Cuba Educational Initiative gave valuable feedback along the way.

My parents gave me love and support long before this book began and somehow sustained both through the many years it took to complete. I dedicate this book to Amanda Jo Goldstein, whose presence permeates its every page, and whose graceful intelligence, boundless curiosity, and mischievous sense of humor have made living with and away from these pages such a pleasure.

Acoustic Properties

Introduction

Learning to Listen

The first transnational commercial radio broadcast in the Americas was the 1923 fight between Jack Dempsey—the United States heavyweight champion of the world—and his Argentine challenger, Luis Ángel Firpo.[1] Broadcast from New York City's Polo Grounds, the transmission reached listeners in Argentina, Cuba, and the United States. Audiences in Cuba described the event as a "miracle," and the Argentine author Julio Cortázar later declared that the broadcast marked "the birth of radio," "the most important moment I experienced in twentieth-century history."[2]

Cortázar was not alone in his belief. The *New York Times* worried "that if Firpo were to win, Americans would have to rethink the Monroe Doctrine and its meaning."[3] Late in the first round Firpo nearly wrote this legislative change with Dempsey's body. Shocking the spectators in New York and thrilling listeners in Argentina, Firpo knocked Dempsey through the ropes and onto a reporter's ringside typewriter, which left a gash in the fighter's head. Although the typewriter was out of commission, the broadcast continued. On the patio of Cortázar's family home the neighbors had gathered around his uncle, the only one able to listen, through headphones, to the relayed transmission from Buenos Aires on his minimal "crystal" set: the single radio in town. Pandemonium broke out, Cortázar recalled, "with the spasmodic reports that my uncle received through his ears and gasped through his mouth." But the fight was not over. Despite witness claims that the referee should have disqualified Dempsey for failing to reenter the ring in the allotted time and on his own strength (dozens of hands helped push him back through the ropes and onto his feet), the fight was allowed to continue. In the second round Dempsey knocked Firpo down for good. With him, Cortázar writes, fell "fifteen million Argentines twisting themselves up in various postures and pleading . . . for a break in diplomatic relations, a declaration of war and the fiery demolition of the U.S. embassy."[4] The hemispheric order of things remained intact, but change was in the air—and also on the page.

Cortázar's anecdote, which he titled "El noble arte," registers the changes radio wrought on writing, politics, and the culture at large in the midcentury Americas. The Firpo-Dempsey broadcast led to a surge in radio sales in Argentina, and this massive audience changed the meaning of radio as a medium.[5] One year after Cortázar's uncle turned himself into an amplifier for the sound heard through his headphones, stores began to sell speakers for radio sets. Radio listening was transformed: no longer bound to the isolation of headphones, listeners normalized the scene Cortázar described as they gathered collectively around the radio, especially as a family in the home. Over the course of the 1920s in Argentina, Cuba, the United States, and elsewhere, radio began to consolidate, thanks to legal regulations of the airwaves, standardized design of radio sets, audio norms in production and reception, regular programming schedules, and an entire media ecology of photographs, articles, and images in newspapers and magazines. In short, radio became a *medium*: an object, an infrastructure, and a set of cultural practices.[6] With the emergence of a mass public, radio became an everyday domestic apparatus: it was more necessary, according to polls, than other modern appliances like refrigerators, sewing machines, or stoves.[7] While early practices of radio—hobbyist design or long-distance (DX) transmission and reception—would continue across the twentieth century, the medium we commonly refer to as radio had been defined by the early 1930s as a mass broadcast system.

When Cortázar included "El noble arte" in his 1967 collection *La vuelta al día en ochenta mundos* (Around the Day in Eighty Worlds), he could compress nearly a half century of radiophonic history into that first broadcast and its effect on his writing. We can locate the touchstones of Friedrich Kittler's "discourse network" in this short text not only because Dempsey's body strikes the ringside typewriter—a physical blow that interrupts the typewriter's symbolic production without stopping the broadcast's sonic transmission—but also because of the way Cortázar frames his anecdote. Introduced by a woman rejecting Cortázar's interest in radio in favor of quoting Hölderlin, the text concludes with the layout and design of the book in which "El noble arte" appears.[8] The not so subtle message here—the message of most media theory that takes literature seriously in the early twenty-first century—tells us that in the age of new sonic media like radio the lyrical expression of Romantic subjectivity (Hölderlin) inevitably gives way to the materiality of the page.[9] Like Mallarmé, Huidobro, Apollinaire, Tablada, Marinetti, Augusto de Campos, and others before him, Cortázar makes readers aware of the page as a medium through the disruption of print's standardized layout. The break with print's previous visual standards, so the story goes, awakens us to the changing conditions and infrastructures under which we organize knowledge. We become aware that a book is a medium, too, one with its own particular and limited repertoire of methods to register and order sensuous experience. We no longer look past the page's surface but realize that the medium is the message (McLuhan), that media determine our situation (Kittler), and that medium specificity matters (Hayles).

In this book, however, I am less concerned with the materiality of the page—which reinscribes the visual as the marker of sound—than I am with the ways specific narrative techniques in print and on radio affect one another to shape popular voice and critique property relations in the formation of a new neighborhood of the Americas. I do not aim to ignore the insights of book history and media theory in order to return to the detailed analysis of rhetorical forms as if the page were once again a vanishing medium. Rather, I borrow from the antihermeneutic tradition of media theory after Kittler, as well as what Bernhard Siegert calls the "cultural techniques" of media, while insisting that attention to the new technical practices elaborated by narrative composers finely attuned to sound in radio and print help us understand the intersection of symbolic, social, and material practices in our study of media. In Cortázar's text this would mean recognizing not only the disruption of the page's layout or his critique of older poetic forms but also his use of specialized radio vocabulary—"superheterodyne," "piedra de galena"—his focus on sonic effects, and his description of listening. For just as audio engineers developed new ways to record, edit, and transmit sound, writers—often learning from those very engineers—developed techniques to work with sound according to the specificity of their medium.

I argue that in the radio age, writing, in particular, became a practice of listening: a specific audile technique that sought to influence radio listening practices in turn.[10] As strange as it may seem to say, books can also teach us how to hear. If we analyze narrative with this sense of mediation in mind, we should understand that media theory cannot simply restrict itself to examining self-reflexive gestures that reveal the medium as a medium. It must also look to how narratives produce and shape the practices that make a technology into something more than itself, not only making a medium but also forming a media culture and a media habitus. Sound, in this case, is not the nonsemantic excess at the limit of language or the page but a transmedial phenomenon the detail of whose production I analyze in the mutual encounter and uneven collisions between radio and print.

Until very recently, literary critics have ignored the immense role radio played in the history of the novel. But, as this book helps to illuminate, just as photography helped reinvent pictorial representation in the nineteenth century, so authors from Argentina, Cuba, and the United States employed radio to retune the "unspeakable sentences" of the novel, reconfiguring individual characters into conduits for popular speech, reorganizing narrative time to approach the simultaneity of radio network listening, and transforming realism from an enumeration of things to an amplification of the sound of social space. Incorporating radio as a medium—and not just a theme—into their narrative forms, Félix Caignet, Raymond Chandler, Julio Cortázar, John Dos Passos, Carson McCullers, Manuel Puig, Severo Sarduy, and Richard Wright wrote against the monologic, charismatic, and nationalist orators who reached them over the radio. Sometimes these writers were known to each other; sometimes they were not. But despite the contingencies of acquaintance, it is clear this cohort of writers collectively pioneered a distinctive midcentury form of composition enabled by and enabling for radio. For this reason, my book takes seriously both Jean-Paul Sartre's search for a "literary art of radio" and Frederic Jameson's speculation regarding a "radio aesthetic" in the novel. It does so in order to demonstrate—as in Cortázar's text—that authors in the radio age developed a new kind of writing that was also a practice of listening. They worked with the novel as an aural technology whose intricate arrangement of vocal registers could train readers to listen critically to the new media of their day. Against the official, institutional politics of FDR's "Good Neighbor Policy" and its legacy, these writers shaped the cultural imagination of radio's new social network from 1932 to 1982.

In doing so, they defined a contingent, informal, but politically potent new neighborhood of the Americas.

What I call "narrative acoustics" describes this literary media theory at the intersection of listening practices, narrative form, and audio engineering. As this book will show, these connections between listening and writing took a dizzying variety of forms. To mention just one example from the following pages, in the mid-1960s, the black internationalist Robert F. Williams had a variety of listeners tune in to his broadcasts from Havana, Cuba, where he urged African American listeners in the southern United States to arm themselves against racial oppression. FBI listening posts transcribed these broadcasts without annotating the way his voice changed pitch or volume or commenting on his musical selections. Everyday listeners, on the other hand, wrote to Williams with passionate letters about the songs he played or the political message he conveyed. And hobbyists wrote with equal enthusiasm to mention that they had heard his broadcast from Havana in New York State and wanted to notify him of the strength and quality of the signal. All of these modes of listening, whether juridical, aesthetic, political, or technical, shape the broadcast's content at the site of reception. With examples like this in mind, I recognize a variety of listening modes in this book and focus then on narrative compositions whose attention to sound and radio broadcasting sought to intervene in the construction of the so-called "popular voice."

Thus, from John Dos Passos's New Deal writing to Severo Sarduy's stereophonic anticolonial radio plays to Manuel Puig's deployment of the banned sounds of early radio and immigrant speech in Peronist Argentina, I trace how writers take sound and its radio transmission as seriously as the populist political speakers they contest. When Dos Passos reports from the U.S. national political conventions in 1932, he writes nothing about what the candidates say: he focuses instead on how their voices sound in Chicago Stadium and the mess of microphones, amplifiers, and cables required to transmit them to the broadcast audience. A year later he combats FDR's signature form of intimate radio address—the "fireside chat"—with a writing style that exposes radio's infrastructure and FDR's sonic style in order to make readers aware of the chasm of power that separates them from the president. Dos Passos's listening feeds into his writing, which, in turn, hopes to change the way his readers listen. The "sound," "tone," and "tenor" of Dos Passos's writing here are not metaphors but linguistic tools whose power we could miss if Dos Passos did not train our ears to listen for them. Situating narratives like this within a larger media ecology of sound in the Americas, my study explores radio plays, novels, *radionovelas*, and political oratory on the radio to study how narrative artists struggled to create an alternative form of the vox populi across the hemisphere.

As I will explain throughout this book, radio proved indispensable to these writers and to the three paradigms of populist political power in the midcentury Americas—FDR's New Deal liberalism, the Peróns' working-class socialism, and Fidel Castro's revolutionary communism—because it offered new modes to manage the problem of populist political speech: to speak *for* the people or *to be spoken through* by the people. Evita Perón, for instance, employed her audiophonic training as both a *radionovela* star and a mimic of famous historical women to create a voice that seemed to emerge from the popular melodramatic radio culture of Peronist Argentina.[11] Further proving the importance of radio and representation in populist power, Fidel Castro negotiated his own status in relation to "the people" through two extraordinary speeches. In his 1953 defense *La historia me absolverá* (History Will Absolve Me), Castro invokes the memory of his political mentor Eduardo "Eddy" Chibás, a firebrand orator who shot himself live on the radio in the hope that his death would spark a revolution. Castro claims that he could have occupied a radio station and broadcast a recording of Chibás's on-air suicide to launch *el pueblo* into instant revolution, but he rejects the idea because he worries about the recklessness of the population taking to the streets on their own, inspired by another's voice. When, six years later, Castro uses the guerrilla radio network Radio Rebelde to secure the revolution's victory through his statement "¡Revolución, sí; golpe militar, no!" (Revolution, yes! Military coup, no!), he establishes his own voice as the voice of the people. As these examples begin to show, across this period in the Americas radio remains the chosen medium to explore and harness populist power, but the speakers' position in relation to the people structures their particular populist politics.

Alongside these speakers, the best-selling realist novelists I study over the course of this book mobilize their own strategies to speak to, for, and as the people. In doing so, I argue, they create alternative forms of popular speech as part of the project of "narrative acoustics." Through new formulations of the relationship between author and character—relations that we will see dissolve across this study—these writers respond to and reorganize the means to produce the vox populi.

A focus on the intricate techniques of narrative works on radio and in print helps us recognize these connections. But it can also change our view of the intersectional poetics of radio history and literary theory. When radio historians and media theorists invoke narrative forms like the print novel or even radio plays, they tend to look to these works' content as evidence of social norms without considering their formal complexity. Rarely do these critics notice how broadcasts deploy sound effects, vocal tone and pitch, or music to organize the narrative perspectives that might amplify or challenge the overt content of a given program.[12] Attention to literary form, however, deepens a listener and reader's sense of these narratives' mediation.

Literature, after all, was not a mere repository for the changes wrought in the discourse network with radio's appearance: radio and the novel coevolved. Alejo Carpentier, Félix Caignet, and Severo Sarduy from Cuba; Orson Welles and Archibald MacLeish in the United States; and Julio Cortázar and Nené Cascallar from Argentina were just some of the writers who worked on radio and helped alter its narrative forms. MacLeish joined with the radio director Irving Reiss in *The Fall of the City* (1939), transposing his sense of narrative perspective into and *in through* radio, using microphone placements to stage and negotiate the place of populism, sometimes as a voice from the masses, sometimes as an individual voice dictating to them. In Paris, Carpentier worked with the surrealist Robert Desnos to develop a number of popular and experimental programs in the 1930s, including the adaptation of the thriller *Fantomas* with Kurt Weill and Antonin Artaud, and a production of Walt Whitman's *Salut au monde!* In addition to these celebrated names from literary history, radio dramas drew on the serial format of the nineteenth-century *folletín*, while *radionovela* writers like Caignet introduced the narrator into radio dramas, shifting them away from audio theater or vaudeville routines and toward the layered and dialogic perspectivalism of the print novel in his massively popular programs. The codevelopment of radio and the novel would fashion both media for years to come. We continue to see this influence in Don Delillo's *Underworld* (1997), Pedro Juan Gutiérrez's *Trilogía sucia de La Habana* (1998), César Aira's *El tilo* (2004), and Daniel Alarcón's *Lost City Radio* (2007), among others. Alarcón made the dynamic relationship between print and radio explicit with his radio program and podcast *Radio Ambulante* in 2013, developed to expand the range and reach of the sonic Latin American *crónica* and the sounds of Latin American voices to radio audiences across the world.

NETWORKS, COMMUNITIES, TRANSNATIONAL RELATIONS

The correlations between radio, popular voice, and narrative come together most explicitly in the introduction John Dos Passos appended to the 1938 publication of his novel *U.S.A.*, which he classified as "a radio network . . . [T]he speech of the people." That Dos Passos sought to characterize his novel as a radio network tells us something about the currency of the network for representing the nation from the New Deal to the Cuban Revolution. The National Broadcasting Corporation (NBC) and the Columbia Broadcasting System (CBS) in the United States; CMQ and Radio Habana Cuba (RHC)–Cadena Azul in Cuba; and Cadena Radio Belgrano and others in Argentina made "the network" a familiar concept throughout the Americas. Simultaneous listening to the same broadcast from multiple individual points across a country connected the material radio infrastructure to that mental and affective nationalism Benedict Anderson named "imagined communities."[13] However, compared to radio, what Anderson calls the "nearly simultaneous" experience of printed newspapers in the nineteenth century seems not very simultaneous at all. Radio's new speed transformed the definition and experience of "simultaneous" experience, since now listeners in Des Moines, San Francisco, and Charleston; or Santiago, Santa Clara, and Matanzas; or Córdoba, Rosario, and Corrientes; or New York, Havana, and Buenos Aires knew they heard sounds at the exact same time. This new experience of shared temporality expressed itself in the term that eclipsed "simultaneous": "live" broadcasting.[14] Moreover, radio distinguished itself from print through a new sense—auditory rather than visual—that included audiences far beyond the literate elites and reading publics located in what Ángel Rama named "the lettered city." Listeners, Dos Passos among them, understood the new reality of a network identity, in which an individual's simultaneous listening with unknown others could produce a sense of connection that made radio believable as a conduit for the popular voice. Moreover, the popular voice in these broadcasts was not necessarily a national voice, and while radio historians have enthusiastically invoked Anderson in their studies of the "radio nation," his national model does not quite fit the histories of radio I examine in this book.

Against the application of Anderson's argument about nationalism to radio history, I understand radio instead as *the* transnational medium par excellence. Not only did radio broadcasting's engineering easily flout national borders, but the very national networks I have listed were all built on transnational relations. As I will explain throughout this book, the transnational precedes the national in radio history. NBC's existence, for example, depended on a multinational combination of corporations from the United States and the Caribbean. Cuban radio's official inaugural broadcast was a presidential address in English to listeners in New York City across the wired and wireless system of the U.S.-owned Cuban Telephone and Telegraph Company (later ITT). And Argentine radio was constructed from imported parts from England, Italy, and the United States. Transnational space, of course, was often also a neocolonial or imperial space, although it could also be an anticolonial force. The infrastructure and the medium depended on its users—producers and listeners—to define its meaning in practice.

Building from radio's transnational foundations, the daily practice of radio transmission and reception challenged national borders. In 1929 Cuban listeners petitioned the island's broadcasters to go silent on some evenings in order for the audience to tune in to opera broadcasts from the United States.[15] "Border blaster" stations in Mexico broadcasting to the United States played music and advertising that had been forbidden by U.S. regulators. Uruguayan radio—easily tuned in from Buenos Aires, across the river from Montevideo—became a second home for Argentine speakers or programs banned from the national airwaves.[16] The Cuban revolutionary government sent signals to Venezuela, the Dominican Republic, and Nicaragua, where the dictator Anastasio Somoza complained that he had to listen to verbal attacks "in my own home as if [Radio Havana Cuba] were installed in Managua." And exiled African American activists Robert F. Williams and Mabel Williams broadcast *Radio Free Dixie* from Havana to listeners across the United States with the tagline, "You are tuned to *Radio Free Dixie*, from Havana, Cuba, where integration is an accomplished fact." With these broadcasts and others, radio helped destabilize national territory, reorganizing the limits of communities one could imagine and detaching political affiliation from national boundaries.

THE NEW NEIGHBORHOOD OF THE AMERICAS

I name the new models of geopolitical and cultural relation that developed through radio in the Americas from the 1930s to the 1960s "the new neighborhood of the Americas." More than the "network" or the "community," the neighborhood defines the sphere of influence, or *radio de acción*, that structured political and cultural relations in the Americas during this time.[17] Building on radio's transnational foundations and the newly felt proximity listeners experienced when they heard other voices in the intimacy of their own rooms, Franklin Delano Roosevelt declared in 1932 that the United States' new foreign policy in Latin America would be the Good Neighbor Policy. The "new neighborhood" I identify in this book stands as an alternative to the official, U.S.-directed "Pan-Americanism" established by Roosevelt.

Both an extension and revision of the Monroe Doctrine that the *Times* worried the first transnational broadcast would eliminate, the Good Neighbor Policy and its ambition to foster Pan-Americanism foisted a U.S. project of cultural diplomacy and economic dependency onto a group of highly differentiated nations that nevertheless had alternative traditions of interregional unity. In the nineteenth century, Latin American critics denounced the economic imperialism of Pan-Americanism and the paternalism of the 1823 Monroe Doctrine, which granted the United States authority to intervene in Latin America if any European power threatened the "southern hemisphere." José Martí's famous call for Latin American community, "Nuestra América" (1891), followed by the Uruguayan essayist José Enrique Rodó's *Ariel* (1900), rebutted the declaration of Pan-Americanism that had emerged from business-minded reinterpretations of the Monroe Doctrine alongside the geopolitical notion of "hemispheric" community with the U.S.-led first Pan-American Conference in 1889.[18] To listen to Roosevelt, one would think that the Good Neighbor Policy sought to acknowledge this criticism and to revise the Monroe Doctrine in order to seek out more equal footing with countries like Cuba.

However, the Good Neighbor Policy was less an attempt to treat Latin American countries like partners and more a practical response to economic hardship in the United States. Amid the financial devastation of the Great Depression, FDR's Good Neighbor Policy sought to replace military might in Latin America with increased economic and cultural power. The policy's most direct action was to repeal the Platt Amendment in Cuba, which had legally allowed the United States to invade the island, in favor of a new trade agreement that made Cuba economically dependent on the United States.[19] Thus, the Good Neighbor Policy introduced the rise of so-called "soft power" (although the United States maintained its rights over the military base at Guantánamo Bay). This new brand of hemispheric politics found its home in

Nelson Rockefeller's Office of Inter-American Affairs (OIAA) and the projects of cultural diplomacy in the 1930s and 1940s through figures like Orson Welles, Walt Disney, and Carmen Miranda. As a result, cultural materials, as well as military might, became key instruments in the power struggles to define the Americas throughout the twentieth century.

My argument takes seriously the fact that the neighborhood became a politically powerful organizing principle with material consequences. However, rather than primarily follow official governmental and institutional channels of U.S. cultural and political power in Latin America—as in the tremendous studies from Catherine Benamou, Claire Fox, and others—I discover that the *radio de acción* enabled by the newly radio-networked Americas does not strictly coincide with the radius and aim of U.S. foreign policy. Following writers, activists, and performers across this new neighborhood of the Americas, I identify a series of aesthetic and political connections unrecognized by official histories, and I reveal how the idea of the neighborhood operated to connect the micropolitical problems of local and physically proximate neighbors to the transnational, or "hemispheric," geopolitical neighborhood defined by the Good Neighbor Policy's treaties and trade. For example, when U.S. officials refused to allow the African American writer Richard Wright to film an adaptation of his novel *Native Son* in the United States because of the book's sharp criticism of racist housing policies and neighborhood redlining in Chicago, Wright filmed in Buenos Aires instead. Argentina was only too happy to produce their first English-language film, *Sangre Negra* (Black Blood), as an indictment of racist ghettoization in the United States.[20] The Argentine government had been excluded by the Good Neighbor Policy, and the country's film industry had been decimated by the U.S. refusal to sell them film stock during the war.[21] Thus, the story of the film's production redefined what it meant to be a good neighbor at the transnational, national, and local level.

Wright's example is just one of the many I analyze to show how the neighborhood's scalar model, from the most intimate to the transnational, reveals the connections between these levels. As their national cultures came into contact through radio broadcasts, economic agreements, and political treaties, the writers in this study took up the radio to imaginatively negotiate the fluctuating local, national, and transnational borders of the Americas. Thus, in chapter 4 I study how the *radionovela*, created as a consequence of the Good Neighbor Policy to sell U.S. goods to Cuban consumers in order to fund U.S.

public housing, tied listeners in the domestic space of the home and the nation to the transnational economic agreements and the national-domestic issue of housing in another country. In order to observe how these neighborhoods—as physical structures or conceptual entities—coincide, overlap, and change, I will shift scales from the hyperlocal to the hemispheric and back. Following the logic of the radio network, I will sometimes zoom in on neighbors in adjoining apartments or apartment buildings or houses, at other times step back to examine the sonic neighborhoods of the extended Caribbean, where radio wars between Cuba and the United States challenge the stability of national frontiers, and at still other times zoom out to observe the political-cultural neighborhood of the Americas from which Carl Schmitt developed his theory of the Großraum, which points to the intimacy of imperial political power FDR tried to soften through the language of the "good neighbor." The oscillating movement between these scales will reveal how seemingly local acts of audition, like tuning in to *Radio Free Dixie* in Alabama or overhearing a bolero in Buenos Aires, underscore the deeply imbricated relationship between all three scales of perspective.

This scalar sense of the neighborhood, coincident with the model of the radio network, also provided writers with the means to critique property relations at the local as well as the national level. For just as radio waves passed over national borders, putting into question national sovereignty, they permeated buildings' walls. This physical fact—a condition of radio's specific engineering—is taken up by writers in the radio age who seek to transform sound's long-standing challenge to property law and built form into a political struggle for equitable property relations.

ACOUSTIC PROPERTIES AND SYMBOLIC MATERIALS

Sound and sonic technologies, from the phonograph to the MP3, have posed recurring problems for copyright law, regulated distribution, and liberal conceptions of property in general. Until 1909 in the United States, property had to be seen, which came to mean that sound—previously thought of as evanescent and immaterial—counted as property only when it could be written. While musical notation had fallen under copyright by 1833, one could not copyright the sound of a performance until after the start of the twentieth century.[22] In the meantime, new technologies created new ways for the law to think about sound and property. The perforations on a piano roll, which allowed the automated player piano to play back music without a human present to touch the keys, and the grooves in a phonograph, which also reproduced sound, provided the courts with a form they argued could serve as an analogue to writing and therefore count as property.[23] New sonic technologies, through analogy to older methods of transcribing sound, helped the law create new sonic property.[24]

Unlike the phonograph, however, radio broadcasting did not write, as legal commentators would put it, or leave visible traces. At least it did not inscribe sound onto physical and visible storage devices for playback at a later time. And unlike the telephone, whose point-to-point engineering contained and constrained sonic information, radio technology followed sound's notorious resistance to capture. What Guglielmo Marconi had initially thought was his invention's failure—publicly broadcasting sound rather than encoding it for private point-to-point communication—became radio's defining feature.[25] In daily practice, radio broadcasting's very engineering transformed the local and often domestic listening experience associated with the phonograph into a new social configuration that Jason Loviglio, borrowing a concept from Lauren Berlant, has called "radio's intimate public." And while commercial retailers would eventually domesticate radio broadcasting's public incursion across the threshold of private real property by making radio receivers into handsome pieces of furniture, and commercial corporations and government regulators would also develop a means to "sell the air," the public structure of broadcast continued to challenge the visible demarcations of real estate, reminding listeners of the medium's structural challenge to fixed property lines.[26]

Radio historians have tracked the struggle to commodify sound, and while I follow them to understand the ways politicians, businessmen, and activists used radio to reorganize property relations, my focus on *narrative acoustics* also takes seriously the symbolic reorganization of cultural forms—especially the realist novel—as a means to participate in and attempt to transform the cultural and economic investment in private property. Realism's strong endurance as a genre—by which I mean a recognized conventional narrative form whose symbolic organization of materials regulates the relationship between the singular and the general, the individual and the collective—makes it a powerful place to observe and investigate the ongoing struggle over what property should mean, how it should be shared, and to whom it might belong (individuals, corporations, governments, the natural land, or "the people").[27] And it makes particular sense that these writers in the radio age interested in new property relations would devote themselves to reorganizing the codes of realism: a genre closely linked to the rise of private property in nineteenth-century bourgeois homes and defined by the enumeration of things to create its "reality effect." Focusing on the aural qualities of one of those things—the radio—the writers I study here turn realism away from the eye and toward the ear, away from the visual catalogue of a room's interior and toward an interrogation of property lines by drawing attention to the negative space of a room's acoustics, or to walls made permeable by sound. Thus, radio, with its ability to bring long-distance and public sounds into the intimacy of domestic space, offers them a means to tune in readers and listeners to the social construction of real property and to imagine it otherwise.

Given that enduring genres like realism sediment social life and ritual, even apparently minor changes in the genre's form can index and entail larger transformations in the organization of ideas in the literary and broader cultural field.[28] Thus, we should not dismiss the role best-selling novelists play in influencing policy conversations, even when their writing about housing never makes it into the pages of a court record or congressional transcript. For instance, against the backdrop of the New Deal housing crisis and the debate about public housing in the United States, novels like Carson McCullers's *The Heart Is a Lonely Hunter* (1940) provided readers with resources to think differently about real estate through radio. In an exemplary rewriting of realism under the sign of radio, McCullers's adolescent character Mick Kelly listens to radio broadcasts through the walls and windows of wealthier houses in her neighborhood and then converts the walls of a vacant home into writing spaces where she lists the names of well-known radio figures—from Dick Tracy to Benito Mussolini—laying claim to the house's value as a semi-public place. Representing a technique central to the writers I examine across this book, Mick's "radio graffiti" learns writing from radio listening in order to critique the private ownership of real property. When McCullers's close friend Richard Wright more directly intervenes in the housing politics of his day through his fictional, documentary, and cinematic work, or when Manuel Puig records a man's story to both produce a novel and help him buy a house, I argue that we can begin to see how the symbolic reorganization of realist form in literature can coincide with and meaningfully affect attempts to change the meaning of property beyond the page.

This narrative work with realism, radio, and real estate helps define what I earlier called *narrative acoustics*, and sound technology's destabilization of the concept of property marks one side of what I mean by *acoustic properties*. In addition to the challenge to real, personal, and intellectual property, I mean *acoustic properties* to connote the physical properties that help determine sonic shape: those inalienable elements characteristic of materials that sound engineers or acousticians consider when constructing a sonic environment.[29] As Brandon LaBelle observes, "in acoustical occurrence . . . sound sets into relief properties of a given space, its materiality and characteristics, through reverberation and reflection, and, in turn, these characteristics affect the given sound and how it is heard."[30] Acoustic science, in other words, teaches that sound is a composite event that in fact includes the space we might normally call its context, and that what is *proper* to a thing actually depends on the materials of its sounding. Sound can be separated from its source but not its context of reception. Rather than hear the voice as a signature of an independent ego, acoustics tells us that the sound of a voice depends on its intersection with a number of materials in its environment. Attention to such properties can thwart the possessive concept of property and the self, making us aware of our relational dependence on other animate and inanimate things to define the voice.[31]

Thus, *acoustic properties*, in this book, refers to both sound's long resistance to property and those sonic qualities that challenge possessive ownership of one's voice. *Narrative acoustics*, meanwhile, brings together both meanings of acoustic properties to reimagine property and people through more equitable relations developed in narrative form. Ultimately, these concepts come together in writers' attempts to shape an alternative form of the vox populi—contagious, apersonal, and unofficial—that intersects with a critique of private property in a series of narrative works that construct a new neighborhood of the Americas.

The trajectory I trace in these authors' engagements with radio and other sound technologies is a sophisticated thread of realist writing that eschews certain forms of documentary, ethnographic, and sonic fidelity—typified in the period and in its criticism by dialect—in favor of using the sonic resources of narrative to produce a more exchangeable, flexible, and contagious kind of utterance. This type of utterance attempts to produce for characters and readers alike an experience akin to singing along with a voice that no one owns. John Dos Passos already hints at this project through the chorus of repetitions in his modern epic *U.S.A.* (1938), a novel we saw him describe as "a radio network" and "the speech of the people," and Carson McCullers's attempts to assemble that voice in the character of the deaf-mute Mr. Singer in *The Heart Is a Lonely Hunter* (1940). However, this project makes its most significant leap in Richard Wright's reaction against Zora Neale Hurston's writing in dialect, and the subsequent publication of his novel *Native Son* (1940). My third chapter shows how that novel comes up against the racial limits of this kind of contagious and democratic utterance: Wright uses repeated phrases in interior monologues, free indirect discourse, spoken court briefings, and finally oral dialogue to make the utterance contagious, but he ultimately suggests the audiovisual voice—that is, the voice attached to a visible body—produces too much tonal distortion to complete the circuit between speakers. Only in Manuel Puig's "dictation novel," *Eternal Curse on the Reader of These Pages* (1980), written after, and in some ways as a response to the experimental work with tone in his friend Severo Sarduy's radio play *The Ant Killers* (1978), do I claim that we encounter what I call a *pop epic realism*. The term, which I take from the Spanish word for "epic," *epopeya*, combines the aesthetic theories of epic mimesis proposed by Eric Auerbach and Mikhail Bakhtin with the "pop art" movement's flat aesthetic. Borrowing from the dictation method of tape-recorded interviews rather than the radio, Puig finally elevates listening to a reading *and* writing practice. In *Eternal Curse*—planned for tape but turned into simple dictation—Puig brilliantly undoes property. In this flat-toned novel without a narrator, readers sometimes fail to distinguish between one speaker and another.[32] And in this form, part novel, part play, "written" in English by recording an interview with a friend, Puig, I claim, arrives at a mimesis of mediation, an imitation of the language others create by listening to mass media. In doing so, he follows Wright's injunction against dialect and arrives at a uniquely transnational text, written in a foreign language (English) in the voice of another, set in a symbolic centerpiece of the United States: New York City's Washington Square. Thus, media technologies—from print to radio to tape—emerge as listening technologies that writers employ to intervene in debates about the popular voice and the possibility of common property in order to imagine new social relations, contingent and yet equitable, a plural popular voice of overlapping sounds: noise and signal at once.

The chapters I detail here set out to explore the ramifications of these experiments in text and sound. Part 1, "The New (Deal) Acoustics," is comprised of three chapters. Chapter 1, "'On the National Hookup': Radio, Character Networks, and *U.S.A.*," examines the technological, political, and historical context of John Dos Passos's *U.S.A.* (1938) in relation to the early transformations of radio and acoustics, and the rise of the network model of broadcasting. It makes an argument for Dos Passos's new form of writing as deriving from his intensive listening, and his attention to how network radio provided a new infrastructure for culture, politics, and subject formation in the New Deal periods of the 1930s.

Chapter 2, "The Sound of the Good Neighbor: Radio, Realism, and Real Estate," examines how the formal and political terrain composed in Dos Passos's trilogy was reformulated by writers like Raymond Chandler and Carson McCullers. Combining what I call a "housing problem" in realist representation with the literal housing crisis of the Great Depression, I see both Chandler and McCullers turn to the radio as a means to reimagine property rights. Particularly in McCullers's writing, a wall can become a social medium rather than a dividing line, as characters listen to the radio through their neighbors' real property: the walls or windows of their houses. I begin the chapter by examining Orson Welles's 1942 radio program *Hello Americans* and its formalist innovations as an aesthetic corollary to FDR's policy of the "good neighbor," turning then to Chandler and McCullers to see how this foreign policy played out in the domestic spaces of the United States. The argument here looks forward to chapters 4 and 5, which study in greater detail how the Good Neighbor Policy affected Cuban politics and culture, and how Cuba helped shape the Depression era in the United States.

In chapter 3, "Struggling Words: Public Housing, Sound Technologies, and the Position of Speech," I argue that Richard Wright's opposition to Zora Neale Hurston's writing and his controversial enthusiasm for Gertrude Stein's writing in black dialect can help us hear his work between 1937 and 1940 as an attempt to detach representations of African American speech from an ethnographic investment in authenticity, and to imagine a non-subject-centered form of speech. I argue that Wright turns against the radio, which he had used as his main narrative device in his posthumously published first novel, *Lawd Today!* (1938/1963), and borrows instead from the phonograph as the operative medium through which he works to elaborate the possibility of shared speech, an idea intertwined with shared housing in "Long Black Song" (1938) and *Native Son* (1940). In the chapter's coda, I explore how the later filming of *Native Son* in Buenos Aires links this national critique of race and real estate in U.S. neighborhoods to the New Deal's foreign policy of the good neighbor.

Part 2, "Occupying the Airwaves," comprises two chapters that focus on writers and broadcasts situated in Cuba that help us reconsider the underlying transnational character of the radio culture we encountered in the book's first section. Chapter 4, "Tears in the Ether: The Rise of the *Radionovela*," explains how the *radionovela*, akin to the "soap opera," was developed as a new narrative form in order to sell the products of the Good Neighbor Policy to Cuban domestic listeners. The chapter takes up the questions of chapter 2 and reformulates them to show how the national housing crisis within the United States was somewhat ameliorated by the exploitation of Cuba's markets, increasingly dependent on the United States because of trade agreements established as part of the Good Neighbor Policy. Part of this story discusses the interconnection of the U.S. and Cuban radio networks, and the radio mogul Goar Mestre, whose later exile to Argentina anticipates the book's final part. The chapter analyzes the work of the Cuban writer Félix B. Caignet, author of the most celebrated *radionovela* in Latin American history: *El derecho de nacer* (The Birthright). His melodrama is the necessary "other" to classical realism and the masculine tales of epic battles and revolutionary speeches I call, following Anke Birkenmaier, "the discourse of the passions." The chapter concludes with an analysis of how the Cuban revolutionaries, who publicly denounced melodrama, used the rhetorical forms of the *radionovela* to build their own political rhetoric.

Chapter 5, "Radio's Revolutions," takes up the work of the African American civil rights activists Robert F. and Mabel Williams's radio program *Radio Free Dixie*, which they broadcast from Havana to the southern United States in the early 1960s to advocate for armed resistance. I pair the Williamses with the Cuban exile author Severo Sarduy's radio play *Los matadores de hormigas* (The Ant Killers) to elaborate on radio's ability to speak across borders, as well as to revise the critique of property that first emerged with Dos Passos's "speech of the people" in chapter 1, focused on real estate in chapters 2 and 3, and carried forward to the U.S. neocolonial relationship to Cuba, and Cuba's revolutionary nationalization of businesses, buildings, and homes in chapter 4. Sarduy's play, which he describes as an attempt to "decolonize the voice," enters into the realm of experimental aesthetics to investigate radio's potential to radically reimagine the voice's relationship to the individual and property.

The two chapters of the book's third and final part, "Hand-to-Hand Speech," turn to the Argentine author Manuel Puig as a writer who reformulates the transnational circulation of radio, realism, and real estate in the other sections in order to compose a literature of the "commonplace." Chapter 6, "House Taken Over: Listening, Writing, and the Politics of the Common Place in Manuel Puig's Fiction," takes up Puig's novels as the epitome of writing in the radio age. Not only does his novel *Boquitas pintadas* (Heartbreak Tango) pit the popular radio exports of Cuba (the bolero) and Argentina (the tango) against each other and position radio at the very center of a real estate dispute that links the domestic sphere to state institutions like the police force, but Puig's work more generally emerges from the combination of melodrama, experimental fiction, and realism studied in the previous chapters. (Puig, who cites McCullers and Chandler as influences, fuses their New Deal realism with his experience as a student at Italy's famous neorealist Centro di Cinema Sperimentale.) The chapter observes how Puig takes up the products of the New Deal—golden-age cinema and radio, the novels of Chandler and McCullers—as well as the revolutionary politics of the 1960s Americas and Argentina's own particular radio history to create a new form of novel and novel writing that, in its most extreme cases, turns Puig into a mere recorder and editor of other people's stories. He thus challenges what it means to be an author, to "have a story," and own anything simply by converting writing into an act of listening. The chapter's concluding section reads Puig's novel *Maldición eterna a quien lea estas páginas* (Eternal Curse on the Reader of These Pages; 1981), which he composed by interviewing and taking dictation from a U.S. graduate student, to argue the provocative ethics of property Puig builds into the novel's compositional and contextual structure.

Chapter 7, "The Ends of Radio: Tape, Property, and Popular Voice," caps the age of radio and the novel as a fifty-year period from 1932 to 1982 (the year of Puig's own tape-recorder novel *Sangre de amor correspondido*, or Blood of Requited Love). The chapter examines Puig's experiment with tape to conclude the intersection of the popular voice, housing, and listening that I have traced throughout the book. If Dos Passos used radio to transform his characters into conduits, and the writers of the New Deal celebrated radio listening as a way to reimagine realist form and the failures of private property, Puig's decision to turn himself into a recorder and publish the transcript of his conversation with a Brazilian bricklayer who would buy a house with the book's profits emerges as the capstone to the new neighborhood of the Americas and the radio age. The chapter's coda offers an analysis of Ricardo Piglia's 1992 novel *La ciudad ausente* (The Absent City) as a novel that looks back at the history of radio in the Americas I have elaborated, and explores the interrelationship between the novel, radio, and tape to argue for a future of storytelling that would transform readers into carriers for stories, rather than subjects telling their own stories. Thus, my argument here once again revises Dos Passos's "speech of the people," away from a cynical fear of the information age and toward a more ambivalent but open understanding of our own moment's newly networked world.

PART I

The New (Deal) Acoustics

CHAPTER 1

"On the National Hookup"

Radio, Character Networks, and U.S.A.

In a section titled "On the National Hookup" in his travelogue *In All Countries* (1934), John Dos Passos details the social and acoustic engineering that electrified Chicago Stadium at the 1932 Democratic National Convention. Describing the scene that remained invisible to radio listeners, Dos Passos observes NBC pages "coaxing the speakers into poses from which they could be heard," maneuvering "the two big white disks above the speakers' platform (the ears of the radio audience)," which "delicately caught every intonation of the oratory."[1] Careful to catch how the whole event sounds, even after the amplified orchestration has ended, Dos Passos listens on to "the proud suave voice of the National Broadcasting Company . . . filling . . . jaded ears from every loudspeaker, enumerating the technical agencies that had worked together to obtain the superb hookup through which they broadcast the proceedings of the Democratic Convention of 1932."[2] Confronted with the pageantry of the political convention—the wired and wireless sound—Dos Passos's satirical tone, noted in the "proud suave voice" and the "jaded ears" that receive it, tunes readers in to the role of acoustics in national politics and social life.

Listening to Franklin Delano Roosevelt deliver his promise of a "new deal" to U.S. citizens in Chicago that year, Dos Passos imagines listeners attuned not to the politicians' content, which goes largely unmentioned in his reportage, but rather to the overall *tone* of "the national hookup" and the newly entrenched forms of listening that knitted together the audience in front of the loudspeakers in Chicago and on the radio "in all countries."[3] What Dos Passos seizes on here, at the dawn of the New Deal era and the Golden Age of Radio in the United States, is the transformative intersection between structural changes in acoustics and politics. Over the course of the decade he would transmit these changes through innovations in literary style that would affect U.S. novelists writing at the end of the New Deal—including Carson McCullers and Richard Wright—but also the political theory and fiction writing of Jean-Paul Sartre, the novels of Luis Rafael Sánchez, and others.[4] Standing beneath the loudspeakers in Chicago Stadium, Dos Passos amplifies how radio, along with a number of changes in engineering and building materials dubbed the "New Acoustics," had already begun to alter the tone of U.S. and world culture.

Apparently minor pieces of reportage, the articles Dos Passos published about the political conventions of 1932 and Roosevelt's later radio oratory condense the major structural, sensory, aesthetic, and political concerns that drive the New Deal's most significant literary event: the 1938 publication of his trilogy of novels in a single bound book he named *U.S.A.* That book, a thousand-plus-page epic he called "a radio network," not only uses the radio to reimagine what a novel could be but serves as a corrective to practices of a mass listening public throughout the Americas. Like the Estridentistas' poetic experiments to reorganize listeners' understanding of radio during the Mexican Revolution in the 1920s (Dos Passos had translated Manuel Maples Arce's *Urbe* as *Metropolis* in 1929), or Alejo Carpentier's incorporation of his radio work of the 1930s into novels that obliquely challenged the U.S.-backed dictatorships in a Cuba Dos Passos knew well from his travels throughout the Caribbean and Latin America, Dos Passos's writing underscored the transnational relations that structured radio networks.[5] Part of his long novel's challenge to those readers who were listeners was to recognize that a "national hookup" depended on many nations and that the micropolitics of a speaker's tone in Chicago Stadium links up with, and perhaps offers the best access into, the knotted network of relations between speakers, listeners, and national territories that Dos Passos depicts in *U.S.A.* If we pay attention to the technical changes and lessons he learned and tries to teach his readers in 1932, and if we trace the differences between those articles and his novel, we can begin to hear how "the speech of the people" in *U.S.A.* arises from the entanglement of specific innovations in acoustic design and theory, and the cultural production that helped shape how these changes were heard. In the following pages I examine the acoustic engineering at the beginning of the New Deal and then read Dos Passos's work with listening during this period in order to demonstrate that the critical writing practice he derived from his careful listening led to the production of a novelistic style impossible to hear but capable of making readers listen to what radio could be. Ultimately, I will argue that the muted tone of the people's speech in *U.S.A.* paradoxically critiques the new homogenization of sound design under the New Deal through the adoption and adaptation of that very same sound. The novel, and the reportage that precedes it, are exhibit A in how we can understand the ways writers responded to and consequentially intervened in the new social experiences of sound that radio made possible.

THE NEW ACOUSTICS: MUTING THE SOUND OF SPACE

Halfway across the country from Chicago Stadium, the connection between the wired room and the wireless audience found its paradigm in the architecture, acoustics, and name of New York City's Radio City Music Hall. Opening on December 27, 1932, Radio City marked the apogee of an engineering era in the United States dedicated to the electrification of sound. The historian Emily Thompson observes that "by 1930, new tools, new techniques, and a new language for describing sound had fundamentally transformed the field of acoustics. 'The New Acoustics' was proclaimed."[6] Radio City, located on the ground floor of 30 Rockefeller Center, became the electric cathedral for this new form of sound.[7] In addition to the loudspeakers outside the hall that brought the evening's entertainment out into the street, inside the hall engineers had "learned to create electrically a spatialized sound" to the effect that "the sound of space was now a quality that could be added electrically to any sound signal in any proportion; it no longer had any relationship to the physical spaces of architectural construction."[8] The New Acoustics' "clear and focused" sound offered "little opportunity to reflect and reverberate off the surfaces of the room in which it was generated."[9] Without reverberation, "*the sound of space was effectively eliminated* from the new modern sound."[10] Whereas the "old" acoustics relied on the particular architectural form and shape of the walls, ceilings, and other available surfaces, the New Acoustics used microphones, insulation, and loudspeakers that rendered the hall's unique nooks and crannies acoustically insignificant. Thus, even adding an echo effect removed the sound of a particular built space. Confronted with these changes, the engineers required a new mathematical equation, as the acoustic formula developed by Wallace Sabine for the reverberations of Boston's Symphony Hall in 1900 no longer made acoustic sense inside the walls of Radio City by 1932.[11]

Those "two big white disks above the speakers' platform," which Dos Passos identifies as "the ears of the radio audience," typify the acoustic engineering manufactured to remove the difference of built forms and suppress their interaction with the human voice and ear.[12] Along with the New Acousticians who sought to homogenize sound in buildings by removing space from their equations, radio engineers and aesthetic theorists valued the same shift in listening and production. Rudolf Arnheim's simply titled treatise *Radio: The Art of Sound* (1936) insists that in broadcasting "*resonance is eliminated*, out of a very proper feeling that the existence of the studio is not essential to the transmission and therefore has no place in the listener's consciousness . . . The listener rather restricts himself to the reception of pure sound, which comes to him through the loudspeaker."[13] As I will explain throughout this chapter, this type of technical and aesthetic engineering, which critics have pointed out went against radio's impressive ability to capture higher-definition sound than the preelectric phonograph, coincided with radio's adoption in the United States as a domestic appliance and a device made for intimate gatherings.[14] Eliminating space at the site of production, Arnheim's theory and the technological changes in Radio City reveal a cultural turn that embraced an acoustic world unencumbered by spatial difference and manufactured to create a sense of proximity with political and aesthetic consequences.

A TRANSNATIONAL HOOKUP: THE NATIONAL SPACES BEHIND NBC'S SOUND

Dos Passos's writings from the 1932 conventions work against the cultural and political ideology that emerged from these changes in sonic engineering. Drawing attention to the "proud suave voice of the National Broadcasting Company" in Chicago Stadium, he connects the sonic shape of those local utterances to their delivery across the national network. He thus intertwines the micropolitics of the sounds of social space, which I will discuss in greater detail in the following section, with the macropolitics of the erasure of territorial space that inheres in the national network.

Identifying how the New Acoustics and NBC depend on extranational affiliations might seem especially odd given that the headquarters for both of these transformations in sound were located across from each other on the single city block of Midtown Manhattan's West Fiftieth Street. Part of New York City's Rockefeller Center, Radio City Music Hall looks onto 30 Rockefeller Plaza, where the Radio Corporation of America (RCA), owner of NBC and the film studio RKO Pictures, would finish moving in by 1933. Radio City, the name for the new home of the six-year-old NBC, and its neighbor, Radio City Music Hall, the shrine to the New Acoustics, formed the seemingly hyperlocal infrastructure and engineering hub for what Dos Passos dubbed "the national hookup."

Yet, it was not without reason that Dos Passos included "the national hookup" in a collection titled *In All Countries*. The metropolitan site of national broadcasting's origins in a single city block in New York City consolidates and masks a number of multinational radio networks whose purchase led to the construction of the RCA building and the rise of a "national" network.[15] RCA, and therefore NBC's early existence, depended on one of the candidates Dos Passos heard in 1932, and the companies of two men whose capsule biographies he would include in his first book from the *U.S.A.* trilogy: "The Emperor of the Caribbean," or Minor C. Keith, president of United Fruit; and "The Electrical Wizard," or Thomas Edison, the founder of General Electric.[16]

In 1919, then assistant secretary to the navy Franklin Delano Roosevelt prevented Edison's company General Electric from selling to the British Marconi Corporation the rights to a radio wave–producing device known as the Anderson Alternator. Roosevelt, along with others in the navy—including Secretary of the Navy (and FDR's future ambassador to Mexico) Josephus Daniels, and then captain, electrical engineer, and later director of Naval Communications, admiral, and RCA board member William H. G. Bullard—feared that the United States would lose control over radio communications in the United States to companies from Germany and Britain.[17] These worries arose as early as 1916, when Roosevelt urged the U.S. government to counter British and German telegraph companies' incursion into the Americas by expanding the U.S. sphere of influence over radio communications throughout Latin America.[18] Just a few months earlier, Roosevelt and Bullard also recommended that the government subsidize the construction of United Fruit's Tropical Radio Company's station at San Juan

Batista in Mexico in order to improve communications with ships traveling through the western Caribbean between Texas and the Panama Canal.[19]

Although the plan to subsidize United Fruit's Tropical Radio Company was ultimately rejected, they would be one of four companies to make up the Radio Corporation of America (RCA) in 1920, when the U.S. government decided to formally subsidize a commercial corporation with radio interests. With Roosevelt's intervention, instead of GE leaving the radio business at its very dawn, it became the largest radio business in the United States. In exchange for giving up rights to radio in Europe, the U.S. government purchased the American Marconi Company and its subsidiary, the Pan-American Telegraph Company, which was itself on the verge of establishing the first trans-American radio broadcast from New York to Buenos Aires. (It would take another three years for the broadcast of the Firpo-Dempsey fight to draw listeners across the Americas.) A year later GE would be joined by Westinghouse, as well as AT&T and the United Fruit Corporation, to form RCA. AT&T and United Fruit sold their shares of the company back to RCA during the 1920s, and GE was forced to abandon its interests in RCA in 1930 due to an antitrust suit brought by the government, by which time RCA, with ownership over NBC and RKO Pictures, was already signing a lease at Rockefeller Center.[20]

The "*national* hookup," in other words, was founded on a multinational network of radio stations that required governmental aid as well as multiple corporate agreements. The rise of NBC's radio network is due, at least in part, to the United States government's desire to gain commercial and military interests in Latin America. Thus, NBC's prehistory offers something of an international and infrastructural corollary to the sonic erasure of space across the street from 30 Rockefeller Plaza.[21]

Dos Passos would take these affiliations into account in the composition of *U.S.A.* The trilogy's second book, *1919* (1932), a year that not only marks the Treaty of Versailles but also Roosevelt's intervention in GE, begins in Buenos Aires, the site of the first Pan-American radio connection, and follows a character named Joe Williams as he sets off up the Atlantic coast of South America and then through the Caribbean on a banana boat. The story of Joe Williams, brother of Janey Williams, a stenographer to the publicist J. Ward Moorehouse—a character to whom I will return—marks a turning point in the trilogy, redrawing the map of *U.S.A.*, "a radio network," to include

Buenos Aires, Cuba, and the Caribbean, as well as Mexico and parts of Europe. These incursions into other national territories insist that the country's story cannot be told without an appreciation of its imperial ties through business and war, and the radio networks that helped facilitate and depended on both to create the fiction of a nation.

NEW DEAL ACOUSTICS: THE SOUND OF SOCIAL SPACE AT THE 1932 CONVENTIONS

The larger spatial politics of the radio network are always implicit if rarely mentioned in the sounds of a given broadcast speaker's individual utterance. In his writings from 1932, Dos Passos complicates radio's apparent "elimination of resonance" and what later theorists have too quickly assumed to be the medium's "disembodiment," as he listens to the distinctly embodied radio address of that year's U.S. Communist Party's presidential candidate William Z. Foster.[22] Recovering from a heart attack in Moscow, Foster speaks over the radio to an assembly in the New York CPUSA headquarters, and the medium helps amplify those two sonic aspects Arnheim claims it vanquishes: "His speech is going to be broadcast from his bedroom. The feeling of farawayness and emptiness is enormously intensified. Is it that we're ten thousand miles from Moscow? When his voice starts coming over, the accent and intonation of a native American workingman fills the hall for a moment with warmth. Hathaway has to finish reading it for him; his voice is American, too."[23] Although unencumbered by the explicit image of the body, Foster's voice comes across the wireless freighted with the gravitas of bodily illness. His fragile bodily condition, in turn, accents "the feeling of farawayness and emptiness," precisely the distance and resonance meant to vanish with the New Acoustics and Arnheim's ideal announcer. When combined with Foster's tone, this radio voice, carrying the signals of its absence and the sounds of a body in decline, makes felt the room's necessary vacancies, "the wide empty platform" from which the broadcast is heard, and the negative space in which the voice resounds. In more theoretically familiar terms, affect arises out of a constitutive absence in the radio voice rather than a naïve extension of presence. The sudden "warmth" in the hall does not abolish the feeling of distance.

The room's spare qualities amplify the phenomenal difference produced by Foster's broadcast, yet this amplification derives not from any acoustic engineering intent on the erasure of space but from the interaction between a technologically mediated voice and its reverberation in a mostly empty space. In other words, against the homogeneous neutrality of electronic information, the "clear and focused" sound promised by the New Acoustics, Dos Passos's rendering of Foster's broadcast voice fills the space with the noise of tonal difference ("the accent and intonation of a native American workingman"). In choosing to marshal the *tonal* aspects of political speech instead of the basic semantic content, Dos Passos's representation of these scenes of audition calls for an attention to acoustic difference. And as these voices come to us not in audible recordings but filtered through the written word, this listening practice necessitates a reading practice able to attend to the sound of (social) space.

In this context, Foster's voice comes close to what Asif Agha has called "*enregistered voices*."[24] Explaining how Mikhail Bakhtin's notion of "social voices" implies less a specifically vocal attribute than a specific social position, Agha comments that "a register's forms are social indexicals in that they index stereotypic social personae (viz., that speaker is male, lower-class, a doctor, a lawyer, an aristocrat, etc.) . . . [T]hus every register has a *social range*, a range of figures performable through its use."[25] In Foster's case, Dos Passos emphasizes the oral aspect of register, his "accent and intonation," to index his social class, national heritage, and his political position. Perhaps more importantly, Dos Passos decides for the reader into which category Foster's voice should be placed—an "American workingman"—instead of representing his speech and leaving it to the reader to decipher Foster's social register from the content of his words, or their formal rendering in the report. Strangely, through his attention to the sonic aspects of register Dos Passos delimits the social range of their possibility but also expands what Agha calls the "*social domain*," the group for whom such indexes are recognizable.

In preselecting for readers how this voice should be recognized, the text also risks closing down the possibilities of transferring this voice to others. While Hathaway can ventriloquize Foster's speech, almost contagiously assuming the same tone as he steps into the speaker's position, the interpretive work performed by the narrative, naming the tone as "American," forecloses the opportunity for many readers, listeners, and potential speakers to follow Hathaway's lead. As some later authors will intuit, readers can in some sense overhear the text, but they encounter accented obstacles in any attempt to claim Foster's voice as their own. If an attention to tone helps position Foster's voice in social space, the narrative also includes a kind of descriptive inflexibility that fails to recognize the potentially liberating collective attributes of the invisible voice. Precisely the challenge between identifying the tonal attributes of a voice and attempting to secure the potential utopianism of an invisible and, therefore, seemingly transferable voice frames the desire Dos Passos and writers after him pursue. The search for such a contagious form of speech will lead Dos Passos to one of his main stylistic modes in *U.S.A.*

By explicitly converting a voice's accent and intonation, the typically nonreferential aspects of speech, into the content of his account, Dos Passos asks his readers to incorporate their ear as much as their eye into their reading practice and to conceive of reading as a type of listening. As I suggested in my introduction, we might call writing that both references an acoustic situation, such as a voice in a hall, and carries the markers of oral speech's nonreferential sounds in order to help tune the reader's ear, *narrative acoustics*. This writing and reading practice would amplify rather than mute the sound of space. It would therefore learn from acoustic engineering as well as provide an initial response to some of its more reductive effects.

BUILDINGS THAT SPEAK: AMPLIFICATION AND AURAL TECHNOLOGY IN CHICAGO STADIUM

The intellectual attention to acoustic engineering and the stylistic investment in a speaker's acoustic properties to sketch his political portrait were not lost on the orators Dos Passos discusses. Indeed, one of William Foster's political rivals in the 1932 election, the incumbent president of the United States, former secretary of commerce, and engineer Herbert Hoover, set out to expand and regulate radio in the United States throughout the rise of the New Acoustics in the 1920s. Despite his familiarity with radio's political and technological functions, Hoover was not entirely comfortable with the technology. In a 1924 radio address he admonished "the deadly inexpressive microphone" and called for "a method by which a speaker over the radio may sense the feelings of his radio audience. A speaker before a public audience knows what hisses and applause mean; he cuts his speech short or adjusts himself to it."[26] From his position behind the radio microphone Hoover elaborates another side of the disembodied voice:

the silent audience and the lack of a communicative circuit or "positive feedback" in radio speech.[27]

However, what intrigues Dos Passos in 1932 as he reports from the same Chicago Stadium where he will listen to Roosevelt speak less than a month later is not the problem or advantage of disembodied speech but instead the embodied asymmetry and closed circuit established between the man he calls "the Hoover manikin" and the New Acoustics. As he watches Hoover deliver his speech at the 1932 Republican National Convention, Dos Passos notices a very different acoustic relation than that which governed Foster's radio address: "A dry phonograph voice comes from the loudspeakers that hang from the shadowy ceiling in the center of the hall. Through the glasses I can see the mouth barely moving. The expressionless face when he turns our way is like the face of a ventriloquist."[28] In this representation of a voice severed from the body by acoustic engineering, Dos Passos depicts the now banal trick of sonic media, a version of Michel Chion's "acousmatic" voice, "sounds one hears without seeing their originating cause."[29] Distinct from the radio's apparently disembodied voice, the mechanized voice in this scene has a body present but fails to coincide with it. Asynchronous, the voice issues not from Hoover's mouth, its expected "originating cause," but a moment later from the loudspeakers, where the acoustics perform the mundane magic of the vaudeville ventriloquist, transforming the animate into the inanimate, the president into a manikin.[30] In a variation on Foster's radio voice resonating on the spare platform in New York, now the wired room, and not the man, seems responsible for the voice. Hoover looks on, mutely, as Chicago Stadium speaks.

If Dos Passos sounded out the political tone of Foster's voice through attention to acoustic properties in the absence of the body, with Hoover the same attention disassociates the voice and body to the point that the "dry phonograph voice" becomes more proper to Chicago Stadium than to Hoover himself. Furthermore, while the phonographic analogy mechanizes Hoover's tone, it also dates the voice to an earlier technological moment and therefore helps index Hoover's oratorical ineptitude and his admitted discomfort with the new acoustic media. We might think of the result in the acoustic vocabulary developed by Chion, as we hear the building itself as a version of the *acousmêtre*, "an acoustic agency . . . present and audible and effective within the visible scene, but . . . not *seen to speak*."[31] As mentioned above, the surprise in Chicago comes from the displacement of the acoustic agency, or acoustic character away from its potential human source, and redirected onto the building and its audio technology. Thus, the essay more fully destabilizes the identification between a speaker and his or her voice, not only echoing the gothic genre's reversal of subject and object, for which the phonograph was emblematic, but more fundamentally challenging the difference and sustainability of either position.[32]

In other words, listening to the sound of space in these wired rooms, speech no longer makes sense as an individual expression. This newly mediated form of speech participates in an interactive circuit that includes the apparatus—the radio, phonograph, microphone, and/or loudspeaker—the speaker's bodily mediation (the tongue, throat, and teeth), and the built forms that help mold the sound's acoustic shape. In Dos Passos's work, as well as that of the other authors to whom I will turn later in this book, aural technologies like the radio tend to mediate the relationships between the other actors in the circuit, amplifying the often inaudible tension between social actors and the spaces in which they speak. As a consequence, voice exists in a strange dialectic: it becomes the alienable property of a supposedly inalienable aspect of a speaker's identity.

ACOUSTIC HISTORY: THE POLITICAL LEGACY OF VENTRILOQUISM

In order to negotiate how these lessons from the acoustic realm might function in represented speech, we should not forget that ventriloquism serves as the operative figure to explain the electrical acoustics' lack of synchronization between Hoover's voice and body. Indeed, Dos Passos's reference to ventriloquism turns the scene to the cultural prehistory of modern sound media, such as the telephone, phonograph, and radio. Attention to this earlier moment can better establish how Dos Passos's interest in technology connects to a literary, cultural, and political interpretation of that technology. In doing so I will show how the literary and political world makes these technologies live in particular ways. Ventriloquism, in this case, emerges as a necessary corollary or antecedent to whatever acoustic and social changes might occur with the technological innovations of the 1920s and 1930s.

Echoing critical commentary on the technological construction of the female voice in film, as well as Arnheim and the New Acousticians' desire to erase the sound of space through acoustic engineering, Steven

Connor's history of ventriloquism asserts that the practice of throwing voices changed dramatically in the United States during the eighteenth century "from a condition of spiritual malady to a form of expertise and entertainment; as it became associated with the *control of space* rather than with the invasion of the body, it became a male accomplishment rather than a female malady."[33] Ventriloquism, in Connor's reading, appears to anticipate, and perhaps even participate in an earlier moment of the same discursive network that associates the acousmatic voice with male control in sonic media from radio broadcasting to cinematic voice-over.[34]

But what of the "speech of the people," the dangerous democratic possibility Dos Passos associates with the radio? Unlike Connor's more familiar secularizing and liberal story of modernity, Nancy Ruttenburg tunes in to a strain of speech that counters the totalizing desire of disembodiment associated with later ventriloquism and its legacy. The literary and theological hybrid she names "the voice of democratic personality" is a voice we might understand as the forebear to the vocal current Dos Passos seeks in his political portraits and the "speech of the people" that will help define *U.S.A.* Ruttenburg's description of this seventeenth- and eighteenth-century practice of enunciation dependent on the figure of ventriloquism serves as an apt if unacknowledged historical antecedent for Dos Passos's attempt to imagine a new narrative acoustics meant to sound out a different form of utterance running counter to the "clear and focused" channel of the New Acoustics.

Acknowledging "democratic personality" as the precursor to Dos Passos's populist voice reveals the particular political utterance that emerges with the "new media" of the radio as a reformulation and extension of a longer political and literary phenomenon. Ruttenburg writes that democratic personality develops out of the religious domain of the Salem witch trials and the eighteenth-century Great Awakenings as an "uncontainable" and "conspicuously transferable" voice through which those subjects previously ignored or flatly denied the right to speak because of their race, class, gender, or age spoke freely when "possessed by the spirit." To speak as if spoken through, either by a divine agent, or a numinous "people," as Walt Whitman later claimed to do in his poetry, performs what Ruttenburg names "the practice of humble self-enlargement (the publicly performed disintegration or 'undoing' of the preternaturally knowledgeable self which authorized disruptive popular speech)."[35] For Ruttenburg, such public speech disrupts the association between the speaking body and the spoken word, enabling a more expansive rhetoric less limited by self-reference and the strictures placed upon specific bodies—a flexibility achieved because of the fact that the voice in this performance claims its source comes from elsewhere.[36] Thus unencumbered, democratic personality makes itself available for diverse appropriation, an appropriation more easily secured because the utterance, which was "often inarticulate if radically persuasive," depended less on what it said than how it was delivered.[37] In this manner, a new form of popular public speech that empowered socially marginal subjects to condemn a society's most powerful members spread quickly, independent of the expertise of literacy or any specific body.[38]

With this history in mind, Dos Passos's depiction of Hoover's speech resounds as either a vicious mockery of popular speech's betrayal by presidential oratory or an indication that the dissociative properties induced by technological change might themselves help reinstate the grounds for a renewed, if revised, version of such speech. What I described earlier as the "contagious" tone picked up by Hathaway as he finished Foster's address points to a minor example of how shared acoustic properties might begin to form a collective utterance in the spirit of Ruttenburg's "voice of democratic personality." And even the odd way that the acoustics in Chicago Stadium seem to turn Hoover into an inhabitant of his own voice—to occupy an almost impossible role as author and character—begins to elaborate the democratic potential of sonic media. Hoover's ventriloquistic act in the wired room, therefore, appears within a longer literary and political history regarding the struggle over the popular voice in the United States, where the triangulation of authorship, character, and acoustic attention suggests a means to read how Dos Passos inscribes listening into his writing practice.

THE RADIO VOICE: FDR'S "INTIMATE PUBLIC"

Back in Chicago Stadium in 1932, Dos Passos pondered the "unassuming speech" of New York governor Franklin Delano Roosevelt, a man he described as overwhelmed "by the giant muddling awesome blurs, a hundred times amplified, on the radio and screen."[39] Two years later, the second book of his *U.S.A.* trilogy completed, he listens to Roosevelt speak again, this time over the radio, and includes what he hears as the final piece of the section "On the National Hookup." Entitled

"The Radio Voice," the brief essay re-creates a scene of a family in a barn listening to the now President Franklin Delano Roosevelt speaking over the radio: "Then there is a man leaning across his desk, speaking clearly and cordially to you and me, painstakingly explaining how he's sitting at his desk there in Washington, leaning towards you and me across his desk, speaking clearly and cordially so that you and me shall completely understand that *he sits at his desk there in Washington with his fingers on all the switchboards of the federal government*."[40] As with the previous descriptions of Foster, Hoover, and then governor Roosevelt, we hear *how* the speaker intones his words ("clearly and cordially"). However, unlike those earlier scenes, the narration does not stop with descriptive vocal adjectives. Instead, Dos Passos quickly shifts into represented speech, bringing tone into the speaker's content for the first time in these sound portraits. Most notably, rather than the marked pathos of "farawayness" encountered in Foster's radio voice, the reader confronts an insistent, repetitive voice intent on demarcating space in order to interpolate its listener, thereby concealing the power divisions between listener and speaker. The simple choice of pronouns, the use of the second and first person, enables every individual in the massive audience to hear each "you" addressed to him or herself alone. The massive address of "broadcasting" becomes an intimate moment, like the point-to-point "narrowcasting" of a telephone conversation or face-to-face speech.[41] The audience, reinforcing the success of the speaker's intimate tone, confides: "[W]e feel we are right there in the White House. When the cordial explaining voice stops we want to say, 'Thank you Frank.'"[42] As if completing the imaginary circuit, the audience feels familiar enough to thank "Frank." The evocative sensation of "farawayness and emptiness" conjured up by the weightiness of Foster's radio voice vanishes, now replaced by the intimacy and presence of a single room and a familiar name.

The mass audience's sense of personal closeness with "Frank" signals what has been called the logic of "radio's intimate public."[43] As Jason Loviglio writes, this "new cultural space created by radio broadcasting in the 1930s" found its political embodiment in FDR, "the *best modulated* radio voice in public life," whose "Fireside Chats invited listeners into a privileged realm of mobility that enabled them to feel as if they had crossed the boundary separating public and private, backstage and onstage . . . These authorized transgressions—public speech in 'intimate' spaces—were part of the unique allure of network radio."[44] Roosevelt's radio voice, from this perspective and, to a certain degree, from that of Dos Passos's narrative, appears to partake of the technical innovations of the New Acoustics and transfer their lesson to the realm of political theater and social life, folding the social distance and difference of specific positions of power into an intimate context and thus mimicking the erasure of space generated at the engineering level through material insulation, higher-fidelity microphones, loudspeaker placement, and more precise volume control. Indeed, Roosevelt and his advisors considered his tone of such importance that they outfitted him with a dental bridge in order to close a slight gap in his teeth responsible for an occasional whistle in his speech.[45] Loviglio understands this action less as an "erasure" than as an invitation "to think about the shifting boundaries of public and private space and speech at the start of an era in which these terms had come to be understood as permanently in 'crisis.'"[46] This more generous reading suggests Roosevelt's radio voice encouraged listeners to self-consciously transgress the border between public and private and in so doing recognize the ongoing construction of these social spaces and thus their fluidity. A more suspicious listener, Dos Passos hears these invitations only as sonic seductions meant to deceive the public while politicians continue to exploit them through the tricks of public relations firms manufacturing intimacy.

Dos Passos recognized that Roosevelt's keen political manipulation of micro-acoustics and radio broadcasting molded the medium's meaning. Distinct from the fascist oratorical styles criticized by Theodor Adorno and Max Horkheimer and the amplified zealotry of American radio orators like Aimee Semple McPherson, Huey Long, and Father Coughlin, Roosevelt's radio speeches carry with them their own particular micropolitics of acoustics and language intended to sculpt a place in the American imaginary.[47] In response, Dos Passos, and authors after him, struggled to counter Roosevelt's political and cultural influence by redefining his radio voice and the radio medium's function in their essays and novels.

We can hear how the work with tone exhibited in Dos Passos's essay counters Loviglio's idea that friendly "Frank" sought to increase listeners' awareness of social space. Indeed, Dos Passos's narrative performs the work Loviglio attributes to Roosevelt. Like any good caricature, the excerpt from "The Radio Voice" exaggerates its subject's most recognizable features: the speaker's direct address ("you and me"), his clear enunciation ("speaking clearly and cordially"), and self-reflexive contextualization ("at his desk").

And repetition, another hallmark of Roosevelt's rhetorical style, becomes the means to unstitch that style and expose its seams. For as the droning cycle proceeds, the narrative voice that began the sentence before slyly slipping beneath the radio voice—as if the initial voice's volume has been reduced in the mix—reemerges at the very moment that the narrative reveals whose hand sits at the controls in Washington. Dissonance distinguishes the voices, one intimate ("you and me"), the other expository ("then there is a man"), and the technical skill required to sound out their difference also exposes its own hand on the narrative controls. Descriptive tone, the adjectival categorization of how speech sounds, carries through here to a syntax that expresses tonal difference by writerly means. Narrative acoustics disturb the intimacy between radio speaker and listener, reinserting the sound of social and discursive space into this written rendering of a radio voice.

The free indirect discursive mode responsible for this toggling between registers has been called "unspeakable."[48] However, the minute formal attention to tone and acoustics that characterizes each of what I am calling the "sound portraits" in "On the National Hookup" calls for these passages to be heard as well as read.[49] Listening, in these essays, becomes a model for writing and reading, the means to "de-acousmatize" the speaker behind the desk, to give him back a body and a powerful sociopolitical position as the man "with his fingers on all the switchboards of the federal government." Narrative style tunes the reader's ears to these minute changes in tonal shape that writing can register to help instantiate a distance between the narrative voice and the radio voice: a distance the essay charges Roosevelt with attempting to erase through his intimate tone. Indeed, to revisit the language from earlier in this section, Dos Passos's use of free indirect discourse here comes closest to written ventriloquism. Whereas eighteenth-century speakers oscillated between author and character to seek out a "democratic authorship" in their harmonic interplay, "The Radio Voice" works to unhinge a speaker from control of his characters/listeners through a self-exposing ventriloquism. Again, attention to acoustic properties loosens the speaker from possession of his voice and reasserts the politics of space against the erasure and control associated with the New Acoustics and the ventriloquist history established by Connor.[50] Attention to Dos Passos's technical work with radio listening through the medium of print reveals sound as a transmedial phenomenon, and literature as a space in which to elaborate the conditions that create radio as a medium: the physical infrastructure, rhetorical techniques, and sonic engineering Roosevelt and Dos Passos struggle to influence as the means of political power.

U.S.A.: THE NEW ACOUSTICS AS NARRATIVE

Brought to the grander stage of the novel, the narrative lessons gleaned from listening to the acoustically engineered political soundscape at the beginning of the New Deal help Dos Passos compose what historians have often heard in Roosevelt's voice: "the speech of the people."[51] The phrase closes the two-page preface attached to the first bound edition of the *U.S.A.* trilogy (1938), which begins with an isolated and lonely "young man" but ends with a chorus of voices linked in to the national hookup wherein "only the ears busy to catch the speech are not alone; the ears are caught tight, linked tight by the tendrils of phrased words . . . his mother's words telling . . . his father's telling . . . the speech that clung to the ears, the link that tingled in the blood; U.S.A. . . . a radio network . . . the speech of the people." This lyrical passage's movement from the heard to the spoken word, from the individual family utterance to the radio network identified with the nation sounds slightly off-key when considered alongside the ironic, skeptical ear reporting from Chicago Stadium six years earlier. While the passage recreates the undulating, curling syntax reminiscent of the author's *Manhattan Transfer* (1925) and the aesthetic hopefulness of its modernist contemporaries, the style soon gives way to the book's dominant mode, what Jean-Paul Sartre, writing in the year of the trilogy's publication, refers to as "the garb of populism."[52] Indeed, the "speech of the people" in *U.S.A.* sounds little like the sweep and grandeur of the preface. While Dos Passos remains attentive to tone, he neither converts the modern din of urban overcrowding and noisy machinery into jazz-like riffs of adjectives and nouns nor seeks out the crannies of acoustic difference in protest against the erasure of the sound of space. Instead, in a turn that must have surprised readers of *Manhattan Transfer* and the reportage I have just discussed, the dominant narrative style in *U.S.A.* applies the clear and focused channel of the New Acoustics to represent the people's speech.

It is not that Dos Passos has forgotten how to listen in this novel or uncritically adopted the erasure of the sound of social space. Strangely, the flat, paratactic, additive style that dominates *U.S.A.*'s realist sections applies the listening lessons of the 1932 political conventions and "The Radio Voice" to mute, rather than amplify, sonic difference; however, the ambition to awaken readers to the sound of power remains. Indeed, I hope to show that the construction of such a homogeneous tone responds critically to both the political failures of Dos Passos's previous writing about radio and politics, and some key concerns about character, individuality, and collectivity in the realist novel. The critical power of this new tone—a tone that seems to match the erasure of sonic and social space enacted through the political uptake of what we might now call the New Deal Acoustics—no longer depends on an intellectual critique that would point out the structural conditions of power. Instead, in a thousand-plus-page novel that portrays the decline of active democracy and the rise of oligarchs in control of public relations firms, Dos Passos seems to recognize that his readers are no longer convinced by the old form of critique. As an alternative, he subtly quiets the individuating signals of narrative tone to represent "the speech of the people" as a homogeneous tone, creating a new connection between characters and a new relation between individual and collective that conveys a surprising political charge.

In order to recognize those effects, we need to first understand how *U.S.A.* transforms the free indirect discourse employed to unmask FDR's manipulative erasure of space. While "The Radio Voice" relied on amplifying the distance of competing registers, Dos Passos now turns to indirect speech to make intimate conversation sound the same as any other speech: "He was still trying to clear his throat when her voice came in an intimate caressing, singsong. Of course she remembered him, her voice said, too sweet of him to call her up, of course they must see each other all the time, how thrilling, she'd just love to, but she was going out of town for the weekend, yes, a long weekend. But wouldn't he call her up next week, no, towards the end of the week? She'd just adore to see him."[53] While we're told to hear the anonymous voice as "intimate," as "caressing, singsong," the form's lack of differentiation, the barely punctuated movement from word to word, converts apparent markers of sentiment ("sweet," "thrilling," "love," "adore") into equivalent and exchangeable terms on a chain of empty signifiers. History and politics receive similar treatment: "He didn't worry much about Gertrude in the months that followed because a strike came on at Homestead and there were strikers killed by the mine guards and certain writers from New York and Chicago who were sentimentalists began to take a good deal of space in the press with articles flaying the steel industry and feudal conditions in Pittsburgh as they called them, and the progressives in Congress were making a howl, and it was rumored that people wanting to make politics out of it were calling for a congressional investigation."[54] Responding to passages like these, Sartre explains that Dos Passos's "American journalistic technique" sounds flat because "for Dos Passos, narrating means adding. This accounts for the slack air of his style" (90). Just as resonance is removed from the equation in the New Acoustics, thereby erasing the reflections and architectural syntax through which each sound must move, the additive, paratactic style of *U.S.A.* removes tonal difference and deflates narrative tension by removing the causality of occurrences, rendering them into exchangeable information, like sounds broken from their context and homogeneously reproduced to every seat in the room.[55] And just as any piece of information can be exchanged with any other, any voice can belong to any character; the style sounds "populist" because the narrative speaks from everywhere, it "adopts the point of view of the chorus, of public opinion" (94). Literalizing the preface's statement, USA, the nation, and *U.S.A*, the book, become a radio network, a mass listening audience, speaking through any body.[56]

Dos Passos turns to this particular "populist" tone to respond to a problem in the political imaginary linked to ongoing issues of realist form: how to negotiate and represent the relationship between the individual and the collective without subsuming one to the other. This structural problem, however, is also indelibly linked to another narratological issue: the problem of character. Other readers will argue that Dos Passos has resolved these difficulties through the larger architecture of the book's four narrative modes.[57] The passages above come from the realist narratives that follow the unfolding lives of one of the novel's thirteen main characters; however, between these chapters Dos Passos intersperses the lyrical and autobiographical montage of what he called his "Camera Eye" sections, the public history of his Dadaist collage of newspaper headlines and popular songs in the novel's "Newsreels," and the capsule biographies of celebrated inventors, politicians, businessmen, etc.[58] Each of these modes helps Dos Passos create layered perspectives—personal, public, historical—through which the realist protagonists find a connection, however weak and contingent, to a totality that exceeds each of them alone.

Some events bind the modes more strongly than others. Wesley Beal has noticed that Armistice Day, for instance, filters into each. Toward the close of Joe Williams's narrative in France he goes to celebrate the Armistice at a cabaret "where the music was playing *The Star Spangled Banner* and everybody cried Vive L'Amerique."[59] The Newsreel that begins on the same page opens with a fragmentary news report ("the arrival of the news caused the swamping of the city's telephone lines"), then another that reads "at the Custom House the crowd sang The Star Spangled Banner," before quickly jumping to headlines "(WOMEN MOB CROWN PRINCE FOR KISSING MODISTE), the Marseillaise (*Allons les enfants de la patrie / Le jour de gloire est arrivé*), and Cole Porter's lyrics from *Kiss Me, Kate* (*Kakakatee, beautiful Katee / She's the only gugugirl that I adore*)." This collage of song and news gives way to the Camera Eye, whose anonymous narrator begs, "Hey sojer tell me they've signed an armistice   t e l l m e t h e w a r s o v e r they're taking us home   latrine talk the hell you say."[60] Finally, in a capsule biography dedicated to Woodrow Wilson, the narrator states, "Five million men stood at attention outside of their tarpaper barracks every sundown while they played *The Star Spangled Banner*," before he declares, "Almost too soon the show was over."[61] Repeated mentions of the national anthem connected to the Armistice link the modes together, but their strikingly different forms of registering the event, from the chaotic assemblage of songs and news reports to the idiolect of the Camera Eye's "latrine talk" and its speaker's hope for the war's end, to the meager mention of the Armistice in Wilson's biography—"almost too soon the show was over"—inflects the felt interpretation of the Armistice with conflicting class registers and perspectives on the event. Some have seen this assortment of styles—what Bakhtin would call "heteroglossia"—as depicting the atomization, fragmentation, or disaggregation of the novel's social world, as well as its attempt to connect these particulars into a grander social totality. And in a phrase to which I will return, at least one critic claims that the novel organizes these two poles through an alternative "network aesthetics."[62] All of these arguments fit neatly into what we have come to expect from the modernist novel, and yet their emphasis on difference seems bound to that previous use of critical narrative form I claimed Dos Passos discarded as he created his new narrative tone. Indeed, the homogeneity of tone in *U.S.A.*'s realist mode would stand in direct contrast to the dialogic organization of multiple registers and perspectives where we're told democratic form is supposed to reside in the novel genre.[63]

So how does Dos Passos's homogeneous tone address the problem of the individual and totality through the more local problem of character? In a perceptive comment misguided only in its conclusion, Georg Lukács writes that "Dos Passos' characterizations are transferable from one individual to another . . . [B]y exalting man's subjectivity, at the expense of the objective reality of his environment, man's subjectivity is itself impoverished."[64] What Lukács understands as a weakness in Dos Passos's characters—the easy transference of their attributes—accurately identifies part of the novel's formal response to the problem of character but undervalues its method. Rather than follow Lukács's critical realism in the search for a typical hero who can transform history by shaping fragments into a totality upon which one can act, Dos Passos creates characters with affective affinities, what an early reviewer called a shared "set of sympathies," to connect characters through what Sianne Ngai names "the affective-aesthetic idea of tone."[65] What Dos Passos has achieved in his use of free indirect discourse in the examples I cited above, and those that proliferate across the trilogy's realist mode, is a flat tone that makes all characters sound and feel the same. If we continue to read with Lukács (or Edmund Wilson, or a host of other critics), we hear this tone as the impoverished standardization of experience.[66] However, if we retune how we read character and instead hear this same homogeneity as Sartre would, we encounter a critical edge in this flat tone that Sartre insists will "impel us to revolt."[67]

What is the difference between understanding Dos Passos's tone as a passive capitulation to the culture industry and experiencing it instead as a mode that impels you to revolt? The easy answer is that the difference is between "understanding" and "experiencing," or between an intellectual and affective reading. As I stated earlier, Dos Passos's use of free indirect discourse in "The Radio Voice" was meant as an intellectual strategy that would identify and awaken readers to the false intimacy of FDR's rhetorical and sonic techniques in the fireside chat. By contrast, the homogeneous tone of *U.S.A.*'s realist narratives makes the individual character coincident with totality to instantiate a visceral political reaction in its readers. But this binary between feeling and understanding also overstates the division. Knowledge and affect are codeterminate. Dos Passos's tone will "impel us to revolt" because, in Sartre's words, "it will not take you long . . . to decide that you *cannot* use this tone in talking about yourself."[68] What Sartre identifies as Dos Passos's tone is a particular use of free indirect discourse that Simone de Beauvoir calls "cruel" and says allowed "[Dos Passos's] main characters" to be, "at one and the same time, drawn as detailed individuals and as purely social phenomena."[69] Thus, the feeling of revolt comes from a specific tone that seeks to draw a relation between the individual and totality through the question of character. More specifically, as Michael Lucey puts it, Beauvoir learns from Dos Passos a new means to represent "existing simultaneously in the first- and third-person modes, existing simultaneously as narrator (first person) and as character (third person)."[70] In other words, Dos Passos's homogeneous tone, which allows characterizations to transfer from one character to another, enacts a resolution to precisely what Ruttenburg identified as *the* problem of "democratic personality" as far back as the eighteenth century. In the face of a culture industry that turns every statement into "a statement to the press," that forces standardized messages and clichés into the mouths of its populace, Dos Passos impels us to revolt through a tone whose very imitation of the language of power makes readers bristle against that confinement and feel compelled toward revolution.

RADIO'S CHARACTER NETWORK

Converting his characters into conduits, Dos Passos draws on techniques developed in the novel and radio, but he eschews more spectacular formal innovation for a more nuanced notion of how broadcasting the voice of the people transformed subjectivity. In order to more closely understand the particularity of Dos Passos's innovations with tone, it helps to turn to Sartre again, but this time to see how he adapted Dos Passos's lessons into his own narrative form to arrive at a very different technique. In *Le sursis* (The Reprieve; 1945), the second book of Sartre's own trilogy *Les chemins de la liberté* (Paths to Freedom), which he modeled after *U.S.A.*, he turns to the radio in what he claims is an attempt "to profit from the technical research on simultaneity that had been carried out by a certain number of novelists, such as Dos Passos and Virginia Woolf."[71] In an opening chapter in which multiple characters in Prague, Paris, Marseille, and Munich either listen to the radio or have their voices broadcast over the radio, Sartre links what Alex Woloch calls "character spaces" (the amount of narrative attention devoted to a given character) with an entire "character system" (the network of relations between characters) in an attempt to recreate the simultaneity of radio listening. The chapter's most ambitious moment shows how the novel's discourse can organize a new notion of the people's voice through radio:

> Dr. Schmidt understood that interpreters were once more needed.
>
> Hitler approached the table, and the clear, resonant voice rose into the translucent air; on the fourth floor of the Massilia Hotel a woman on her balcony heard it, and she said: "Gomez, come and listen to the Negro singing, it's delightful" . . . The Negro sang on . . . Chamberlain said: "As regards the first clause, I have two objections," and the Negro sang: *Bei mir, bist du schön*; which means: "You're the fairest in the land."[72]

The movement in this passage from the translator, Dr. Schmidt, to Hitler, to the "Negro's voice" identified by the anonymous woman, then back to Chamberlain before the song is identified, resounds with political complications. Sartre presents readers with Hitler seemingly sharing a voice with a popular U.S. swing composition originally written in Yiddish, sung by a banjo player, the "Negro" singer on the streets of Marseille, who sounds like the eponymous protagonist walking the same streets in Claude McKay's 1929 novel *Banjo*.[73] Furthermore, the Spanish names that appear here, in France, in the narrative's 1938, index the Republican struggle against Franco's Falangists. Thus, the voice(s) stage(s) an encounter between fascism, socialism, the diasporic language of Yiddish, and the non-European "folk" tradition as a collective speech.

Stylistically, it is the radio voice's movement, the almost vertiginous passage across character spaces, that prepares this final moment of what we might call, in a variation on the work of Bakhtin, unified heteroglossia. For unlike the quick scene cutting from modernist classics like the "Wandering Rocks" episode in Joyce's *Ulysses* or even the more open and amorphous movements in Woolf's *Mrs. Dalloway*, in the "clear, resonant voice" Sartre achieves a passing consolidation of voices, wherein this single sound seems owned by everybody and no one. Sartre's "voice of the people" is a collision of ideological differences that can only exist together through the printed juxtaposition of voices in a narrative reordering that audiences can never actually hear as present to each other through the radio.[74]

Such a representation of radio voice departs from the homogenized tone of Dos Passos's work. However, it deepens the question of the radio's influence on *U.S.A.* and how we are to understand it as a "radio network." While Sartre believes he borrows his experiments with radio and simultaneity from Dos Passos, it is difficult to disentangle whether

Dos Passos drew his own narrative experimentation from innovations in radio narrative or whether he was influencing those new developments. The most obvious antecedent for Sartre's novel could be Dos Passos's Newsreels, whose technique Dos Passos himself could have borrowed from what Neil Verma has called the "kaleidosonic" organization of narrative space in NBC's 1930s radio program *The March of Time*.[75] That program narrated a day's news stories by juxtaposing events as varied as "Queen Mary teaching young Princess Elizabeth the Highland Fling, Harvard President James Conant writing a letter to decline an endowment from the Nazis . . . Herbert Hoover debating Henry Wallace, [and] an Irish marquis dying of a family curse."[76] The presentation of these events as all residing on the same plane allows *The March of Time* to "ritualize equality," producing "the feel of simultaneity," as the kaleidosonic aesthetic "leap[s] from one mike to another, 'objectively' arraying the world before us, with everything equidistant and accessed across just two dimensions" in order to "make us certain that we know 'the people.'"[77] Although Dos Passos developed "newsreels" like *The March of Time* as one of the four modes in *U.S.A.*, he chooses instead to locate "the speech of the people" in the odd homogeneous tone that impels revolt. Rather than a radio voice replete with competing voices, as in Sartre, Dos Passos mixes radio's lessons with realism's long work with authorial and character perspective to create the "slack air" style Sartre identifies as the voice of "public opinion."

Key to the construction of this voice is a different notion of "simultaneity" and the construction of a certain tone that bears on the relationship between narrator and character. Rather than the simultaneous combination of ideologically distinct perspectives, such as Hitler and Banjo in Sartre's *The Reprieve*, Dos Passos creates a different notion of the character network. Dos Passos's thirteen characters often appear in multiple chapters—both those that carry their own name and feature them as protagonists, and those that treat them as minor characters in chapters that focus on their compatriots. This version of the character network is Dos Passos's second response to the problem of character as the relation between individual and totality. Moreover, he combines these character networks with his particular homogeneous tone to create the dull, flat style of characters free of psychological interiority or any belief in self-expression:

> Dos Passos reports all his characters' utterances to us in the style of a statement to the Press. Their words are thereby cut off

> from thought, and become pure utterances, simple reactions that must be registered as such . . . Little does it matter, thinks the satisfied chorus, what Dick had in mind when he spoke that sentence. What matters is that it has been uttered. Besides, *it was not formed inside him, it came from afar.* Even before he uttered it, it existed as *a pompous sound*, a taboo. All he has done is to lend it his power of affirmation.[78]

In other words, the characters in these novels do not possess their utterances as expressions of an interior world brought forth into the social world. Neither does the narration enter into their private thoughts through interior monologues. They are less hybrids than conduits for certain speech forms that they attract. Similar to the "atmospheric realism" Eric Auerbach finds in Balzac's notion of milieu, Dos Passos creates the people's voice by narrating every individual as if he or she were the instantiation of the public. Their speech, as Sartre puts it, comes from afar, like utterances floating through the air, like voices heard over the radio. The characters need only listen and then step forward to repeat their line.

RADIO AND REAL ESTATE IN *U.S.A.*

A clear example of the character as conduit occurs toward the end of the trilogy's first book, *The 42nd Parallel.* A narrative voice begins the paragraph in free indirect discourse with the observation that "The sinking of the *Lusitania* had made everybody feel that America's going into the war was only a question of months."[79] By the end of the paragraph this bland statement finds its way into the mouth of the real estate agent J. Ward Moorehouse, who declares, as Sartre says, "in the style of a statement to the Press," that "*he thought America's entering the war was only a question of months.*"[80] Thus, Moorehouse, whose name indexes his original profession in real estate and his economic ambitions, speaks with the voice of the people, his naïve ventriloquism in debt to his position in discursive space rather than any conscious, individual self-expression. Transferring democratic personality from a religious register to a technical feat, the neutral tone that allows speech to slide so easily between bodies in these novels seems to have shed the possessiveness of authorship.

Yet, as I have stated, in doing so the critical function Dos Passos gave to listening in "The Radio Voice" has vanished. In a less generous reading than I have offered, one might say that in *U.S.A.* listening becomes the foundation for a cynical revision of Ruttenburg's "voice of democratic personality," as mere publicity and mimicry, echoing the betrayal of popular speech in Hoover's mute ventriloquism, and the performance of an intimate populism to mask paternalistic governance in Roosevelt's radio voice. "The speech of the people" in *U.S.A.* might belong to nobody, but the political potential of "conspicuously transferable" speech in eighteenth-century American religious gatherings might also be heard, in Dos Passos's novel, as a sign of that speech's political absenteeism.[81] For while Dos Passos develops a technical response to the domination of the culture industry, he also ignores real questions of difference that were already complicating popular radio culture in the United States through minstrelsy programs like *Amos 'n' Andy*.[82] Sidestepping questions of ethnic and racial difference—an issue we will find that later writers, like Richard Wright, will not forget—Dos Passos pursues a narrative mode whose simultaneous interiority and exteriority transforms personal experience into collective experience without attending to the pluralism within that collective.

In grounding, for a moment, the speech of the people in J. Ward Moorehouse at the conclusion of *The 42nd Parallel*, we hear again the conflict surrounding the voice throughout Dos Passos's writings in the 1930s. It is a conflict between (1) a belief in the possessive property of the voice: the voice as identity; (2) the acoustic properties, the tonal shape of the voice in space, as it interacts with several media and no longer seems to "belong" to one particular object; and (3) the architectural forms that participate in determining tone, and, as in Hoover's speech, seem to take full possession of the voice. Keeping in mind Moorehouse's original profession as a real estate agent, we might begin to think of the connection between the idea of possessing a voice, the acoustic properties of a voice, and the architectural forms that participate in shaping the voice as a keen attention to "real" estate, or real property. In this sense, it is worth recalling Michael Denning's observation that *U.S.A.* was partially conceived as an attempt to bring the building lessons of what Dos Passos called Frank Lloyd Wright's "new clean construction" to a narrative style that "writes straight."[83] For if the people's speech in *U.S.A.* seems structurally akin to voices on a "radio network," it also coheres with Dos Passos's comment that "buildings determine civilization as the cells in the honeycomb the functions of bees."[84] In other words, *U.S.A.* sounds, in the end, like the voice NBC broadcast from Chicago Stadium in 1932, that is, like a building engineered to displace the voice from any single body and out into a network signaled by a neutral, homogeneous tone that erases even the space between characters.

RADIO, REALISM, AND REAL ESTATE AT THE END OF THE NEW DEAL

Dos Passos's sound bites from the 1932 conventions, written at the end of an era of radical acoustic change, the beginning of a period of tremendous political and economic upheaval, and in the middle of the trilogy he eventually thought of as both a piece of architecture and a testament to "the speech of the people," can therefore be heard as helping to inaugurate a new literary acoustics, or at least a new attention to acoustics and its political and formal importance in the New Deal novel. What remains a dense, often obscure connection between rethinking discourse in the novel through radio acoustics and real estate in *U.S.A.* becomes an explicit theme in the New Deal acoustic narratives at the end of the 1930s and the close of the New Deal project to imagine the "good neighbor." Authors as varied as Raymond Chandler (*Farewell, My Lovely*; 1940), Carson McCullers (*The Heart Is a Lonely Hunter*; 1940), and Richard Wright (*Native Son*; 1940) exhibit a shared investment in listening attentively to changes in aesthetics, architecture, engineering, and social space, addressing the New Deal soundscape with subtle alterations in literary form and content. Exploring how these writers reworked Dos Passos's careful attention to acoustics and politics, one can see how a New Deal Acoustics was consolidated in 1940 as a critique of the radio's failed promise to deliver the speech of the people, a failure coincident with the government's inability to properly house the people with whom FDR was apparently conducting "fireside" chats.

CHAPTER 2

The Sound of the Good Neighbor

Radio, Realism, and Real Estate

Two defining cultural events revealed radio's aesthetic and political impact on the United States in 1938: Orson Welles and *The Mercury Theater of the Air*'s infamous Halloween night broadcast of *War of the Worlds* on CBS, and the publication of John Dos Passos's *U.S.A.* trilogy. Already a national celebrity when he appeared on the cover of *Time* magazine in 1936 to celebrate the publication of the trilogy's final book, *The Big Money*, by 1938, when the trilogy appeared in print as a single bound copy, Dos Passos was heralded by Sartre as "the greatest writer of our time."[1] However, *U.S.A.*'s central ambition and narrative challenge—to represent "the speech of the people" as a radio network—found new formal responses when writers worked with sound on the radio. Against the backdrop of FDR's "fireside chats," or tirades from 1930s zealots like Father Coughlin and Aimee Semple McPherson, radio writers and sound engineers in the United States sought to create new national and transnational versions of the vox populi.

Print novelists were not far behind. In this chapter I will explore how both radio writers and best-selling realist novelists in the United States struggled with the ambitious aesthetic and political project Dos Passos completed in 1938, and developed transmedial techniques to work with sound to represent the voice of the people. These writers picked up on the oblique connection between radio, realism, and real estate I described at the end of the last chapter, but further specified this relation as a challenge to private property and a hope for housing equality. The desire to represent the possibility of a communal life—from the discursive difficulty of sharing speech to the infrastructural need for shared housing—motivated these writers' turn to the radio: a medium whose networked connection to listeners held out the utopian promise of a space they could inhabit together. While radio writers, directors, and sound engineers developed new representational strategies to depict disparate listeners as "good neighbors," realist novelists found in the radio—the thing that talks back—an object that would disrupt the genre's dependence on furnishings for its reality effect in order to focus instead on a room's negative space. Their opposition to the defining things of realist literature, circulated through the gothic genre's talking objects, will lead to a realism keenly attuned to the problem of buildings, walls, and property: the problem of real estate. Realism, in turn, will become a genre that struggles not just to give voice to the people but to house them as well.

RADIO'S TRANSNATIONAL NEIGHBORHOOD

The year after the publication of *The Big Money*, the Pulitzer Prize–winning poet Archibald MacLeish, whom FDR would soon appoint as librarian of Congress, broadcast his play *The Fall of the City* on CBS. The play found a new sonic way to think through the national representation of the people's voice. MacLeish and the show's director, Irving Reis, found that working with sound allowed radio narrative to create a more conflicted, and perhaps more sophisticated representation of radio and the people's voice than what Dos Passos could achieve in print. In Neil Verma's reading of the play, MacLeish uses the "word"—meter, intonation, and diction—to help situate listeners, while Reis employs the "microphone"—ambiance, acoustics, and volume—to meditate on the contradictions of New Deal populism in the play's final scene.[2] The "word," or the narrator's commentary, positions us on a balcony overlooking a crowd, while the "microphone" makes it sound as if we are positioned among the masses. Depending on how we listen, we hear a "pessimistic liberalism" that associates the narrator's authority as the paternalistic voice speaking for the people, or a populism hopeful about the voice of the multitudes. An enormous success among popular listeners and critics, the play not only transformed its announcer, Orson Welles, into a national star; it also trained him in the techniques he would employ to reimagine popular speech through the radio over the course of the next five years.[3]

Welles, who recorded a voice-over penned by Dos Passos for the 1937 film *The Spanish Earth*, would continue Dos Passos's aesthetic innovations with a difference, as he reimagined the U.S.A. as a radio network through narrative *on* the radio that explored a less nationally focused voice. Recalling the ambiguity of his character's position in *The Fall of the City*, these artistic experiments were directly connected to political infrastructure, more concretely linking the people's voice in *U.S.A.* to the government body, and to the government's involvement outside national borders. These two connections should not be surprising given that in addition to his film and theater work and his radiophonic adaptations of *Heart of Darkness, Dracula, War of the Worlds*, and other literary classics, Welles became an official employee of the New Deal propaganda campaign as a "goodwill ambassador" to Latin America. Appointed to the post by the RKO studio executive and U.S. head of Inter-American Affairs, Nelson A. Rockefeller, Welles toured Latin America to film a never completed project, posthumously released as *It's All True*.[4] Salvaging some of the material that had been rejected by RKO for political reasons—the U.S. and Brazilian governments, as well as RKO executives, disagreed with Welles's depiction of the African diaspora, represented by his planned sections on the "story of the blues" and "the story of the samba," as the unifying thread between North and South America—he returned to CBS in 1942 with a radio program entitled *Hello Americans*.

The program's message of Pan-Americanism made historical sense at CBS. The network's founder, Samuel Paley, had begun his career in the United States as a lector in a Chicago cigar factory, following the Cuban tradition of reading novels, magazines, and newspapers to the cigar rollers.[5] When he began his own cigar company, from which he soon made millions, Paley purchased the Columbia Phonographic Broadcasting System to advertise his La Palina cigars, and his son, William S. Paley, transformed the consortium of radio stations into one of the country's major broadcasting networks.[6]

The network founder's appropriation of Latin American material for capitalist ends set the backdrop for Welles's propagandist mission. Always the promoter, Welles informed Nelson Rockefeller that a radio program was the best way "to sell South America to North America,"[7] but he chose more diplomatic language on the program itself. To the audience listening to *Hello Americans*, Welles noted, "It is important for the people of this hemisphere to get better acquainted, and the Mercury [Theater] has been given the job of helping out with the introductions." For Welles, the means to introduce North and South America to one another was to provide an aesthetic corollary to what Franklin Delano Roosevelt had defined as the policy of the "good neighbor."

When FDR delivered his first inaugural address on March 4, 1933, he spoke of a New Deal for the citizens of the United States, but also of a new approach to foreign policy: "In the field of world policy I would dedicate this nation to the policy of the good neighbor—the neighbor who resolutely respects himself and, because he does so, respects the rights of others—the neighbor who respects his obligations and respects the sanctity of his agreements in and with a world of neighbors." Broadcasting his message across national and international radio networks, Roosevelt's "good neighbor policy," its language strikingly resonate with later notions of technological community like the "global village," imaginatively shrank the world to neighborhood proportions. Aided by radio's multinational reach and the relatively new ability of listeners in Des Moines, Denver, and Decatur to hear the voice of the president, as well as radio programs from Tijuana, Toronto, Shanghai, Buenos Aires, Paris, and elsewhere, the concept of a worldwide neighborhood seemed possible. The audience's resulting sense of connection with the world also extended the United States' sphere of influence, anticipating the country's role as the unofficial leader of the "neighborhood watch."

Borrowing FDR's concept, as well as the narrative logic in *U.S.A.*, Welles and his crew in *Hello Americans* depict the Americas as a unified radio network, pasting scenes from multiple historical periods and geographic locale into juxtaposition. Explaining on the program that "you can twist the dial of history and listen in wherever you like," Welles used sonic techniques borrowed from *The March of Time* radio broadcasts' "kaleidosonic" technique I mentioned in chapter 1, in an attempt to place the cultures of the United States and the countries of Latin America on the same plane, creating a sonorous "neighborhood" for the radio audience. For instance, in one scene in the final episode, actors depicting U.S. senators from 1816 argue over a diplomatic mission to South America. When Henry Clay says, "It was the doctrine of kings that man was too ignorant to govern himself," the line is repeated in another voice, that of Simón Bolívar, now reading Clay's words to his men to liberate Latin America from the Spanish Empire. This simple repetition of a single phrase by two differently accented voices imaginatively unites the Americas through a shared antipathy to European royal imperialism. Such juxtapositions are typical of the series, portraying the easy circulation of ideas about equality through sonic splicing. As a propaganda piece intended for English-speaking audiences in the United States, the show cleverly retains some sense of U.S. leadership, even as it pretends to forge historical bonds between the regions. Ignoring a whole series of U.S. imperial pursuits in Cuba, Panama, and elsewhere, the scenes retain the horizontal logic of North American words inspiring, if not dictating, South American action, as the Latin American Liberator imitates and borrows from the North American senator.

In between these moments Welles inserts another variation on the neighborhood motif. Commenting on the congressional debate, he states that resistance to Pan-American diplomacy has been the same since 1816: "The differences they talk about are the ones you'll find in any one of our large cities, but people seem to get along alright. Any New York subway car is an international express."[8] Through this variation on the U.S. ideal of multiculturalism—a subway car instead of a melting pot—Welles returns FDR's foreign policy doctrine of the "good neighbor" to metropolitan life within the United States, thereby linking national and international concepts of neighborliness. Welles thus reminds the listening audience—whose radios not only brought voices into their rooms but blared sounds through their walls and out their windows—of their roles as neighbors and, implicitly, the importance of housing in constructing a neighborhood.

Adamant that his audience understand the power and potential danger of radio's capacity to convert long-distance signals into neighborly voices, one of the final episode's last scenes depicts a family listening to the radio. Transporting his audience to "a clay hut somewhere in South America," Welles's narrative voice introduces the audience to a small family in which the son, Juan, builds radios at a new factory, and his father, José, listens to the device. Once the broadcaster's voice begins, the listening audience within the frame—Juan, José, and his mother—share the same narrative space as the audience listening to *Hello Americans*. A broadcast voice introduces Franklin Delano Roosevelt's inaugural address, then Roosevelt's recorded voice announces his Good Neighbor Policy, and finally static interrupts Roosevelt's voice, and a German-accented voice in English takes over the air to complain about "the republic of the Jew Franklin Roosevelt." When José comments that the broadcast seems strange, his son Juan answers:

"Oh no, they're all like that. Señor Schmidt at the factory told us it's uh, it's because we're closer to Germany." At this point, a musical curtain falls, and Welles's voice returns to reassure his audience that "Juan has learned. Today he works in another factory and he's a member of the Latin American Confederation of Workers."

The passage is remarkable for the ways in which it distinguishes between voices—José, his son Juan, the U.S. broadcaster, Roosevelt, and the German broadcaster—and both reflects and creates in sound the imagined proximity implicit in Roosevelt's idea of the Good Neighbor. In these scenes, the radio broadcast incorporates itself as a medium and reflects on its own conditions of production, but this second-order reflexivity is not the final message. The technique also reveals how "Señor Schmidt" and his countrymen take advantage of the geographical confusion made possible by the radio's transcendence of space and national borders. In the 1938 broadcast of *War of the Worlds* such confusion between imagination and reality had launched listeners into panic, but now Welles steps in to calmly and condescendingly reassure the audience that Juan "has learned."[9] He has learned, that is, to join the left-leaning Confederation; less than a decade later this would be an untenable statement on U.S. radio.

The sonic splicing Welles employs here to represent the neighborhood of the Americas borrows from Dos Passos's literary techniques, as well as the formal innovations Welles himself had developed in radio and film, to present a more optimistic portrait of "the good neighbor" than the haphazard and inadvertent community of circumstance Dos Passos structured into his trilogy.[10] Although their attitudes toward the policy differ, it is remarkable how both Welles and Dos Passos turn to radio at a political moment concerned with the good neighbor to develop a narrative solution to what became the central problem of realist representation and domestic politics for the writers at the end of the New Deal: the problem of too many people.

In the remainder of this chapter I want to suggest that the representational problem Welles, MacLeish, Reiss, Dos Passos, and others all turned to the radio to solve was tied to a domestic problem—the need for housing, and specifically public housing—that went to the heart of the policy Welles placed center stage: the policy of the good neighbor. While the Good Neighbor Policy used radio's new sense of felt proximity—a transnational version of "radio's intimate public"—to make foreign policy seem domestic, in the wake of Dos Passos's writing, Raymond Chandler, Carson McCullers, and Richard Wright took up radio to reimagine the borderlines of local domestic property. These writers took seriously FDR's political construction of the metaphorical "neighborhood" as opening the door to a symbolic response with material consequences. Thus, they sought to reorganize the representational logic of the realist novel—*the literary* genre of the home—to address the housing problem within the United States' physical neighborhoods. They did so through a "narrative acoustics" that turned away from the realist enumeration of things—personal property—to listen instead to the negative space of vacant buildings—real property. This renewed attention to listening seized on radio's sonic challenge to national borders in FDR's "neighborhood" and turned back to domestic buildings to listen to radio voices escaping through windows or permeating walls as signals that contested local property lines. Walls, under this new logic, became social media rather than isolating partitions. And radio's intimate public now became a call for public housing. While Richard Wright—the subject of chapter 3—makes the material connections between these issues explicit, the writers in the rest of this chapter help us recognize how radio was used to rethink realist design, and how such symbolic reordering obliquely participated in the housing reforms that made up the domestic side of the Good Neighbor Policy. The novelists' interrogation of these codes will help us better understand how first Wright, and then writers and politicians in Cuba and Argentina, connected changes in housing to the transnational politics of the neighborhood of the Americas.

The voice of the people remains a central concern here, but it is linked, especially in McCullers's writing, to the representational and social issues I mentioned above: realism's population and housing. As Alex Woloch has argued, "the realist novel is structurally destabilized not by too many details or colors or corners, but by *too many people*."[11] Indeed, in the decade after the New Deal novelists I discuss in this chapter, the realist novel's most dominant critics insisted that realism's development depended on the increased inclusion of more and different types of people: Eric Auerbach's definition of the novel's history as the ever-expanding representation of more varied social classes of people, Leo Spitzer's argument that the novel genre begins with the eruption of realist "perspectivism" in the encounter between Sancho Panza and Don Quixote, and Georg Lukács's insistence on critical realism as a representation of social "totality" all theorized the attempt to reconcile "the image of the people" with a housing crisis that failed to find a literal place for them.[12] Dos Passos's work with radio had attempted to resolve this formal problem, and Sartre's radiophonic montages in *Le sursis* would apply these technical means to create in print a form akin to Orson Welles's "dial of history" on *Hello Americans*. However, Sartre's work and other novelistic experiments with radio form seemed to drive away readers even as they sought to find space for people to coexist in their novels. The best-selling novelists I study in this chapter, on the other hand, seek out formal means that restructure social space in novels that will be read by a popular audience.

Furthermore, while the metaphor of the radio network seemed to help in the struggle to find a position between the individual and the collective, the metaphorical, structural, and cynical solution Dos Passos presented through the radio network in *U.S.A.* had erased the sound of specific spaces and with it a key marker of the social differences and inequities that were the obsessive target of New Deal politics. In response, New Deal realists connected the radio, and its unruly status as an object that speaks, to the real property that radio broadcasting's very engineering challenged. Drawing attention to the radio as a thing that brings other voices into domestic space and thus unsettles the boundary between private and public, these writers countered the coincidence between the government's Communications Act and National Housing Act of 1934 with a new realism that connected radio and real estate to call for a more equitable experience of the American "neighborhood."

NEW DEAL REALISTS: BUILDINGS AND THINGS

Michael Szalay has named some of these writers "New Deal modernists," but in order to understand the social meaningfulness of their challenge to form we should instead recognize them as realists who revise that tradition through the construction of a vacancy, or the building of an empty room. These writers engaged FDR's neighborhood concept by way of generic changes that advanced various plans for housing reforms. Fiction writers, they did not engage in direct, documentary statements about legislative housing issues but rather sought to entail new social configurations through a shift in narrative codes. This was not a new approach in realist fiction. Rather, it drew on the sedimentation of symbolic practices in the realist novel to make the change in realism readily felt. From Frederic Jameson's "Realist Floor Plan" to Sharon Marcus's "apartment house plot," literary theorists have identified the central role housing and interior design play in the history of the realist novel, and realism's endurance as a genre makes even small changes in its codes carry hefty symbolic weight. Contributing to this theoretical terrain, across this chapter I will show how the empty spaces in New Deal novels from Chandler and McCullers signify the construction of a new public, one attuned to how voices and volume pull individuals into other people's lives. Spaces vacant of things reveal walls that enable a more intimate association between neighbors because of and not despite impediments to sight. More than any other medium, radio broadcasting helped model this form of association, converting overhearing and eavesdropping into a socially acceptable daily practice. In linking literary form and radio—the most popular commercial and political medium of the moment—these writers bent their craft toward the problem of public housing.

The legislative and literary connections between these seemingly separate issues—radio and housing—are apparent when one turns back to the conversations guiding New Deal reforms. The National Housing Act (the Caperhart Act), which created the Federal Housing Administration (FHA) in order to provide federal insurance for home rehabilitation loans and mortgages for newly purchased homes, was part of the same New Deal platform that pushed through the radio Communications Act of 1934.[13] The policy was updated subsequently by the Housing Act of 1937 (the Wagner-Steagall Act), which created the United States Housing Authority (USHA) to pay government subsidies to local public housing agencies.[14]

Key to the public housing debate were transformations in aesthetic design that corresponded with the changes realist authors called for and instituted in their writing. The most influential voice in developing public housing in the United States, Catherine Bauer, soon became USHA's director. In her 1934 study of public housing in Europe, *Modern Housing*, Bauer sought to establish the platform for the housing policy and legislation she drafted for the government of the United States. Published the same year as her mentor Lewis Mumford's *Technics and Civilization*, Bauer's housing ideas should be thought of as part of the technological developments of the age.[15] Moreover, Bauer, who insisted on the connections between ideas and materials, establishes the ground to consider why formal changes in other fields, like realist literature, should matter to questions in architecture and urban planning. Cynthia Hammond makes clear that Bauer's ideas of design were essential to the politics of public housing:

> Thus Bauer explained to her readers—potential inhabitants—that such details as the smooth and unarticulated arched doorway, the functional ceramic fittings of the bathroom, the green and social spaces afforded to view by the generous provision of light-admitting windows, were the carefully crafted hallmarks of this new vernacular. This vernacular and the life it was to support were essential components in a comprehensive design philosophy, one that located users—inhabitants, not owners—as the vital essence of architecture's modern mission. As part of the basic human right that Bauer and others had identified as "minimum standard" housing, such apparently minor details as doorframes and play spaces were in fact cornerstones in the larger battle to improve the lot of the poor or working person during times of considerable economic hardship, to give them a central place in modern, urban, democratic life. *Thus the absence of ornament was not only the liberation of the individual, but a homecoming of the community.*[16]

New Deal realism and New Deal housing coincided in their belief that new design was needed and that a change in form could lead to a change in living. Significantly, Bauer's hope to bring Bauhaus-influenced architectural and housing models to the United States as part of public housing coincided with Willa Cather's call to reform realism through the construction of a vacancy: "throw all the *furniture* out the window!"[17] This mission, I argue in the remainder of this chapter, defined New Deal realists' attempt to redesign the novel to transform the real politics of real property.

THE DETECTIVE'S EAR: THE SOUND EFFECTS OF REAL PROPERTY

Frederic Jameson's essay "The Synoptic Chandler" offers a hint as to how we might think of New Deal narratives from Chandler, McCullers, and others as diverse representatives not just of a new realism but also of a "radio aesthetic which has no equivalent in the earlier novel or silent cinema."[18] Such an aesthetic would address Alex Woloch's concern about the problem of too many people as an extension, rather than refutation, of Roland Barthes's claim that realism is produced as an effect by piling up too many contingent details. The radio apparatus's strange status as an unruly object that literally talks back, bringing the extradomestic news into the family room, points away from its apparent owner and seems to welcome those voices and people bourgeois walls were erected to keep out. In this sense, the radio goes beyond or further literalizes even those "things" Emily Apter, Bill Brown, and Elaine Freedgood have identified as burdened with histories suppressed and concealed when read as merely contingent objects.[19] Radio, in other words, destabilizes its own objecthood as it makes audible the negative space of a room, the acoustic dimension that reminds readers of meaning that extends beyond the walls that partition the visible.[20]

A detective novelist acutely aware of how the contingent detail meant to produce a reality effect can slip into the realm of the clue or red herring, Raymond Chandler challenges the motivated link between the discursive enumeration of details and reality.[21] As his detective Philip Marlowe comments: "The guy that sees too much detail is just as unreliable a witness as the guy that doesn't see any. He's always making half of it up."[22] Marlowe's suspicion of "too much detail" corresponds to Jameson's claim that Chandler's novels distinguish themselves from the details of the Flaubertian room intended only "to stand in for the sheer massive contingency of reality itself" or the Balzatian furnishings in which "the object-world was meant to give a metonymic signal (like a wild animal's den or an exoskeleton)."[23] If the objects in Chandler's novels fail to fulfill either of these realist functions, however, we are left to ask what narrative purpose they might serve. For Jameson these objects strangely participate in the "construction of a vacancy, an empty space," where "whatever the objects mean . . . they also outline a space of a specific type."[24] Although Jameson remains silent as to how this alteration in realism might occur, we can understand Chandler's shift from furnishings to the rooms that contain them as a change in perception, from the eye to the ear. When Marlowe enters a building looking for clues, he hears something else entirely: "I went farther into the room and stood peering around and listening and hearing nothing except those fixed sounds *belonging to the house and having nothing to do with the humans in it*."[25] Disrupting realism's inherited design code, Chandler moves away from the intimate possession of things and toward the aural attention to spaces. Realism here has less to do with how people possess and are possessed by their things than how buildings produce their own sounds.

Marlowe's recognition of the house's autonomous sounds begins to tune us in to how Chandler develops "a kind of substitution of an architectural language for that of individual characters: it is not so much that these 'people' in Chandler are their spaces, as that these spaces in Chandler are 'characters' or *actants*."[26] Thus, in Marlowe's encounter with the empty room, the character's personal property, the details realist discourse previously employed to index a character's identity or establish the effect of reality, has vanished, replaced by a house that *possesses* its own sounds. In more specific terms, we might say that with the evacuation of realism's things—that personal or "portable property" key to so many recent studies in realism—real estate, or immobile and "fixed" property, comes to life. Rather than turning the novel into a house, Chandler's method exacerbates the housing problem in fiction by pointing to real estate's increasing power over individuals.[27] Perhaps it is no surprise that one of his characters asks his celebrated detective and narrator, "Are you in the real estate business, Mr. Marlowe?"[28]

In shifting the realist attention from things (private property) to housing (real property), Chandler tips the old gothic fascination with haunted houses toward the more mundane recognition of how housing policy, and its failures during the New Deal, forced people out of their homes.[29] This change in realism had been anticipated by Henry James in *The American Scene* (1905), albeit to much different effect. Writing of James's prosopopeia as a related fusion of the gothic and realism, Bill Brown notes that "James's ventriloquism seems . . . to confer rights in the mode of conferring voice in order to grant objects some belated yet originary right to narrate their own stories, and thus to express the longing to have their rights realized. Determined as their fate seems to be, they nonetheless retain, in this animating vocalization, some agency, however ghostly."[30] Whatever need for rights that buildings might have had in James's turn-of-the-century realism, by the end of the New Deal the problem was just the opposite: buildings had rights denied to the people. From Faulkner's Sutpen's Hundred to Chandler's office spaces to the tenement buildings of Chicago's South Side in Richard Wright's *Native Son*, New Deal novelists' borrowings from the gothic were meant to illustrate how far people had become things subjected to the laws of real estate.

Chandler's *Farewell, My Lovely* is one of a group of novels published in 1940 that mark this transformation and make plain the constellation of radio, realism, and real estate I have outlined above. Chandler's novel centers on a burlesque dancer in Los Angeles who tries to transform her life through radio but ends her life hoping someone will recognize her voice on a lonely stage in Baltimore. Radio, in this case, becomes the object through which the novel tests the borders of genre, gender, and nation. Tracing the path of Velma Valento from a burlesque dancer to radio broadcaster to (as Mrs. Grayle) the wealthy wife of a radio-station owner, the novel turns against the visual regimes of the realist novel and detective fiction to reveal instead how radio networks both threaten the bounds of real property and invisibly consolidate new landed power.

From the novel's opening chapter, real property and visual violence cohere. We begin with a homecoming of sorts, as the recently released convict Moose Malloy returns to find the bar where he had worked now under new management. More specifically, the neighborhood's racial demographics have changed from white to black. Furious with his feeling of displacement, Malloy spews racist epithets at the bar's clientele and then murders its new owner. Chandler sums up the extraordinary violence as an attempt to restore whiteness to the bar's proper name: *Florian's*. On the other hand, standing beneath the bar's neon sign, Malloy pointedly refuses to confer proper names on the neighborhood's African American residents. Calling them "shines," "smokes," or "dinges," Malloy represents a racist figural logic that transforms people into things based on the color of their skin. The racist murder at the novel's opening thus prefigures the problem with the whole *structure of seeing* that governs the detective genre: the synecdochal reading that moves from a physical attribute to a stereotype and classification.

The linguistic dexterity that critics have long celebrated in Chandler's writing depends on a similar logic. Observing Malloy walk down the street, the detective-narrator Philip Marlowe says he "looked about as inconspicuous as a tarantula on a slice of angel food."[31] Marlowe turns Malloy, already a "moose," into another animal here, and in doing so, modifies the scene's dynamics of color. The stark black-and-white image reverses the racial attribution of the scene, positioning blackness as threatening but associating Malloy with that blackness. Marlowe's language does not escape the dehumanizing violence Malloy practices. Whether commenting on that violence or sharing its ambitions, Marlowe watches Malloy attack one of the African American men in the neighborhood and repeatedly notes that Malloy kicked "it."[32] Indeed, Marlowe's depersonalizing language carries over to the other men in the bar, who, he says, "chant" and "chatter" but never speak.[33] Thus, the descriptive work in these opening chapters highlights Marlowe's verbal ingenuity as the linguistic corollary to the spectacle of racial violence that silences the African American population beneath the sign of Florian's.

The bar's neon sign reveals not only the transformation of real estate but also something of the archaeological permanence of real or immoveable property. While Malloy and Marlowe practice new forms of physical and visual violence in these scenes, they also recall Florian's previous incarnation as a burlesque bar. As I mentioned above, Velma Valento used to dance at Florian's, making her living under her patrons' gaze. Malloy, the bouncer meant to protect the bar, helped assure that gaze was untroubled; thus, he secured a different kind of power through looking that we later learn Velma felt as a particular kind of visual violence. Thus, Florian's becomes a topos through a palimpsest of scenes where looking does damage to bodies.

Although the novel's narrative voice often empathizes with this linguistic and physical power, the plot slowly begins to turn away from the visual regime through the invocation of radio. In doing so, it subtly shifts from the nineteenth-century realist codes built on things toward a newly architectural understanding of power. And yet, when we first encounter the radio in this novel it seems to stand for exactly the kind of metonymic signal Jameson attributes to Balzatian realism. Stepping inside the home of the late Mr. Florian's wife, Marlowe's gaze is drawn to "A large handsome cabinet radio [that] droned to the left of the door in the corner of the room. *It was the only decent piece of furniture the place had. It looked brand new.* Everything else was junk."[34] Since Mrs. Florian's neighbor identifies her as someone who "ain't neighborly. Plays her radio loud late nights. Sings. She don't talk to anybody,"[35] her investment in a new radio should not seem surprising. Moreover, in one of her less reticent moments, she speaks to Marlowe, and "her voice came from her mouth sounding like a worn-out phonograph record."[36] Chandler's rendering of Mrs. Florian goes beyond metonymy, as she comes to embody the technological reproduction of sound. The author's simile literalizes the link between the sonic playback and the woman's identity: recalling Dos Passos's description of Hoover, she seems more sonic technology than human subject.

As if to assure readers of this identification between Mrs. Florian and the radio, its silence eventually signals her death. Returning to the house later with his fellow detective Randall, Marlowe notices that "the radio was off."[37] The acoustic signal changes the atmosphere in the room, and yet, at first, it seems we have returned to the previous scene. Repeating the narrative's earlier appraisal, Randall comments: "That's a nice radio . . . Cost money. If it's paid for."[38] Then he plugs the radio in, and when its sound fills the room, the object's noise evokes a miniature narrative within the house's confines: "The light went on at once. We waited. The thing hummed for a while and then suddenly a heavy volume of sound began to pour out of the speaker. Randall jumped at the cord and *yanked it* loose again. The sound was *snapped off sharp*. When he straightened his eyes were full of light. We went swiftly into the bedroom. Mrs. Jessie Pierce Florian lay diagonally across her bed . . . She had been dead long enough."[39] Randall's actions with the radio mimic what is eventually determined to be Mrs. Florian's brutal death by strangulation, bringing woman and machine into a more intimate identification. While the object's initial silence and its sudden, overpowering liveliness help indicate the immediate circumstances of her death, the radio also goes on talking while Mrs. Florian lies lifeless; the scene perversely emblematizes Marx's commodity fetish. This seeming victory of the object world does not so much rewrite the earlier realist tradition as push it to its extreme: identified with the radio, Mrs. Florian herself becomes an object within the room.

As the narrative moves on, however, the "brand new" radio's peculiarity in this room points less to Mrs. Florian than to an entire system of subterfuge in Chandler's Los Angeles. The radio's presence indexes a network in which Mrs. Florian's murder by radio helps reveal the alternating dynamic between public, private, and transnational life in the broadcast voice. As the novel's climactic scene makes clear, Mrs. Florian was in fact blackmailing the former burlesque dancer Velma Valento, who left the late Mr. Florian's now defunct bar in order to work at radio KDFK in Beverly Hills, where she met and married the station's owner, Mr. Grayle.[40] In the expository style typical of detective story revelations, Marlowe describes this biography of Mrs. Grayle/Velma to Velma/Mrs. Grayle, explaining that "on the way up a shabby old woman recognized her—probably heard her singing at the radio station and recognized her voice and went to see—and this old woman had to be kept quiet."[41] The "shabby old woman" described here is Mrs. Florian, and to make sure she is "kept quiet" Mrs. Grayle enlists the help of another employee at KDFK, Mr. Lindsay Marriott.[42] More precisely, Mrs. Grayle, the now-wealthy wife of a radio-station owner, acts through Marriott to control Mrs. Florian through real estate: he "made her monthly payments and owned a trust deed on her home and could throw her into the gutter any time she got funny."[43] Thus, the initial realist marker, the radio, now indicates Mrs. Florian's main means of retaining her home. It is an emblem for the blackmail agreement that assures Mrs. Grayle will continue to pay her. Likewise, anxious to gain some leverage against the wireless that binds her to a former life, Velma/Mrs. Grayle uses real estate to counter the threat indexed by the radio. Lastly, in order to track down Moose Malloy and bring him to identify Velma, Marlowe heads to a gambling boat, with "two stumpy masts just high enough for a radio antenna."[44] Although called the *Montecito* in the novel, the ship is an obvious stand-in for the first "pirate radio" station, RKXR, based on board the SS *City of Panama*, which broadcast from "international waters" in 1933.[45] With these plot points and this context in mind, we can return to the brand-new cabinet radio, "the only decent piece of furniture" in Mrs. Florian's house, and understand the radio's association with its owner as a savvy realist red herring that also contains and organizes the layers of private, public, and transnational listening the radio network assembles.

In Chandler's novel, radio represents the means by which Velma sought to escape her former life and, at the same time, the medium through which Mrs. Florian heard Velma's voice and pulled her back toward that life. The public broadcast enables a breach of her private life, a breach that also threatens to transform her public image and personal identity, to change her name back from Mrs. Grayle to Velma Valento. Chandler's novel thus counters Arnheim's fantasy of the disembodied voice unencumbered by spatial difference, which ignores how a voice's acoustic markers function within social space. In Velma's case, her radio voice both releases and binds her to a social class: even as she sips martinis in her mansion with Mr. Grayle, she fears Mrs. Florian's listening will bring her back to her barroom days caught working under the eyes of men paying to watch her.[46] Ultimately, Chandler situates the radio between freedom from the visual regime typical of the detective genre, and the fear that the radio listener will restore that regime to power.

Carefully manipulating social and acoustic space, as well as the overlap between reading and overhearing, the climactic conversation between Marlowe and Velma/Mrs. Grayle occurs with a third party hidden out of sight. Marlowe welcomes Velma into his bedroom and claims: "We're all alone here. Nothing either of us says has the slightest standing against what the other says. We cancel each other out."[47] Yet, he knows that Moose Malloy, Velma's former lover, the ex-bouncer at Mr. Florian's bar, and the recent murderer of Mrs. Florian, can overhear the entire conversation from his position hidden in Marlowe's closet. Marlowe, in a feature typical of detective fiction, helps produce Velma's admission while pretending to listen to her alone. Against the possibility of social mobility through the networked voice we encountered in the last chapter, the sound of Velma's voice—specifically, Marlowe's recognition of that voice—threatens to limit her social range and return her to an immobile position on the burlesque stage. As with her broadcasts on KDFK, however, the pretense at anonymous speech ignores the overhearing audience. As occurs repeatedly in Chandler's novels, in this scene of apparently intimate conversation, Marlowe's bedroom becomes his office space.[48] More precisely, the room functions as a microcosm of the radio experience, bringing the broadcast booth and the invisible listening audience all into a single space.

For a brief moment, the reader is also an eavesdropper, brought into the closet with Malloy and forced to inhabit the character's space as we listen along with him.[49] When Moose emerges from the closet and speaks, leaving the reader behind, his words and his acoustic recognition imply the same threat posed by Mrs. Florian: "'I thought I knew the voice,' he said. 'I listened to that voice for eight years—all I could remember of it . . . *You* turned me into the cops. *You*. Little Velma.'"[50] Faced with this hulking reminder of her past, with the anonymous listener stepping into the broadcast booth, Mrs. Grayle shoots and kills Moose at the very moment he pronounces her former name.

But the novel's coda also makes impossible the simple conclusion that voice is an inexpungible marker of identity. In an ending that reads almost like a eulogy for her radio voice, a police officer finds Velma three months later in the dressing room of a Baltimore nightclub where she sings. As she admits to the cop before taking his life and then her own: "I thought I had a voice that would be remembered. A friend recognized me by it once, just hearing it on the radio. But I've been singing with this band for a month—twice a week on a network—and nobody gave it a thought."[51] Evading the simple transposition of object and voice, Chandler's narrative resists the phonocentric maneuver of replacing the thing as a marker of one's identity with the voice. As the policeman responds to Velma, "'I never heard the voice.'"[52] Rather, in moving from one coast to another, Velma realizes the lost links in "the national hookup," radio's regionalism, and the social and geographical limits of her radio voice. Having finally escaped her past, she nostalgically turns to the radio half hoping for a connection to be heard. And the radio, the object first established as the key to one of the novel's realist interiors, becomes instead a tool for exploring the scope and limits of recognition and self-transformation in the nonvisual medium of sound.

Writing in the radio age, at a time when, as Emily Thompson's history of the New Acoustics shows, buildings already speak a particular electronic language, Chandler invokes an alternate narrative acoustics, modified in the aftermath of the New Acoustics, wherein the project to make space heard and felt attains a political as well as aesthetic purpose. By the late 1930s, it was no longer enough to give voice to buildings, as in the gothic novel or the later narratives of Henry James. While in James such writing might remain a "thought experiment," for the writers at the end of the New Deal the new rights of buildings and the potential to change the face of "housing" becomes a central concern. From the projected rights of buildings in Bill Brown's reading of James to the acousmatic voice of Chicago Stadium described by Dos Passos, to the buildings that become characters in Jameson's reading of Chandler, the New Deal realists of 1940 should be understood as sketching the transformation of real property as it slowly gains a voice and gains possession of itself. Notably, for Chandler, or for the Dos Passos who comments that "buildings determine civilization as the cells in the honeycomb the functions of bees," it is the radio that talks back, the buzzing signal that challenges the inert furnishings of realism, emptying the room of things and filling it, instead, with sounds and voices that give the buildings a sense of animation, helps render them *actants*. It is radio that seems to speak through walls, thus challenging the notion of real property, at the same time that it seems to convert listeners into objects. It is radio, in other words, that pricks up the ears of New Deal novelists and enables them to rethink the political connection between generic codes, acoustics, housing, and social space.

HOUSING PROBLEMS: NEW DEAL PUBLIC NEIGHBORS

As I explained earlier, the housing problem in these novelists' work with radio and narrative form pertains to a fundamental change in thinking around public and private space in the United States during the New Deal years. While I described Brown's and Jameson's comments about buildings gaining rights and assuming the status of characters in mainly aesthetic terms, the new building codes and building projects of the Housing Acts of 1934 and 1937 also created new legal rights for housing under the watch of USHA. As Lawrence Vale has described this period, "the stormy history of public housing may . . . be seen as the confluence between the *public neighbor* and the new practice of the *public neighborhood*."[53] For Vale, the category of the "public neighbor" derives from the residents of English almshouses, "needy people whom community leaders felt an obligation to assist or, since the early nineteenth century, to reform."[54] The public neighborhood, initially meant to house working citizens, soon took on the identity of this older social type, partially due to the fact that, in Vale's words, "there is a problem when one's neighbors become seen as 'the public' rather than as a specific set of individuals and families of known qualities."[55] Soon after their creation, public houses stigmatized their inhabitants, and much like residence in a hospital indicates illness or incarceration in a prison bespeaks criminality, residents of public neighborhoods found their identities chosen for them by the houses in which they lived. In this sense also, buildings became "actants," determining the identity of their inhabitants.

It is into this dense network of public housing, realist codes, and radio that Mick Kelly steps as she crosses the threshold of a newly finished construction site and enters an empty building in Carson McCullers's *The Heart Is a Lonely Hunter* (1940). Away for a moment from the bustle of her parents' house, which they run as a boardinghouse to help pay the mortgage, she enters

> the new, empty house . . .
>
> The rooms smelled of new wood, and when she walked the soles of her tennis shoes made a flopping sound that echoed through all the house. The air was hot and quiet. She stood still in the middle of the front room for a while, and then she suddenly thought of something . . .
>
> . . . Mick drew the big block letters very slowly. At the top she wrote EDISON, and under that she drew the names of DICK TRACY and MUSSOLINI. Then in each corner with the largest letters of all, made with green and outlined in red, she wrote her initials—M.K. . . .
>
> . . . She stood in the empty room and stared at what she had done. The chalk was still in her hands and she did not feel really satisfied. She was trying to think of the name of this fellow who had written this music she heard over the radio last winter . . .
>
> . . . She hummed one of the tunes, and after a while in the hot, empty house by herself she felt the tears come in her eyes. Her

> throat got tight and rough and she couldn't sing any more. Quickly she wrote the fellow's name at the very top of the list—MOTSART.[56]

In this "new, empty house" Mick enters into the quite literal "construction of a vacancy," the unfurnished room made iconic in Willa Cather's call—republished in 1936—to her fellow realist novelists: "How wonderful it would be if we could throw all the furniture out of the window."[57] The acoustic resonance of Mick's steps echoing "through all the house" slowly shifts into inscription, as Mick attends to the walls, turning her radio listening into writing as she composes a strange but semantically charged list: MOTSART, EDISON, DICK TRACY, MUSSOLINI. The phonetic spelling of Mozart, the name and music "*heard over the radio*," supplies the acoustic and narrative organization to these proper names. Mozart is heard on the radio; Edison's phonograph provides the material support for the radio; Dick Tracy was perhaps the most famous radio detective; and Mussolini notoriously employed the radio to consolidate popular support for his fascist government.[58] Similar to, if perhaps slightly more oblique than, the narrative acoustics practiced in Dos Passos's writing, radio emerges here as the absent signifier that structures this diverse semiotic chain.[59] The chosen proper names point to radio's acoustic property: namely, its apparent transcendence of property in the open system of address that is *broadcast*. Thus, what begins as sound in empty space becomes a translation of the exhausted voice—"she couldn't sing any more"—into a scene of writing radio on blank walls; she writes to reclaim, in some minor sense, private "real" property for common use. Mick's graffiti, her word art ("mots art"), in other words, emphatically inscribes the coincidence of radio, emptied realism, and the critique of property that I have argued characterizes New Deal acoustic narrative.[60]

In dispensing with furnishing for radio graffiti, this scene literalizes the "radio aesthetic" I associated with Chandler's writing, and the increasingly fluid borders between public and private spaces in New Deal narratives that take up the radio. If, as Brown argues, Henry James's enduring achievement was to test "the limits of realism by evacuating the genre of . . . its material possessions" and move the novel toward a protomodernist attachment to things,[61] and especially buildings, that speak, McCullers might be thought to pick up where James left off. However, she writes with a difference, in that she presents us with a building not yet furnished by any single owner but nevertheless marked everywhere by transient subjects. These markings, what Susan Stewart might call "crimes of writing," enact a very different "non-proprietary possession" than that which Brown argues for on behalf of James. As each "writer" in McCullers's building marks their place they do not so much take possession of the real estate but challenge the claim to "real" ownership and worry little about "personal" property. Not quite or not yet a domestic interior, the new, empty house brings the domestic space over into a more public, or at least contested and unstable, arena. While posted signs attempt to prevent children from entering, preserving the space's sense of private ownership, a notably free indirect discourse reports that "some tough boys wee-weed all over one of the walls and wrote some pretty bad words. But no matter how many Keep Out signs were put up, they couldn't run kids away until the house had been painted and finished and people had moved in."[62] The boys' territorial markings claim transient ownership in this unfinished space, an open territory to which Mick adds her own personal writing. Likewise, the narrative style in these sentences marks the public/private dialectic; the very terms that describe the boys marking their territory ("wee-weed," "pretty bad words") open the discursive space to the neither wholly interior nor wholly exterior movement of free indirect discourse, oscillating between Mick's adolescent speech and the narrator's register.

If the scene's discursive strategies link the idea of the radio network to a challenge to the borders of real estate, it should not be seen as an uncritical endorsement of radio's social possibilities or a simplistic embrace of a shared public sphere. Indeed, Mick's radio listening not only challenges property lines between neighbors—as when she perches on a tree "in the rich parts of town [where] every house had a radio," or sits on the stairs of her parents' boardinghouse "listening to Miss Brown's radio on the floor below"—but helps her distinguish between what she refers to as her "inside" and "outside" rooms.[63] Those things she encounters over the radio, "foreign countries and plans and music were in the inside room . . . [It] was a very private place. She could be in the middle of a house full of people and still feel like she was locked up by herself."[64] If this intimate reception of radio, which sculpts a private space within her crowded boardinghouse, appears to combine with her scenes of auditory transgression or "accidental audition" to recall what Loviglio names "radio's intimate public," they depart from that form's national ambitions.[65] Radio, in other words, helps Mick develop a foreign interiority. And, in a turn that helps return us to the transnational good neighborhood Welles developed in *Hello Americans*, when Mick comments that "Mister Singer was in both rooms,"[66] the narrative links her "private place" to the novel's central critique of New Deal radio politics.

A deaf and mute man who rents a room in Mick's parents' boardinghouse, Mr. Singer comes to embody radiophonic populism in McCullers's novel. Early in the narrative Mick admits to herself, "Maybe it was true that she came up on these top steps sometimes so she could see Mister Singer while she was listening to Miss Brown's radio on the floor below."[67] When Singer eventually purchases a radio, Mick goes to his room where "[s]he seemed to listen all over to whatever it was she heard . . . She asked him if she could come in and listen sometimes when he was at work and he nodded yes. So for the next few days whenever he opened the door he found her by the radio."[68] Singer, himself unable to hear, sensitizes Mick to the physical sensuousness of listening: "she seemed to listen all over." These accidental, contingent lessons in listening help Mick experience her contact with the world, as well as the development of her own "inside room." Thus, it is particularly devastating when, at the novel's close, she enters Singer's room to listen to the radio and finds that he has killed himself. Shocked at this sudden encounter with death, she reaches out to a single object to move forward: "She did have Mister Singer's radio. All the installments hadn't been paid and she took on the responsibility. It was good to have something that had belonged to him."[69] Through the metonymy of Mick's dual desire, Singer and the radio come to mirror each other: the one produces discourse without listening; the other listens without speaking.

Readers familiar with recent writings in media history and disability studies should not be surprised to encounter a deaf man metonymically related to radio. After all, radio's antecedent technologies—the telephone and the phonograph—developed through experiments with deaf subjects, beginning with the studies of Alexander Graham Bell's father and continuing on through Bell and Thomas Edison.[70] Important for the alternative they offer to media history's infatuation with war or capitalism as the only motivating forces behind technological invention, these media histories also help us recognize how McCullers positions Singer in a network that comments critically on the specific political functions they took up during the New Deal. Indeed, we might say that Singer embodies the problem of feedback that Hoover and Brecht complained about early in radio's history. As a listener—a lip reader—who does not speak, Singer foregrounds the problems associated with radio's one-way street of communication. Perhaps more surprisingly, he also comes to represent the radio voice.

Mick is not the only visitor who comes to Singer's room. However, if Mick spends most of her time there listening to the radio, the others speak over the radio, using Singer as a sounding board. Jake Blount, the alcoholic union organizer; Biff Brannon, the cross-dressing barman; Doctor Copeland, the avowedly Marxist African American doctor; and Mick, the adolescent tomboy each "would come and talk in the silent room—for they felt that the mute would always understand whatever they wanted to say to him. And maybe even more than that."[71] Unable to talk back, Singer can only listen and record his guests' various problems and confessions in his thoughts. Attempting to explain the book to readers of the *New Republic*, Richard Wright wrote, "the nearest I can come to indicating its theme is to refer to the Catholic confessional or the private office of the psychoanalyst."[72] Fittingly, in Singer's silence the other characters encounter consolation, as they imagine he "would always understand."[73]

However, if Singer seemed a priest or psychoanalyst to Wright, the novel situates him more closely to the populist ideal of the New Deal, an empathetic everyman with whom everyone can identify. In a passage that seems lifted from the tradition of folk storytelling, rumors spread, and soon

> [t]he Jews said that he was a Jew. The merchants along the main street claimed he had received a large legacy and was a very rich man. It was whispered in one browbeaten textile union that the mute was an organizer for the C.I.O. A lone Turk who had roamed into the town years ago and who languished with his family behind the little store where they sold linens claimed passionately to his wife that the mute was Turkish . . . The rich thought that he was rich and the poor considered him a poor man like themselves. And as there was no way to disprove these rumors they grew marvelous and very real. Each man described the mute as he wished him to be . . . Owing to the fact that he was a mute they were able to give him all the qualities they wanted him to have.[74]

Singer's silent status as a blank slate ties together an entire community of particulars who can see themselves in him, a single entity, because they *hear* nothing different. An inversion of the disembodied voice, the embodied mute mimics the populist utopianism radio theorists have claimed for the broadcast voice: the neutral voice upon which anyone can project their desires.

With this realization in mind, we might hear Singer's intimate "conversations" with the other characters as a reaudition of the "fireside chat," especially when we learn that "for all of them together he had bought a radio and put it on the table by the window."[75] However, if the narrative's earlier sections with Mick associate radio with new notions of communal space, to read Singer as an ironic Roosevelt, or any ideal populist radio orator, reveals radio's limits, and the fundamental conflict in the project to imagine "the people." For just after Singer purchases the radio as a Christmas present for his individual friends, chaos erupts in his room:

> One night soon after Christmas all four of the people chanced to visit him at the same time. This had never happened before. Singer moved about the room with smiles and refreshments and did his best in the way of politeness to make his guests comfortable. But something was wrong . . . Each person addressed his words mainly to the mute. Their thoughts seemed to converge in him as the spokes of a wheel lead to the center hub . . . Each one seemed to be waiting for the others to go. Then on an impulse they all rose to leave at the same time. Doctor Copeland went first and the others followed him immediately. When they were gone Singer stood alone in the room.[76]

More emphatically than anywhere else in the narrative, Singer appears to embody here the near perfect inversion of the broadcast voice: the single channel converging outward to numerous particulars. The failure when these various listeners/speakers come into contact indexes the medium's fractured capacity for collectivity. Singer's awkward Christmas party gives the lie to the utopian imaginings in Loviglio's "intimate public." Like Velma's "de-acousmatization" in Chandler's *Farewell, My Lovely*, McCullers's novel teaches the power of radio's *invisible speech*. It reveals the advantages in Mick's listening without seeing, and the "humble self-enlargement" she experiences as radio inspires her to write on walls.

In Mick and Singer, as well as the rest of the narrative's ensemble cast, McCullers sounds out the unfinished business of the New Deal as the country once again turned toward war. The empty house, in Cather and Chandler, but also in Faulkner's *Absalom! Absalom!* (1936) and

James Agee and Walker Evans's *Let Us Now Praise Famous Men* (1940), shakes with the voices on the other side of the radio dial and tests out the terrible, if sometimes liberating, spaces between them. The boardinghouses and homeless nomads in McCullers's work testify to a housing problem that radio's new public spaces in the ether only obliquely begin to answer. And in Singer, the mute who listens, we confront something missing in Roosevelt's radio voice and his role as the great communicator. Thus, these realist writers, attuned as they are to literature itself as part of a sonic media system, uncover radio's lack of a *listening technology*, a correlative to the various speaking technologies, those devices that help sever speaking from listening. The communicative blockade induced by Singer's muteness, and ultimately responsible for his suicide, points to the frustrations of the permanent listener and gestures toward the darker side of a reifying culture increasingly attuned to things that talk. On the other hand, in Mick we also find a radical listener: a character who tunes in to broadcast's disruption of property and learns a writing practice based on listening (MOTSART: "word art") in order to write against real estate. At the end of the New Deal, McCullers's novel gives a new voice to democratic personality, and more importantly, a new ear ready to listen through walls to hear other voices in other rooms intent on sharing a struggle for a public housing more attuned to the rights of its inhabitants than the rights of buildings.

CHAPTER 3

Struggling Words

Public Housing, Sound Technologies, and the Position of Speech

Level it down / for him to build a house / on to build a / house on to build a house on / to build a house / on to build a house on to . . .

—W. C. WILLIAMS, "The Defective Record"

The same year Carson McCullers published *The Heart Is a Lonely Hunter* she moved into a boardinghouse on 7 Middagh Street in Brooklyn Heights owned by her friend, *Harper's Bazaar* editor George Davis.[1] Her fellow boarders in the building Anaïs Nin later christened "February House," eventually included W. H. Auden, Benjamin Britten, "Gypsy" Rose Lee, Christopher Isherwood, Louis MacNeice, Jane and Paul Bowles, and, in 1942, another young writer who published his debut novel the same year as McCullers: Richard Wright.[2]

In August 1940 Wright had written a review of McCullers's novel in the *New Republic*, praising the author for her ability, "for the first time in Southern fiction, to handle Negro characters with as much ease and justice as those of her own race."[3] In response, George Davis invited Wright to Thanksgiving dinner. Although Wright would not move into the house until two years later, the Irish poet Paul Muldoon imaginatively conflates Wright's later stay in the house with the 1940 dinner in a section, "Carson," from his long poem "7 Middagh Street":

> I had just left Reeves / and needed a place to stay. As must
> Wynstan, // dear Wynnie-Pooh, who's given to caution / the rest
> of us every / time we sit down, be it to jerky / or this afternoon's
> Thanksgiving dinner, every / blessed time, "We'll have crawfish,
> turkey, / salad and savoury, / and no political discussion"— //
> a form of grace / that would surely have raised an eyebrow / at

> even the Last Supper, / never mind a household where no time ago / when the Richard Wrights moved in the super / moved out, unwilling, it seemed to draw and hew / and tend the furnace for fellow Negroes.[4]

Historically inaccurate but poetically biting, Muldoon's poem ventriloquizes McCullers to parody Wynstan Hugh (W. H.) Auden's noted aesthetic autonomy. Against Auden's attempt to suppress "political discussion," Muldoon's Carson depicts the pressures of race and real estate in the boardinghouse. Recognizing the exigent living conditions at Middagh Street—the roomers "*needed* a place to stay"—Carson's narrative voice mocks a "grace" whose blind decorum would refuse to recognize how others' access to the domestic space—the Last "Supper"—is threatened by the discrimination of a "super" who would turn his "fellow Negroes" out into the cold. The heavy echo of "supper" in "super," as well as the latter's refusal to "tend the furnace," directs attention to the connection between real estate, power, and the basic survival offered or denied by the metaphor of the hearth. While Muldoon again bends history in the last line—the historical super was reportedly upset that the "Richard Wrights" were a biracial couple—the poem underscores the precarious housing situation linked to racial tensions at the end of the New Deal: we find the notion of the "good neighbor" particularly soured when one's "fellow" turns one away on racial grounds.[5]

"Carson," the poem, unites McCullers and Wright through a shared concern with housing problems, in a historical boardinghouse where they would become neighbors and friends. This affiliation carried into their writing. Recognizing the formative role of housing and segregation in both authors' works, Julieann Veronica Ulin claims McCullers's 1961 novel *Clock without Hands* "offers a tribute to Wright and to his struggle to find a home in the United States."[6] Yet, one does not have to wait until the 1960s to find both authors taking on the housing crises of the New Deal. The writers' mutual understanding of the problem of fair housing as the problem of democracy already permeates their debut novels from 1940: McCullers's *A Heart Is a Lonely Hunter* and Wright's *Native Son*.[7] Just as Mick Kelly enters the vacant building in McCullers's novel to reclaim private property for shared use through her "radio graffiti," Bigger Thomas, the antihero of Wright's novel, occupies abandoned tenements in Chicago's segregated "Black Belt" real estate district, and "fight[s] to re-enter those buildings, to live again."[8] And as we will see later, both apply lessons learned from listening to sonic technology to fine-tune their challenges to the power of buildings. In the wake of Dos Passos, who understood himself as a "frustrated architect," Wright and McCullers read more like activist urban planners, anxious to create a space in literature their characters can equitably inhabit. Their novels struggle on behalf of citizens' rights to housing and against a commitment to uphold the borders of real property that would raise the rights of buildings over the rights of their potential inhabitants.

More than any other New Deal fiction writer engaging real estate as the intersection of a narrative problem and a social crisis, Wright immersed himself in the public policy, urban planning, and architectural concerns that drove legislative debates. In 1938 and 1939, after setting aside his first, unpublished novel, *Lawd Today!*, and in the midst of writing *Native Son*, Wright wrote articles on housing and rent strikes for the *Daily Worker* and went to work for the Federal Writers' Project in New York City, composing the "Portrait of Harlem" for the *New York Panorama* (1938) and contributing to sections on Harlem and the Harlem River Houses for the Federal Writers' Project's *New York City Guide* (1939).[9] For Wright, these historical and sociological surveys deepened his ability to render the social space created in the confluence of architecture, urban planning, and real estate. According to J. J. Butts, work on the guidebooks immersed Wright "in the exploration and representation of urban space," leading him and his fellow authors to build "concrete examples of the politics of New Deal spatial representation, showing us how rhetorical battles over particular spaces contributed to national debates over community and justice."[10] This training especially affected Wright, whose correspondence, essays, journalism, stories, and novels throughout his career adamantly return to the theme of real estate and race, deploying a sophisticated knowledge of spatial representation to contribute to the ongoing struggle to build, define, and defend unbiased public housing in the United States.[11]

Wright contributed to these linguistic and legal battles in his work for the 1939 federal *Guide*, which situates Harlem as part of a New Deal progress narrative where federal housing projects join with cultural community to create a concrete response to a history of injustice.[12] Describing the establishment of what it calls "Negro Harlem" as an African American neighborhood, the guidebook states, "Though they did not succeed in escaping economic insecurity and race discrimination, the Negroes were able to build within the metropolis a city of their own, a cosmopolitan Negro capital which exerts an influence over Negroes everywhere."[13] In this city within a city, the guide singles out the Harlem River Houses as the key monument, dedicating a chapter to the buildings among "Major Points of Interest," alongside Rockefeller Center and the Museum of Modern Art. The guide describes the apartment complex, the first public housing project of its kind in the United States, as the architectural response to a social need: "Built in 1937 by the Federal Administration of Public Works, Housing Division, the project is a recognition in brick and mortar of the special and urgent needs of Harlem, and the first large-scale modern housing community made available for low-income Manhattan residents at rents they can afford."[14] The Harlem River Houses emerge here as a concrete utopia, a hopeful alternative to the real estate conditions in the privatized and ghettoized "Black Belts" Wright would condemn in *Native Son* just a year later.[15] As I will demonstrate in greater detail below, these historical conditions reveal that novel's attack on real estate as a powerful cry for public housing. In broader terms, when read in relation to Wright's contemporary and ongoing fictional work, the article on the Harlem River Houses serves as the instantiation of a belief in Bauer's "modern housing" and a reminder of the hopeful construction that could arise from a novel's well-leveled critique.

While Wright would also criticize the failures he saw in public housing, the critical edge in his writing complements the work of the New Deal's most prominent urban planners. The writings of Lewis Mumford—quoted in the 1939 guide for praising the Harlem River Houses as "superior to any comparable area of residential apartments in the city"[16]—along with Catherine Bauer's *Modern Housing* (1934) were fundamental in representing public housing to a public unsure what such housing would mean. Additionally, Charles Abrams, who headed the New York City Housing Authority from 1934 to 1937, argued that public housing—the very name of which invokes the social life of buildings—depended as much on language and representation as it did on building materials and blueprints. New Deal planners and New Deal novelists like Wright converged in a mutual ambition to deploy a new design style, on one hand, or a new narrative style, on the other, to effect a change in the personal values of a building's inhabitants or a book's readers.[17] Anticipating the work of Henri Lefebvre, David Harvey, and other cultural geographers and theorists of social space, both the planners and novelists sought to engage the dynamic interaction of language, place, policy, and shelter.[18] Both Butts's and Ulin's readings affirm the complementary relationship between Wright texts like *12 Million Black Voices* or his unpublished novel "Black Hope," and the planners' works, such as Abrams's *Race Bias in Housing* or his later *Forbidden Neighbors: A Study of Prejudice in Housing*.[19] When the Harlem River Houses opened in 1937, these mutual ambitions converged: planners like Abrams fought Robert Moses over the definition of "slum"; Bauer and others struggled against private interests challenging whether government-funded housing at affordable prices fell under the rubric of a "public purpose"; and Wright, the novelist, issued a "*Blueprint* for Negro Writing" in which he declared, "Surely this is the moment to ask questions, to theorize, to speculate, to wonder of what materials can a human world be built."[20]

In this context where housing needed language and literature required a blueprint, Wright's fictions go on to reveal not just language in general, but the voice and vocal technologies like radio and the gramophone, as the primary materials mediating speech and architecture in his critique of race and real estate. In even greater detail than his contemporary novelists Chandler and McCullers, Wright's work fuses policy debates about housing with the technical innovations and cultural histories of sound technology to reimagine the borders of real property through a sophisticated mediation of the voice in print. The confluence of housing and technology in this case renovates the fictional world, just as those fictions help transform that "human world [to] be built."

More specifically, across a number of works from 1936 to 1940, Wright joins these concerns in closing scenes that depict the human voice as intricately connected to real property. Both *Lawd Today!* (1938/1963), which structures its story around a radio broadcast, and "Long Black Song" (1938), whose plot hinges on a graphophone recording, end with female voices lamenting their husbands' destruction of their real property in response to white power. *Native Son* (1940), in a revision of this template, closes with a struggle to share a voice in a dialogue about the ambition for public housing across race and class lines. Returning to, and reworking the intersection of the voice, technology, and real property in these years, Wright's compositions struggle to bring the advancements in sound technology to bear on the literary, legal, and social legacy of African Americans' historical status as voiceless property.

In order to respond to the ongoing influence of this racist history, Wright, I argue, takes up different sonic technological media—the radio in *Lawd Today!*, the graphophone in "Long Black Song," and the voice-over in *Native Son*—to explore how alternative strategies of mediating the voice might reconfigure racial property lines in the history of U.S. housing and literature. The transition through this media ecology tells a specific story about voice, race, and property in which Wright searches for a nonessentialist form of speech and identity to build toward a collective community.

Radio, which has been the model for this collectivity in the previous chapters, presents impassible problems for Wright's narrative project. Unlike in the fiction of his contemporaries, in Wright's work radio exclusively bears the voice of white authority. While it remains a powerful emblem of the disruption of public and private space—as in Chandler's and McCullers's novels—the utopian *communitas* of that disruption is replaced by a sense of invasive subjection and alienation. What was the "speech of the people" in Dos Passos sounds only like the voice of white power in Wright's fiction.

While the trilogy of terms—radio, realism, and real estate—that governed chapter 2 remain important here, radio ultimately fails to serve Wright's critique of housing. The gramophone, graphophone, and similar sonic recording technologies, on the other hand, uniquely carry the history of the "fugitive" black voice and the imbrication of sound technology and slavery that Wright needs to tell the story of race and real estate.[21] In an argument to which I will return, Stephen Best has explained how the copyright of human voice recordings borrowed its legal logic from the Fugitive Slave Act's definition of a human self as an alienable quality of (inalienable) human *being*. The result of this association has turned the gramophone and attendant technologies like the graphophone, in U.S. legal history, into examples of new forms and domains of property—such as the ownership of voice—strongly linked to slavery's transformation of the human person into a commodity to be owned. Wright brings these peculiar notions of property and personhood to the doorstep of real property in "Long Black Song," where he finds a medium more historically appropriate than radio to represent the history of racial bias in housing through the voice.

Finally, in *Native Son*, an uneven synthesis of these media—one linked to Standard English and white power, the other to the black vernacular—opens the possibility of a shared voice united with the project of shared housing. Voice-over, which borrows from radio and early sound recording, models the potential for characters to assume other voices. Although rarely mentioned in the text, the technology enters into the novel's discursive form in an attempt to share an utterance across a racial divide without adopting a color-blind approach that would conceal lived contradictions and erase a past of racial violence. Having challenged literary segregation by transforming the fugitive slave narrative—the apparent *urtext* of African American literature—into the antecedent for the period's popular genre of noir, Wright invokes the "fugitive" voice as the potential mode to imagine a means to respond to the ways in which real, personal, and human property segregate possible communities along racial lines.[22] That this experiment fails testifies to the unfinished character of this project in fiction and everyday life alike. Yet Wright's texts do more than merely point to the continued absence of racial equality in the democratic project. Among the slums and ruins of the Great Depression's landscape, Wright leaves behind a construction site on which later generations may build.

THE NATIONAL HOOKUP (A SLIGHT REPRISE)

Wright's first attempt to confront these issues in his fiction appears in his posthumously published novel *Lawd Today!*, which he began in 1935 with the working title "Cesspool"—invoking the slum conditions in Chicago and New York—and set aside in 1938. The book's final scene features Jake Jackson, the misogynistic and abusive mail sorter at the novel's center, as he returns home drunk and wrathful and throws an ornamental glass elephant through the house's front window. While his wife, Lil, cowers in the corner and begs him to lower his voice for fear of disturbing the neighbors, he shouts, "GAWDDAMMIT, I WANT 'EM TO HEAR!"[23] His rage increases, but rather than the visual cliché of "seeing red," the narrator informs us, "There was a heavy deafness in his ears and he had to shout in order to hear himself. 'LET 'EM HEAR! LET 'EM HEAR!'"[24] Deaf with rage, he corners Lil, who, in an act of desperate self-defense, stabs him with a shard of the broken windowpane: "Her voice was high, hysterical, in one breath, on a dead level."[25] Sobbing, she looks on at her apparently unconscious, and perhaps dead husband as "darkness roared in his brain."[26] The windows destroyed, and the house's interior now open to the world, the final sentence listens to the Chicago winter's most notorious sound: "Outside an icy wind swept around the corner of the building, whining and moaning like an idiot in a deep black pit."[27]

Recognizing the racial politics in this operatic denouement—which will echo throughout the endings of "Long Black Song" and *Native Son*—depends on tuning in to the novel's extraordinary orchestration of voices. Listening to the rest of Jake's day in this noisy novel, we hear a street-corner preacher, the songs of a "back to Africa" parade, the disembodied voice of a post office foreman, the mechanical call-and-response between mail workers and their boss, the friends' collective and shared speech in the post office, and the musical shouts in the brothel Jake visits just before he returns home to Lil. Among this polyphony, however, one voice seems to structure the meaning of the others: the radio voice.

When Jake shouts at Lil in capital letters, his voice looks like the radio broadcast that opens the novel. The day's broadcast, from "STATION WGN, TRIBUNE SQUARE, CHICAGO," focuses on a single subject: "THIS IS FEBRUARY TWELFTH, ABRAHAM LINCOLN'S BIRTHDAY!"[28] From the moment Jake awakes with this voice infiltrating his dream as a voice calling for him "from the top of the steps"—a voice he says that "*sounds just like my boss*"[29]—the radio voice establishes itself as the voice of authority. Recurring over the course of Jake's day, the Lincoln broadcast opens each of the book's three sections and makes up part of the soundscape in each chapter. At various moments the broadcast implicitly comments on the novel's action. When Jake and his friends pass time at a drugstore before heading to work at the post office, the radio sounds like a scolding conscience: "AND THE WHOLE NIGHT THROUGH LINCOLN PACED WITH HIS HANDS KNOTTED BEHIND HIS BACK, MUMBLING MOURNFULLY: WHY DON'T THEY COME?"[30] Similarly, as the men eat heartily at the brothel, the radio blares, "NOW SHERMAN'S TROOPS . . . WERE ROBBING IT OF EVERY EDIBLE THING."[31] And, in a different register, when the group watches the "Onward to Africa" parade, "From their rear came the sound of a radio: 'WE ARE COMING, FATHER ABRAHAM, THREE HUNDRED THOUSAND MORE.'"[32] This last invocation of a song originally published in response to Lincoln's call for Union volunteers, transforms the lyrics' original meaning as well as that of the parade. Heard against the radio's historical reminder, the parade sounds out the ongoing history of the Great Migration: one legacy of the Civil War, and a testament to that war's failure to fully emancipate African American citizens from social inequality on racial lines. Furthermore, the parade's Garvey-influenced theme—encouraging African Americans to emigrate to Africa—changes the destination for these "thee hundred thousand more" from the northern United States to Africa. Not only does the radio voice's added sonic layer provide an additional narrative perspective, it invades and comments on Jake's life without leaving any space for him to respond.

More than just an ironic note in the novel's already busy soundscape, the capitalized radio discourse takes on a more specific meaning when mimicked elsewhere in the text. When Jake and his friends arrive at the post office where they sort the nation's mail, "a voice" commands, "LINE UP FOR DETAIL."[33] Throughout the rest of the chapter, this voice remains unassigned to any particular body. Found in the second of four chapters that make up the novel's second section, "The Squirrel Cage," this disembodied voice emerges after Jake's confrontation with the post office bosses in the section's first chapter, the mechanical and military-like call-and-response between the foreman and the workers in chapter 3, and Jake and his friends' chorus of unattributed voices in chapter 4. The capitalized and disembodied voice uttering commands—the emblem of the acousmatic voice of authority—represents the unification of the bosses and foreman that bookend it.[34] Its monologic form stands in opposition to the series of unattributed voices in dialogue among Jake and his colleagues meant to represent, as the narrator puts it, that the workers' "common feelings were a common knowledge."[35] Yet, lest one interpret this latter form of unified dialogue as an optimistic antidote to mechanical labor in lockstep obedience to a voice of authority, the narrator describes the workers' voices in ominous tones, "like dead bodies floating swollen upon a night sea."[36] Rather than a liberating alternative, even this chorus ultimately sounds like the fatal sound effect of forced labor.

Against Jake and his friends' voices, the radio voice continues to dominate the soundscape. In Jonah Willihnganz's pitch-perfect reading of the novel this capitalized voice of authority resounds as the voice of radio and the voice of white power. Pointing out how the novel aligns "radio and the black experience of white power," Willihnganz definitively links the voice of Jake's white employers with the radio voice's disembodiment, call for obedience, and compulsion to endless labor.[37] It is especially important that this voice emerges in the post office—the center of the nation's oldest information network—and describes, Willihnganz argues, Jake's lack of "power to engage the voices that hail him or that he serves."[38] Thus, the radio voice stands as a reminder of Jake's exclusion from the national network of communication and power.

In contrast to the radio, the novel's other voices, such as the post-office dialogue described above, stand in for a folk vernacular. However, as Willihnganz rightly observes, they always remain circumscribed and outside the system of power. At the post office "their dialogue is observed almost anthropologically by the whites, as a curiosity, but in no way does their dialogue challenge the whites."[39] The careful orchestration of the structural difference between the voices of African American characters and the radio voice associated with white voices returns us to the novel's closing scene.

This work to represent power also corresponds to Wright's sense of print as a specific sonic medium. When we look again at Jake's capitalized demand to be heard, we can now hear this typographic shift as an ironic sign of his vocal disempowerment, as well as his attempt to step into a position of power and violence previously associated with white authority and the radio. On one level, the capitalized voice ironizes Jake's demand to be heard. He speaks with the volume of that voice, but without its effect and without its tacit assumption that everyone is already listening. Despite Lil's worries about the neighbors, there is no sign that his voice carries outside the house. His mimicry only further underscores his difference from the radio voice and his exclusion from the network of power it represents.[40] In this impotent declaration, Wright sounds the key to which he will return in his other works from this period.

On the other hand, the capitalized voice wreaks a physical destruction that literalizes the structural violence aligned with the voice of white power elsewhere in the novel. In a gesture that will also be repeated in Wright's other fictions, Jake's "struggle for self-sovereignty" not only results in his own death but also depends on his brutal attempt to silence his wife, Lil. Although Lil escapes with her life—barely more fortunate than other female characters in Wright's oeuvre—she remains little more than a sound without speech ("her voice was high, hysterical").[41] With Jake most likely dead, and Lil whimpering beside him in a house reduced to little more than its frame and exposed to the fatal cold of Chicago's winter, the capitalized voice that at first appeared ironic now sounds sincerely deadly.

Looking at this house undone by the radio voice's embodied invasion, one can hear a microcosm of New Deal literary acoustics. When the wind sweeps across this destroyed home, "whining and moaning like an idiot in a deep black pit,"[42] it is hard *not* to hear the echo of another novel about race, voice, and real estate published in the middle of Wright's own composition: William Faulkner's *Absalom, Absalom!* (1936).[43] That book closes with Jim Bond, "a hulking young light-colored negro," whose "idiot face" howls inconsolably as his white great-grandfather's estate goes up in flames.[44] The destruction of that house, and the howl that accompanies it, signal the funeral pyre for a nightmare built from an entire system of racial hatred and economic exchange that transformed people into things. Jim Bond's name, with its obvious connection to property and the shackles of slavery, also helps sound out the pun in his mixed-race grandfather's name, Charles Bon: the "good" not so much a moral state as an item of exchange.[45] Thus, when Wright personifies—turns a thing into a person—the wind outside Jake's blasted house as a "moaning . . . idiot," we can hear Jim Bond—the legacy of slavery and literature's attempt to wrestle with its meaning—haunting the text.

When one couples this last voice with the radio voice, one can tune into *Lawd Today!*'s intricate revision of the New Deal's literary engagement with radio, race, and real estate. While the allusion to Faulkner's text assures that we recognize the ravaged home as the consequence of Jake's embodiment of the radio voice, that latter voice, as I implied in the title to this section, borrows from Dos Passos. In an elegant comparison of Wright and Dos Passos, Willihnganz argues: "For these writers, radio helped reveal the way the subject was becoming, in Foucault's phrase, 'an object of information, and never a subject in communication.' Radio helped them imagine the modern subject as a kind of dead medium, a relay losing its capacity to generate its own signal, a disappearing voice."[46] J. Ward Moorehouse served as such a "relay" in my argument in chapter 1. In *Lawd Today!* Wright fully grasps the lessons of Dos Passos's *U.S.A.* but revises the "voice of the people" to show how racial difference crosses the wires of the national hookup. When Jake voices the radio, demolishes his domestic space, and yields to the sound of a "moaning . . . idiot" outside his house, the narrative brings into collision the two polyphonic masterpieces of the New Deal—Faulkner's *Absalom, Absalom!* and Dos Passos's *U.S.A*—to level the ground for Wright's own reimagining of radio, race, and real estate in U.S. literature.

As it turned out, radio would not suit this project. The year Wright abandoned *Lawd Today!* he published a story in his collection *Uncle Tom's Children* that would take up another sonic technology to think through the same problems raised in his then unpublished novel. Radio, with its associations with white power and authority, could not serve as the medium for the more equitably shareable voice his work increasingly thematizes as the corollary to shared housing in his "desire . . . for writing a communal life."[47]

THE BLACK PHONOGRAPHIC VOICE

The same year he published his "Blueprint for Negro Writing," and a year before he permanently set aside *Lawd Today!*, Wright issued a provocative and still controversial 1937 review of *Their Eyes Were Watching God*, in which he criticized Zora Neale Hurston's novel for what he called its "minstrel technique." According to Wright, Hurston was responsible for inventing a "quaint" rendition of African American life whose tone, he argued, speaks only to a white audience that considers itself a "superior race."[48] In response, Hurston scoffed at Wright's own use of dialect in his 1938 collection of stories, *Uncle Tom's Children*: "Since the author himself is a Negro, his dialect is a puzzling thing. One wonders how he arrived at it. *Certainly he does not write by ear unless he is tone-deaf*."[49] While James T. Farrell defended Wright's "remarkable . . . handling of dialogue" and his "demonstration . . . of the possibilities of vernacular," most critics since have sided with Hurston, questioning, in particular, the way Wright's ear was drawn to another writer in black dialect: Gertrude Stein.[50]

Beyond some of the more facile binary distinctions in African American studies dividing the legacies of Hurston and Wright, their conversation helps tune us in to how the challenge of tonal mimesis shaped the field of African American writing and, as Michael North observes, modernism more generally.[51] According to North, African American writers in the early half of the twentieth century tended to avoid "black dialect" because of its links to "the white minstrel tradition . . . a constant reminder of the literal unfreedom of slavery" and a representative of "a most intimate invasion whereby the dominant actually attempts to create the thoughts of the subordinate by providing it speech."[52] However, in a study that hears white imitations of supposed black dialect as a main engine for modernism's linguistic inventiveness and social resistance, North seems troubled by Wright's insistence that it was Gertrude Stein's language in *Three Lives* that "caused him to hear 'English as Negroes spoke it,'"[53] or Wright's claim that "Miss Stein's struggling words made the speech of the people around me vivid."[54] Surprised by Wright's acceptance of Stein, coupled with his outspoken rejection of what he described as Hurston's "minstrel technique," North wonders, "Perhaps Wright valued the way Stein's 'struggling words' struggled against one another, producing the tension that [Henry Louis] Gates [Jr.] identifies as the 'masking function of dialect,'" "its self-conscious switch from white to black or, more properly, from standard English to the black vernacular."[55] Unable to accept that an African American writer might hear an innovative approach to representing African American speech in a white author's prose, North turns to an African American literary critic in order to recuperate white minstrelsy for Wright by resignifying such "speaking through" as a politically viable "masking function."[56]

Wright's valorization of Stein's writing does not need to be excused by recuperative arguments, nor should it be heard as a "tone-deaf" endorsement of Stein's own modernist "minstrelsy." Instead, when we hear Stein's "struggling words" as her contemporary reviewers heard them, as "*a stubborn phonograph*," her influence helps elucidate a tradition of constructing blackness in relation to what Katherine Biers identifies as the "black phonographic voice." This phrase also helpfully collects recent statements by Stephen Best, Lisa Gitelman, Mark Goble, Brent Hayes Edwards, Fred Moten, Bryan Wagner, and Alexander G. Weheliye.[57] Of these various and in some cases opposing definitions, I think it most helpful to borrow from Best's and Wagner's arguments in order to describe the particular historical logic in which I conceive Wright's enthusiasm for Stein, and the means by which he employs acoustics to imagine new social relationships that would undo historical bonds forged through specific legacies of property.[58]

For Stephen Best, "fugitive sound" denotes "sound's evanescence and the troubling fugacity of the voice."[59] Oddly, the invention of the phonograph, which seemed to "capture" this "fugitive" sound, induced a crisis in copyright law in the United States. The jurists heard the recorded voice as an inalienable aspect of personhood that the phonograph had rendered into an alienable commodity. The recorded voice thus echoed an earlier notion of the "fugitive," for, in nineteenth-century law in the United States, "reference to inalienables in market terms (terms such as 'exchange,' 'transfer,' and 'alienation') . . . carries with it the 'specter' of slavery."[60] As Best argues, the law's attempt to categorize the voice as either alienable property or an inalienable expression of the self borrows from and furthers the reconfiguration of personhood begun in the Fugitive Slave Laws. The fugitive slave, in the law's thinking, stole his master's alienable property, the slave's labor, even as he fled with his inalienable being. In dealing first with fugitive slaves, "property that, paradoxically, behaved like a person—'thinking property,' in Aristotle's memorable formulation,"[61] and then with "fugitive" voices, American jurisprudence came increasingly to recognize the previously inalienable qualities of personhood as alienable things.[62]

If Best primarily investigates the trope of the fugitive phonographic voice in legal terms, Bryan Wagner underscores the cultural means by which the phonograph creates an "authentic" black folk voice through its very "fugitivity." As Wagner puts it, alienability produces inalienability, as "[a]lienating the voice from the body . . . creates rather than disrupts speech's capacity to stand for subjectivity."[63] For ethnographers who already heard this constructed image of a homogeneous black voice as fugitive because it resisted their established systems of notation, "the phonograph offered a new explanation for why the black voice sounded not only disenfranchised but disembodied, as if it came from nowhere."[64] In its "hisses, pops, and clicks, the warped passages where the acetate yielded to summer heat,"[65] and through its distortion and disembodiment, the phonograph's acoustic infidelity included the notational resistance already attributed to black performance. This fit between technical infidelity and the purported elusiveness of black voices heightened the apparent cultural authenticity of the black phonographic voice. In Wagner's words: "From the point of reproduction, the black voice's primary effects became indistinguishable from their technological condition of possibility, and this led to a situation where, for the first time in its history, the music could be commonly considered as folklore on the grounds that it was indexed directly to the individual consciousness of its producer . . . The aura is made, not destroyed, by the phonograph."[66] Hence, a technology Best understands as inducing changes in law that made personhood increasingly thing-like becomes, in Wagner's reading of an archive that stretches from the close of the nineteenth century through the Lomaxes' recordings of Lead Belly (Huddie Ledbetter) in the 1930s, the technological means to create a cultural idea of the authentic "black folk voice." No less a critic than Richard Wright confirmed this insight in a 1937 review of the Lomaxes' recordings of Lead Belly, commenting, "it seems that the entire folk culture of the American Negro has found its embodiment in him" ("Ledbetter").[67]

Heard alongside the tradition detailed by Best and Wagner, we can begin to understand Wright's preference for Stein over Hurston not as a "tone-deaf" mistake but as participation, whether conscious or not, in a specific tradition of hearing and constructing blackness through attention to the specifically mediated voice of the phonograph rather than the ethnographic ideal of the vanishing mediator represented in Hurston's work with African American dialect. Wright's interest in the repetition, the *stubborn phonograph* style of Stein's prose, full of hisses, pops, and clicks, his sense that with Stein's words he can "tap at will into the vast pool of living words" and "hear the speech of my grandmother who spoke a deep, pure Negro dialect"[68] and "the language of my entire race"[69] elaborates a reading practice that *listens* to writing in such a way as to hear the black folk voice in the phonograph, and the phonograph in the black folk voice.[70] It is a reading practice that, in Wright's case, became a writing practice, and a means to think through how the legacy of an African American voice's interrogation of acoustic "property" relates to a critique of New Deal housing politics.

"LONG BLACK SONG": THE PHONOGRAPH AND REAL PROPERTY

Wright's most explicit narrative treatment of these complicated exchanges between mediated voice and property—both real and moveable—appears in his story "Long Black Song." Part of the collection mocked by Hurston and praised by Farrell, "Long Black Song" pulls together the dual trajectory of the black phonographic voice: Wagner's conception of "blackness" as dependent on the phonograph, and Best, Biers, Weheliye, and others' notion of the phonograph as helping instantiate a disembodied, "fugitive" black voice. Between the story's opening lullaby in dialect, which fails to soothe a baby's wailing, and its closing gunshots that draw out the opening singer's own cry, Wright positions the phonograph as the emblem of a certain breach in real property that sutures together two formerly, and hopefully different historical times.

In the sonic logic of the story's first section, the opening song, "Go t sleep baby / Papas coming back," is quickly countered by a baby's dissonant cry, "its wailing drowning out the song."[71] The song's failure to calm the infant leads its mother, Sarah, to let the baby choose its own comfort object: a clock:

> The baby crawled after it, calling, "Ahh!" Then it raised its hands and beat on the top of the clock Bink! Bink! Bink!

> . . . [Sarah] fetched a small stick from the top of a rickety dresser . . . "Beat wid this, see?" She heard each blow landing squarely on top of the clock. Bang! Bang! Bang! And with each bang the baby smiled and said, "Ahh!" Mabbe thall keep yuh quiet erwhile.[72]

Song, sung to induce silence, or at least a shift in volume, gives way to the refrain of the beaten clock (Bang! Bang! Bang!) and the satisfied sound of the baby kept "quiet." However, the rhythmic soundtrack also inspires Sarah's longing for her ex-lover, Tom, who is soon to return from the First World War. Thus, the shift in the sound's content recodes the waiting in her opening song as well as the refrain of "bang bang bang" into sounds of desire and wartime machinery.

Soon, Sarah's waiting and the sound's meaning changes again. Interrupting her daydreaming, Sarah hears a "dull throb," "then she heard the throb again . . . The throb grew louder, droning; and she heard Bang! Bang! Bang! There! A car!"[73] Instead of her expected husband, a white salesman approaches the porch, and upon hearing the banging casts his first glance inside the house. When Sarah stuns the man by explaining that the infant is beating the clock, he exclaims: "Well, this beats everything! I don't see how in the world anybody can live without time."[74] Despite Sarah's claim that "we just don need no time, Mistah,"[75] the salesman, like a caricature of Capital and Empire who has come to bring mechanized time to rural folk, insists, "you need a clock . . . That's what Im out here for. I'm selling clocks and graphophones. The clocks are made right into the graphophones, a nice sort of combination, hunh? You can have music and time all at once."[76] Recalling the opening broadcast from *Lawd Today!*, which announced the "CENTRAL STANDARD TIME . . . TING! . . . COURTESY THE NEVERSTOP WATCH COMPANY," whiteness is again associated with voice technologies that standardize time.[77]

The specific technology, the graphophone, underscores Wright's work with time. For the salesman's graphophone—a record player onto which one could record—is notable within the history of technology for its particular standardization of time and pitch. As Jonathan Sterne observes, one of the "major innovations in the graphophone was a device called a *governor* that took the irregular motion of a hand crank or foot pedal and converted it to regular motion for the cylinder's rotations."[78] Since playback speed affected the voice's pitch, the "governor" helped reduce the variability in pitch and thus standardize reproduced voices. While Mark Goble is probably right when he argues that including "an outmoded technology was one way for Wright to register the uneven development of the black South," the graphophone's association with the "governor" also calls forth the "specter of slavery" that Best uncovers in a technology so closely aligned with "his master's voice."[79] The technological standardization of time, linked to the temporality of economic development, reinforces the story's work with the history of slavery, and the cultural histories embedded in the recording technology.

When the salesman starts the gramophone and it begins playing what sounds like a traditional spiritual, the opening scene has been wholly recomposed.[80] Now Sarah is sung to, listening to the spiritual coming over the graphophone-clock, with a new message of waiting: "*When the trumpet of the Lord shall sound . . . and time shall be no more.*"[81] In response to the apocalyptic message of the black phonographic voice, Sarah feels "the rise and fall of days and nights, of summer and winter; surging, ebbing, leaping about her, beyond her, far out over the fields to where earth and sky lay folded in darkness."[82] The graphophone's spiritual, mimicking and replacing Sarah's opening song, transforms her waiting and longing into a religious tone of spiritual redemption tied to the passing of the seasons. Recalling Bryan Wagner's theory of the phonograph's *production* of authenticity, the mechanical turning of the cylinder induces in Sarah an embodied sense of organic time that draws her into temporal folds of darkness.

The same language returns as Wright's narrative has Sarah give herself over to the salesman's sudden aggressive sexual advances: "A liquid metal covered her and she rode on the curve of white bright days and dark black nights and the surge of the long gladness of summer and the ebb of the deep dream of sleep in winter till a high red wave of hotness drowned her in a deluge of silver and blue and boiled her blood and blistered her flesh *bangbangbang*."[83] The operatic scene, with its melodramatic language and climactic reiteration of the governing leitmotif, transforms this simple sound into the overburdened carrier of sensuous and narrative material. It evokes the rhythm of sexual excitement, as well as the underlying violence of the salesman's "seduction" and the overlapping temporalities of seasonal change, while pointing to the child's inane and innocent abuse of the clock, an action already linked to world military history (Sarah's longing for Tom's return from wartime Europe), as well as the impending return of Sarah's husband (connected to the closing lines of Sarah's opening song: "Go t sleep, baby / Papas coming back. . . .").[84] Whatever homogeneous time the clock might once have represented seems thoroughly exploded in this simple repeated sound.[85]

Yet, when her husband, Silas, eventually returns home after the salesman's departure, he recognizes the graphophone as *restoring a past history*—the time of slavery—and establishing a continuity in time that recomposes the "waiting" in the spiritual and the return of the black phonographic voice as undermining his real property. As with the chorus of voices in the post office in *Lawd Today!*, the potentially liberating form of the work song or collective dialogue expresses itself instead as an epiphenomenon of forced labor that only links the song—and threatens to return the listener—to the conditions of its original production. In this machine, meant to faithfully reproduce and capture the voice, Silas identifies his wife's infidelity as linked to slavery's market in persons, a history that does not stand outside of time. Joining the alternate logics of Best's and Wagner's historical arguments about the black voice and the phonograph, we might think of this voice's return as an unwanted and perversely ironic scene of the "fugitive" property come home.[86]

That is, if Best and Wagner identify the black voice emerging from the phonograph and becoming fugitive as it escapes and confounds white European and American musical and legal systems, the "fugitive" voice in Wright's story perversely "returns home" when it enters Silas's house with the intention, to his ears, of pulling him and his wife back into human bondage. Heaving the graphophone box outside his front door to smash to pieces in the dirt, Silas screams: "Ah works mah guts out t pay them white trash bastards whut Ah owe em, n then *Ah comes n fins they been in mah house! Ah cant go into their houses, n yuh know Gawddam well Ah cant!* . . . Fer ten years Ah slaves lika dog t git mah farm free, givin ever penyy Ah kin t em, n then Ah comes n fins they been in mah house."[87] Mechanical reproduction, the source of "blackness" in Wagner's argument, becomes the minstrel means through which the salesman gains access to Silas's real property. Inverting the terms of Best's archival reading, the "fugitive" voice, liberated by the phonograph, helps dispossess Silas of his property. And this breach in real property signals the suturing over of a historical rupture, as the time of slavery, meant to have vanished by Silas's industriousness, returns in words that link his liberating work to enslavement ("Ah slaves lika dog t git mah farm free").

In this story, the eruption of dialect speech—precisely that form Wright criticized in Hurston's writing—inscribes slavery into language and technology. Although Sarah speaks in dialect throughout the narrative, Silas's speech seems to emerge from the graphophone. When, upon the salesman's return, Silas kills him in the same spot where he destroyed the graphophone, "he simply stood over the dead white man and talked out of his life."[88] For Goble, at this moment Silas becomes a stylus "who treats the body of the graphophone salesman as if it were a graphophone in the flesh."[89] The monologue that follows sounds familiar to readers of *Lawd Today!*: "Ahm gonna be hard like they is! So hep Gawd, Ahm gonna be *hard*! When they come fer me Ahm gonna *be here*!"[90] Attuned to the sonic puns, Goble aptly translates the passage, recognizing that "it is only when he talks into the body of the white man that Silas can be 'heard' ('*hard*'), that we can 'hear' ('*here*') his words as they are taken down in dialect and italicized on the page as if to guarantee these double meanings. Only a graphophone, or its human equivalent, will do."[91] In *Lawd Today!* only a radio voice, associated with white power and Standard English, would do as Jake shouted, "'LET 'EM HEAR! LET 'EM HEAR!"[92] In "Long Black Song," it must be a graphophone—the playback device that also functioned as a recording device—that gives voice to black dialect and a violent challenge to white authority as Silas screams, "Ahm gonna be *hard!*" Radio, associated with the white voice of the national hookup, cannot carry the history of slavery Wright locates in the "black phonographic voice." Furthermore, as Goble's reading implies, this latter voice is produced through Silas's violent use of the dead white body, whereas Jake's voice was little more than a hollow imitation of white power, which left him lifeless. And yet, "Long Black Song" does not conclude with an endorsement of this new voice of black power.

Ultimately, the only means for Silas to be heard by anyone other than Sarah in this story about race and real estate is to burn silently in his house. In Silas's violent response to the newly sutured history he encounters in the graphophone, the sound meant to claim freedom from time, the infant's banging on the clock, resounds instead as an encounter with the phonographic voice and the legacy of slavery. At the same time, the narrative repetition of the "bang" creates an alternative mode of structuring time within the story, a time that would not join with the sutured history Silas resists.[93] Amplifying these connections, Silas shoots the salesman upon his return for the graphophone the next morning (bang bang bang) and then holds off a white mob with his gun (bang bang bang). Resisting the phonographic "fugitive" logic of speech and access employed by the salesman, Silas continues to shoot, "bang bang bang," as they set fire to his house, where he "stayed on to burn . . . stayed without a murmur."[94] As Silas burns silently with his house amid the soundtrack of gunfire ("bang bang bang"), his acts once again revise Sarah's opening song and the silence it hoped to conjure. If the banging silenced the baby, now Silas's silence and the bang bang bang of the guns draw forth Sarah's own cry as she flees to the fields in the final sentence "crying 'Naw, Gawd!'"[95] The narrative's fearful symmetry transfers the infant's cry to Sarah's mouth, the bang bang bang of the guns heard now as the cause of rather than the antidote to her wailing.[96]

We might be tempted to hear sound functioning here like an empty, or floating, signifier, ready to accept any one of numerous meanings. However, rather than shedding its contents as it moves, the reiterated sound only increases its burden, carrying one meaning on top of another. By the story's end, "bang" is no nonsignifying sound but indexes a dense network of personal and global desires, temporalities, and economies. And alongside this sound, Sarah's cries outside the house where her husband burns echo once again the voice of Jim Bond in Faulkner's *Absalom, Absalom!* It is this howling protest against the legacy of slavery's property rights that makes up the notes of "Long Black Song."

Weighted down with history and increasingly dangerous, Wright's acoustics challenge the universalism of any apparently disembodied voice and the liberating theories that attend them.[97] The migrating voice in Wright's work tells a story of dispossession different from McCullers's hopes for listening discussed in chapter 2. Although Wright's work was published the same year as Dos Passos's *U.S.A.* trilogy, the voice and buildings in Wright's writing seem a far cry from that trilogy's "radio network . . . [T]he speech of the people." In the collision of particulars within these sounds resides a resistance to the homogenizing logic of the New Acoustics, and the ideology of the presidential radio voice I described in chapter 1. Hardly tone-deaf, "Long Black Song" tunes readers in to Wright's early attempt to compose what he hears as the intricate soundscape of New Deal politics and its relation to racial discourse and real estate.

LISTENING OUT OF BOUNDS: *NATIVE SON'S* ACOUSTIC BORDER CROSSING

When Sarah asks the white salesman in "Long Black Song" why he is selling graphophones, he tells her, "If I can make enough money out of this Ill go to Chicago."[98] Reading Wright's 1940 novel *Native Son* in the light of this earlier story, then, allows us to follow an itinerary of acoustic properties Wright constructs across his narratives, from the repeated radio reports of Lincoln's centennial in *Lawd Today!* (1938/1963) to the falsely reported death of Cross Damon over the radio in *The Outsider* (1953). Once again, the attention to acoustics and housing reminds us that Wright's day job during the composition of *Native Son* involved writing about Rockefeller Center—the brain center of the New Acoustics—and the Harlem River Houses, "the first of the government's large scale housing projects" for the 1939 Federal Writers' Project's *New York City Guide*.[99] Although Wright rarely mentions sound technologies in *Native Son*, these other texts help tune us in to how he embeds their engineering into the formal logic of his narrative's work with voice and property. In the pages that follow I demonstrate how the novel's acoustic attention structures the narrative's exchange of voices attached to social positions in order to undermine belief in a whole, autonomous, and ethnographically authentic or stable subject. Instead, the novel offers a series of voices, borrowed, echoed, and mimicked. It is a system of transfer that reveals voice in *Native Son* as a fugitive voice already thrown, meant to imagine, as Wagner writes about the phonograph, "the possibility of hearing the sounds of others in your own voice."[100] Thus, my approach here shows how literature registers sonic mediation in its execution of literary technique. This media hermeneutic of literariness corresponds to my reading of Dos Passos's use of "unspeakable sentences" to train readers how to listen differently to FDR's radio voice. Literature, across these texts, serves as an education in media.

Whereas the fugitive voice became a carrier for the legacy of slavery and fugitive flight in "Long Black Song," in *Native Son* this vocal mode produces the possible grounds for communal life. Put another way, when we arrive at the final scene from *Native Son* we will encounter how it violently combines the radio's voice of white power in *Lawd Today!* and the black phonographic voice from "Long Black Song" in an attempt to overcome the property lines in language and housing. The novel hopefully reimagines the logic of the "dead relay" that

Willihnganz associates with radio, and Goble's notion that "race is what happens when you become the medium of someone else's instrumentality."[101] In Wright's work, voice technologies *do* transform subjects into relays. However, *Native Son*'s innovation is to universally apply the lessons of racial difference to turn *all subjects* into relays, not to reify and instrumentalize them but to produce new forms of communal subjectivity according to the logic of the fugitive voice.

GOING ON RECORD: THE VIOLENCE OF EMBODIED SPEECH AND COURTROOM LISTENING

Wright's *Native Son* tells the story of Bigger Thomas, a poor, African American twenty-year-old from Chicago's South Side who finds a job as a chauffeur for the wealthy white real estate developer Henry Dalton. After driving Dalton's daughter, Mary, and her Communist boyfriend, Jan, on a tour of the South Side one evening, Bigger is forced to carry Mary into her bedroom after she has passed out from drinking too much. When Mary's blind mother enters the room, Bigger, fearful he will be accused of rape, accidentally smothers Mary. Unaware of what has happened, Mary's mother leaves, and Bigger disposes of Mary's body in the home's furnace, flees to the South Side tenements, where he murders his girlfriend, Bessie, and is eventually captured by the police who bring him to trial, where the court ultimately sentences him to death. The novel, a best seller that made Wright an international celebrity, combines the fugitive slave narrative, noir fiction, and documentary realism to indict Chicago's racist real estate redlining policies and a white New Deal liberalism that pretends at ideal notions of collectivity without addressing the structural inequalities of race concealed in any rhetoric that speaks of class alone.

Perhaps the story's most unique formal maneuver to manage the book's complicated plot and address the unevenness of collective speech, is to incorporate the act of overhearing into the novel's narrative structure. At the end of the book's second section, "Flight," Bigger is arrested on a South Side rooftop, in the "Black Belt" real estate district, on suspicion of murdering Mary Dalton. In the scenes leading up to this moment, we read as Bigger listens: to a newsreel about Mary Dalton in a movie house, to the struggling sounds she makes as Bigger fearfully suffocates her in her room, and, from an apartment adjoining a vacant tenement, to the voices of two men debating his crimes. The plot, in other words, develops Bigger's character through an evolution in listening, from the movie house's passive object of interpolation to the eavesdropper hearing his life become news.

However, when we turn to the novel's third and final section, "Fate," Bigger's unspoken thoughts, "overheard" by readers in the interior monologues of the novel's first two sections, are put on the public record in the voice of his lawyer, Boris Max. In the courtroom Bigger hears his own words in a new tone, and moves closer to an act of public speech through the strange experience of ventriloquism. Earlier in the novel, Bigger thinks, "He had killed twice, but in a true sense it was not the first time he had ever killed. He had killed many times before, but only during the last two days had this impulse assumed the form of actual killing."[102] Later, in his speech before the court, his lawyer, Max, says, "He has murdered many times, but there are no corpses. Let me explain. This Negro boy's entire attitude toward life is a *crime!* . . . Every time he comes in contact with us, he kills!"[103] Unable to speak publicly throughout the novel, Bigger now finds his thoughts spoken aloud by Max.

Or does he? Other critics note that Max repeats the *narrator*'s language, not Bigger's own thoughts. In doing so, Max's speech "functions to deconstruct the very rhetoric that it employs" and "underscores the limitations of the perspective shared by both" the narrator and Max.[104] Both limit themselves with a "symbolic outlook" that transforms Bigger into a "mode of life" rather than an individual.[105] If Max's repetition deconstructs the authorial rhetoric, the voice of authority, it also further tears or folds the narrative fabric that stitches one voice to one body. Echoing the narrator, Max might undermine what these critics call symbolic reasoning. More significantly, however, the novel's eloquence works to undermine attribution, to challenge the notion of property, to ask to whom a voice is proper. The transfer of utterances, beginning with the already unstable linguistic property of free indirect discourse that Max transforms into direct speech in his courtroom defense, and which is then reiterated in Bigger's final speech, slowly strips the voice from any original mouth, depositing utterances across a variety of positions. We encounter speech as listening. Contrary to the critical literature, *Native Son* does not progress to an articulate statement of self-understanding.[106] Its narrative maneuvers, its careful attention to listening as part of speech, work, a priori, to undermine a comprehensive notion of the self, and of the voice as identity. When, in his closing speech, Bigger echoes utterances already attributed to Max and the narrative, and declares, "What I killed for, *I am*!," the statement resounds as a parodic representation of the *cogito* and its implied self-possessive authority. His statement asks us to seek out another way of thinking and being than Cartesian logic and the visual grid it implies.[107]

Nevertheless, critics tend to read the novel's climactic broken dialogue between Max and Bigger as an example of Bigger's linguistic and psychological mastery. Such readings depend on an approach inattentive to the ways Bigger's speech echoes his most unjustifiable actions from earlier in the novel: his murders of his girlfriend Bessie and Mary Dalton. In the terrifying scene in which Bigger rapes and murders Bessie, inarticulate speech and a dreadfully articulate body collide in a terrible clash: "she spoke, not a word, but a sound that gave forth a meaning of horror accepted. Her breath went out of her lungs in long soft gasps that turned to a whisper of pleading. 'Bigger . . . *Don't!*' Her voice came to him now from out of a deep, faraway silence and he paid her no heed. The loud demand of the tensity of his own body was a voice that drowned out hers."[108]

Bigger's literal embodiment of voice—"his own body was a voice"—silences Bessie and seems to suffocate her speech. If Mary Dalton's death seemed an unintentional act, Bigger rapes and murders Bessie because of, not despite, his will. Wright registers the shift in terms of listening and speaking. Whereas Bigger listens and tries to stay silent when he accidentally suffocates Mary in the Dalton house, he becomes an embodied voice that refuses to listen to Bessie—"he paid her no heed"—in the abandoned South Side tenement. In this third telling of the rape plot (the white salesman and Sarah, Bigger and Mary, Bigger and Bessie), an African American man attains a type of voice but only at the expense of an African American woman. Recalling the violent conclusions of *Lawd Today!* and "Long Black Song," in this scene embodying voice, a state of being proffered by many critics as the novel's goal, sounds terribly wrong.

Indeed, we notice the same language return in Bigger's final speech. Speaking with confidence for the first time in the novel, Bigger refuses to listen: "Max opened his mouth to say something and Bigger drowned out his voice . . . 'I didn't want to kill!' Bigger shouted. 'But what I killed for, I *am*!'"[109] Once again "drowning out" another's voice, the narrative echoes Bessie's rape at the moment critics cite as evidence for Bigger's linguistic competence, and his status as a symbol of the freedom of black writers.[110] Attuned to the narrative's acoustic logic, the means by which it repeats strips of discourse across the space of the novel in order to have them resound, often dissonantly, in a new context or a new mouth, we can hear Bigger's statement as a struggle with competing forces in American literary and social life. In *Native Son*, it is no longer enough to speak articulately; in fact, the novel runs counter to the tradition of the fugitive slave narratives of Douglas or Jacobs, wherein the act of autobiography already testifies to the subject's success. Nor does *Native Son* fit comfortably in what Katherine Biers identifies as an American national imaginary that features blackness "as a *failure* of speech."[111] At least not as Biers understands "failure," as an inability to speak articulately. In Bigger's case, articulate speech actually contributes to a failure in the belief in articulacy first registered by *Max*'s linguistic failure—his stutter.

DETERRITORIALIZING REAL PROPERTY: STUBBORN PHONOGRAPHIC STUTTERS

In the final failed dialogue between the jailed Bigger Thomas and his lawyer, Max, we hear Max's voice, "y-you've got to b-believe in yourself, Bigger."[112] It is a stutter that inspires Bigger to speak: "'Sounds funny, Mr. Max, but when I think about what you say I kind of feel what I wanted' . . . Max opened his mouth to say something and Bigger drowned out his voice . . . 'I didn't want to kill!' Bigger shouted. 'But what I killed for, I *am*! . . . I know what I'm saying real good and I know how it sounds. But I'm alright' . . . Max's eyes were full of terror."[113] Bringing together the professional rhetorician's stutter with the previously inarticulate fugitive's existential affirmation, the inversion upsets and reconfigures an entire literary genealogy and the social world in which that literature and its interpretations labor. In order to piece together how these few sentences function as an intricate attempt to wrestle with these tense and knotted strains in American discourse, I want to turn to Max's stutter: the trigger for Bigger's speech.[114]

Why insert the stutter into the speech of this professional rhetorician? What makes Max's words falter, and what does this slight disruption, this "vocal defect," tell us? At first, it seems we have completed the equation drawn between Wright's work with African American voices and Stein's "stubborn phonograph." The Steinian style that Wright praised was phonographic precisely through its tendency to repeat and stammer, to evoke "struggling words." Stein's style resists the fluidity and forward movement of the mother tongue in much the same way as the modernist community Gilles Deleuze claims "shapes and sculpts a foreign language that does not preexist *within* h[er] own language."[115] She puts the language system [*langue*] into "perpetual disequilibrium," wherein "*language itself will begin to vibrate and to stutter*, and will not be confused with speech, which always assumes only one variable position among others and follows only one direction."[116] At first, Max's simple stutter seems less disruptive—if it indicates disequilibrium, it does so in prosaic, basic language ("y-you've g-got to believe, Bigger"), seemingly remaining at the level of the single utterance [*parole*] rather than the formal experiments of Stein, Beckett, and the other modernists favored by Deleuze.

Yet, despite its simplicity, Max's stutter registers a fundamental struggle in and against language. It is akin to Barthes's claim that stammering exists "neither in language, nor outside it; it lets us know that failure might be near."[117] The faltering exposes a territory riven with competing discourses and begs the question, "Which failure?" On one hand, Wright's novel has struggled up to this point to condemn a culture that has fought to silence the African American voice. At the same time, as Nancy Ruttenburg argues, that culture's literary ideals have long valued silence and inarticulacy as markers of authentic national expression.[118] How, then, does a novel address silence as oppression when silence betokens the oppressor's highest value? How does an African American novelist write against articulacy when the tradition's founding texts celebrate the teleology of literacy and self-expression? What kind of failure does Max's stutter indicate?

For Nathaniel Mackey, as well as the black authors he and later critics celebrate, the turn to the inarticulate, the stutter, evinces a knowledge of linguistic and cultural rules that, when broken, open a new space for the utterance. Deleuze's "foreign language" in Mackey's case becomes a specific racial tone, as "the black musician's stutter . . . reflects critically upon an experience of isolation or exclusion . . . [I]t symbolizes a refusal to forget damage done" and stands as a "critique and a partial rejection of an available but biased coherence."[119] Resisting the coherence that arises out of a suppression of racial difference, the stutter is a "two-way witness" to "the need to go beyond the confines of an exclusionary order" and to its "limited success at doing so. The impediments to the passage it seeks are acknowledged if not annulled, attested to by exactly the gesture that would overcome them if it could."[120] This impeded speech is therefore a "telling 'inarticulacy'" whose failure in speech "implicitly indicts a white-dominated social order and the discourse of racial difference by which it explains or makes sense of itself."[121] Writing with a stutter differs from the code switching that North, following Gates, upholds as typical of African American modernism. The stutter does not switch between two languages so much as disrupt and disarticulate a language from within.

If the stutter and inarticulate speech help, as Deleuze and Guattari write, to "minorize" the dominant language, why does Max, the lawyer, and not Bigger, the fugitive, stutter?[122] Indeed, Bigger's speech in response to the stutter, his transformation into a "body [that] was a voice," invokes and revises the most famous stutter in literary history, that of Billy Budd, and his statement: "Could I have used my tongue I would not have struck him."[123] While that famously innocent sailor issued his "stutter and blow in place of the word of self-identity,"[124] Bigger, fusing voice and body in his articulate statement of self-determination ("what I killed for, I *am*!"), morally condemns himself in the eyes of Max. Putting the stutter in the mouth of the lawyer, the narrative invokes the legacy of American literature's paragon of innocent violence only in order to adamantly reject a romantic inarticulacy as well as the teleology of the fugitive slave narrative that imagines literacy as liberation.[125]

Instead, Max's stumbling words help link *Native Son* to a literary genealogy the novel struggles to recompose. Rejecting the identitarian, or, at least, autobiographical tradition as a viable social model and turning down the communist logic proposed by Max, the narrative wrestles with the role of the black vernacular and the problem of address: the two issues Wright identified as particular failures in Hurston's writing. We should recall here that in the courtroom scenes immediately preceding their conversation in Bigger's jail cell, Max echoed the free indirect discourse through which the reader accessed Bigger's thoughts in the novel's first two sections, and Bigger now speaks those words to Max. With this in mind we can hear how their exchange employs the stutter as a signal to disarticulate competing claims to any notion of linguistic mastery. Wright's apparently "tone-deaf" endorsement of Stein's "struggling words" emerges instead as a critique of imagining speech as color-coded, a rejection of the phonographic and ethnographic construction of "the black voice," and a refusal to fit established criteria of articulate liberation. Placing the stutter in Max's mouth challenges the standard placeholder for articulate speech, and transforming articulate speech into the medium through which one attempts to justify murder undoes the heroic bildungsroman narrative others have applied to the novel. The narrative, in other words, places available discourses of African American literature and the more general category of American literature into a state of disequilibrium registered by the fractured dialogue between the stutter and speech.[126]

When Max hears Bigger speak the words Max himself stated in court and then refuses to endorse those words, the stutter's fugitive signal becomes audible. Revisiting Silas's encounter with the graphophone in "Long Black Song," Bigger speaks with a fugitive voice—a voice unhinged from any original source and displaced onto different bodies throughout the narrative. Different than the position of the voice in its earlier instantiations, Bigger now embodies and literalizes the fugitive voice from his position as a literally captured fugitive. Speaking as a fugitive he attempts to return the fugitive voice to Max, uttering the words Max spoke, the words readers first encountered in the free indirect narration of the book's evocatively titled second section: "Flight." Notably, it is through this stylistic maneuver that Wright told Gertrude Stein he could show that "the facts of Negro life constitute a great body of facts of importance about mankind in general."[127] The universal applicability of black experience depends, for Wright, on the fugitive voice he renders in print.

The key to such fugitivity, to a voice that can travel beyond its expected locale, resides not just in free indirect discourse but in the nonsemantic sonic properties of the voice. Attentive to the tone of Max's voice—"*Sounds funny*, Mr. Max, but when I think about what you say I kind of feel what I wanted . . . I know what I'm saying real good and *I know how it sounds*"—Bigger's words attempt to bridge with Max's voice in court: "He had not understood the speech, but he had felt the meaning of some of it from the *tone* of Max's voice."[128] The unrealized hope in Bigger and Max's broken dialogue is that the alienable qualities of the voice—the tone and sound—will fuse with the inalienable notion of the self to produce a contagious utterance that allows each speaker to hear the voices of others in his speech. For Wright, tone becomes the affective carrier for a sympathetic knowledge precluded in the semantic field. Bigger cannot understand Max's words, but he "*feels* their meaning." Insisting on the tone of these voices, rather than their content, the narrative attempts to make Bigger's statement of identity a shareable, transportable utterance, one whose sound would encourage the "conspicuous transference" Nancy Ruttenburg has described as the legacy of democratic personality: a democratic charge that would now reverberate out from African American experience. However, when Max recoils at Bigger's statement—"his eyes were full of terror"—he closes or breaks this potential circuit; the voice will no longer be shared.

This failure to share a voice marks the limit case of the novel's hopeful ambition and its desperate pessimism. Although they speak the same words in the same tone, neither Max nor Bigger prove themselves wholly capable listeners. The fugitive voice is caught in the captured fugitive's failure to communicate with his lawyer, and the lawyer, having spoken in the courtroom, fails to listen in the jailhouse. Moreover, the communicative failure, an inability to share words, will also come to index a failure to reimagine property in terms of equal access. Wright reveals real property to be built by words and bricks, but he struggles to communicate how a change in the concept of the personal utterance could also undo the property lines of the buildings Max's words indicate.

This communicative failure makes explicit the novel's work with a speaker's tone, the transferability of an utterance, and the space of housing. We can call the interdependence of each of these elements "narrative acoustics," and we can find their encounter in Max's stuttering declaration, delivered at the end of his proclamation about Bigger's need to reclaim the houses of the South Side. Pointing to the buildings outside the jailhouse windows, Max asks Bigger to realize that "If men stopped believing, stopped having faith, they'd come tumbling down."[129] Adding that building owners like the real estate mogul Dalton oppress others in order to keep them away from the buildings, Max concludes, "But men, men like you, get angry and fight to re-enter those buildings, to live again."[130] These remarks help collect the novel's challenge to property ownership as racially and class biased. They restate, in miniature, the ideas laid out in Max's court speech, where he spoke of Bigger's youth in the tenement housing owned by Dalton's South Side Real Estate Company, how those experiences indirectly led to Bigger's murder of Dalton's daughter in Dalton's house, and later, of Bessie, in one of Dalton's condemned buildings.[131] While fleeing from the police, Bigger, in fact, did "fight to re-enter those buildings, to live again," as he broke into empty, overpriced rentals in the so-called Black Belt area of Chicago.

Fugitivity and real property collide in this broken dialogue much as they did in Silas's speech after he broke the graphophone. However, unlike Silas's monologue, Bigger and Max's fractured conversation reveals the property lines in language. The failure to share language—a failure anticipated in Max's stutter and confirmed in his look of terror—this disarticulation brings real property and language together, as the failure in speaking undermines the claims for public housing, for housing open to the public, housing made to be shared. The scene struggles with a linguistic and material "deterritorialization" indexed by the stutter but never achieved. Max's inability to share language, and, instead, his continued insistence on it as one's property, as pertaining to one's own position alone, only reinforces those property claims made in the simple signs on empty buildings in the so-called Black Belt real estate district: "Property of the South Side Real Estate Company." Recalling chapter 2's discussion of Fredric Jameson's speculation about a "radio aesthetic," houses—real property—act as characters within the language of Wright's novel. Through these placards, and the entire legal, financial, and cultural system they imply, these real properties "speak" their "rights" in *Native Son*, emerging as characters endowed with more rights than their human counterparts.[132] The late Jamesian extension of realism to give real property a voice comes under brutal critique here. Indeed, Bigger's "flight" through the Black Belt's houses depicts him as mere movable property, a temporary fugitive in the towering vacants of the South Side. Despite his words, Max's look of terror at Bigger's "self-possession" ultimately dispossesses Bigger. When Max turns his back, we are left to conclude that the connection between belief and real property, in the form of Dalton's South Side Real Estate Company, possesses a right to exclude and segregate communities, to force them into the streets or, in Bigger's case, to the jailhouse.

Published six years after the National Housing Act was written into law, *Native Son* testifies to the failure to realize public housing as a social good in the New Deal United States. The new building codes and building projects under the National Housing Act of 1934 and the subsequent Housing Act of 1937 created new legal rights for housing under the watch of the United States Housing Authority, itself created in 1937. As I have attempted to demonstrate over the course of this chapter, *Native Son*'s narrative works against precisely the pejorative association and a priori definition that we saw public housing enforce on its inhabitants in chapter 2 through Lawrence Vale's understanding of "the public neighbor." At the same time, Max and Bigger's "struggling words" strive to imagine a shared property that would include shared housing. While the stutter indexes the phonograph's fugitive voice, it departs from Silas's fight for private property, in order to think through sound as a challenge to private property's limitations. Indeed, when we trace Bigger's fugitive travels throughout the narrative, we realize that what appeared to Michael Denning as the novel's lack of a "unifying narrative" emerges instead as a meticulous ordering of acoustic properties.[133] The closing stutter suddenly reveals the resonances between the narrative's work with the fugitive voice and Bigger's fugitive movement from his family's cramped tenement to the movie house where he first hears Mary Dalton's name pronounced on the newsreel, to Mary Dalton's room, where he accidentally suffocates her to prevent her from speaking his name, to the vacant tenements where he listens to others speak about his crime, to the courtroom where he hears Max speak on his behalf, and finally to the jailhouse where he and Max speak together. An emblem of racial bondage in "Long Black Song," the phonographic stutter becomes in *Native Son* an opportunity—broken and still not possible in 1940—to listen beyond the limitations of real estate.

In this sense, the novel holds out the promise that chapter 1 located in the radio voice—the belief in a democratic personality that would work collectively to challenge proprietary claims on speech. Yet, this hope also ultimately fails in the conclusion to Wright's novel. While we can acknowledge the social and cultural symmetry Wright described in his writings for the Works Progress Administration regarding the public Harlem River Houses and the wired Rockefeller Center, his novel condemns the limits of both projects, revealing that the "speech of the people" and the dream of New Deal public housing founder against the intertwined legacies of private property and racial bias. Rather than a mere expression of architectural determinism, *Native Son*'s final scene, delivered in a jailhouse blocks from the South Side tenements and Chicago Stadium, stands as a careful indictment of the voices from the Republican and Democratic national conventions echoing throughout that stadium in 1932, the voices broadcast outside those walls, and the message proclaimed there of a New Deal for the people of the United Sates and a policy of the good neighbor for all of the Americas.

CODA: THE ARGENTINE AFTERLIFE OF *NATIVE SON*

A decade after the publication of Wright's novel he flew to Argentina to perform the role of Bigger Thomas in *Native Son*'s adaptation to film. Although the decision might seem odd at first, it makes particular political sense for Argentina and Wright himself. As Butts comments, "[Wright's] vision of culture dovetailed nicely with New Deal visions of shared community and the importance of black culture, yet it also contained a critique of national culture as it expressed the hope of new intra- and international alliances."[134] Wright himself placed his decision to head to Argentina in terms of a larger investigation into the connections between white racism and imperialism. He told Roland Barthes: "Having left America and having been living for some time in France, I have become concerned about the historical roots and the emotional problems of Western whites which make them aggressive toward colonial peoples. You can see from this that my travels in to [*sic*] the Argentine, into Africa and Asia even have an autobiographical inspiration. I was looking for explanations of the psychological reactions of whites."[135] As Thy Phu argues, Wright's interview with Barthes reveals how his time in Argentina helps him develop "a racial consciousness of the far-reaching effects of U.S. and European colonialism."[136] I want to specify that claim here by showing how this more global awareness of U.S. race relations derives, in some ways, from the new social, political, and economic alliances associated with the Roosevelt administration's Good Neighbor Policy.

Although intended to usher in a period of friendlier inter-American relations, the effects of the Good Neighbor Policy hit Argentina's powerful film industry hardest. While Orson Welles and other "goodwill ambassadors" traveled to Latin America to establish cultural ties between South and North America, and Hollywood made a concerted effort to curtail pejorative and stereotypical representations of Latin America in their films, Argentina's declared neutrality during World War II made them appear a less than favorable neighbor and led to severe restrictions of imported film stock from the United States. Debilitating the Argentine hold on the Latin American film market, Hollywood gained ground throughout the region. In a move to support their now flailing national film industry and react against the ill-treatment by their "neighbor" to the north, the Argentine government censored and reedited U.S. films and instituted a screen quota in 1944 to limit the U.S. theatrical releases in the country.[137]

At the same time, European directors fleeing World War II headed to Argentina as well as Hollywood. The Belgian-born Jewish director Pierre Chenal established himself within Argentina during this time and returned in 1951 to shoot a film that would reevaluate the terms of the "Good Neighbor." A successful director in France known for his

1935 adaptation of *Crime and Punishment*, and *Architecture Today* (Architecture d'aujourd'hui)—his 1930 study of architectural design cowritten with Le Corbusier—Chenal filmed several Spanish-language films in Argentina throughout the 1940s. His first English-language film in the country, which appeared in Argentina under the title *Sangre Negra* (Black Blood), was, in fact, the first English-language film shot and produced in Argentina, and the first film adaptation of Richard Wright's *Native Son*. With Wright discouraged by the political and social climate in the United States and unable to find a studio willing to release the race-conscious film on which he insisted, he and Chenal found a willing partner in the Argentine film industry.[138]

We might think of the production of *Sangre Negra* as a revision of the Good Neighbor Policy's political and cultural legacy. While wartime Hollywood went so far as to produce Disney's *Saludos Amigos* (1943) and other films of "goodwill" with "Latin" or "inter-American themes,"[139] the cinematic representations of African Americans failed to achieve the same acceptance by U.S. policymakers.[140] Anticipating the association between U.S. Black Power movements and Fidel Castro in the early 1960s, *Sangre Negra* offered the Argentine government and film industry an opportunity to critique the United States' racial inequality as a failure of democracy.[141]

The cinematic version of *Native Son* that was eventually released in the United States bears the scars of this embattled relationship. Initially, *Sangre Negra* looked like a perfect vehicle for "Wright's belief in transnational, interracial collaboration," one with the potential to become "a 'borderless' film hit that would reach the international masses with his searing message intact."[142] The movie's first showing, on a Pan American flight from Buenos Aires to Paris, emblematized the work's successful transnational production.[143] Furthermore, within the film, simple name changes—"Gus" is now "Panama"—subtly broaden the local racial politics to a Pan-American issue connected to U.S. imperialism in the Canal Zone.[144] Finally, Bessie, a domestic servant in the novel, becomes a jazz singer in the film to trade on jazz's status as the music of black internationalism.[145] The noir product of Wright and Chenal's collaboration—a critical and commercial success upon its release in Argentina—appeared destined for international success.[146]

However, U.S. censors quickly sought to dismantle the film's political and artistic potential. Already suspicious of FBI informants trying to undermine the film's production, Wright began communicating in code from Argentina to his literary agent in the United States.[147]

In correspondence between Wright and Chenal, the director's letters register the changes. He shifts from celebrating the movie's reception in Argentina—where it was shown with subtitles—to excoriating U.S. censorship: "They've cut more than EIGHT HUNDRED METERS from our film!"[148] Banned and then reinstated in New York and forbidden in Ohio until the lawyers of U.S. distributor Walter Gould defeated the Board of Censors in court, the film eventually released in the United States received largely negative reviews and bore little resemblance to its original form.[149] As Ellen Scott argues, the New York Motion Picture Division's "six densely filled, single-spaced pages [of eliminations] . . . essentially re-authored the film."[150] The censors cut entire scenes where Bigger touches Mary, removed the word "white" from dialogue, cut all allusions to lynching, deleted scenes of white mob violence, and, intent on weakening the film's leftist critique, cut nearly all of Max and Jan's dialogue so that "seemingly random bursts of speechifying come from characters to which we have barely been introduced."[151] In an attempt to assure these cuts would not be followed in Europe, Chenal sent Gould a list of corrections for the film's international version and made the sound mix correspond to the Argentine rather than the U.S. version.[152]

Chenal's control of the sound mix as a means to resist the censors marks perhaps the final turn in Wright's work with voice technology, race, and property. Wright recognized that motion pictures "had 'sensitized the ear' of millions of Americans," and, in order to keep his writing up to speed with this newly sensitive audience he used an Ediphone—a competitive version of the Dictaphone—to help him write dialogue.[153] This intensive practice to tune his ear and his writing makes the censors' cuts—and silences—even more punishing. At its broadest, *Sangre Negra*'s "censorship suggests that film speech was often the basis for censor boards' deletions when it came to articulating racism."[154] More specifically, the politics of the fugitive voice so central to the novel are muted in the film. Wright's own performance as Bigger in itself was a means to imbue this work with voice throughout the film. Rather than merely make himself the star, Wright as Bigger transforms the authorial role into that of the character and undoes the apparent biographical connection between author and character, precisely by stepping into the character's place.[155] However, the adaption of this fugitive voice stopped there. Eliminating the narrative elaboration of that discursive work, the censors transform Max's speech in court into a twenty-five-second voiceless montage overlaid with the film's dramatic score.[156] Likewise, Bigger and Max's climactic exchange shifts from a conflicted dialogue over the belief in public housing to a paternalistic humbling. The film reduces Bigger's political challenges to a meek request: "Mr. Max. Tell Mr. . . . Mr. . . . Tell Jan hello." The stuttered struggle to share the voice is now the meager equality of calling another by his first name.

The film's ending more strikingly exacerbates the adaptation's impoverishment of the politics of race through voice technology. Turning from Bigger, the camera watches Max walk into the distance toward the jail's exit. Two soundtracks exist for the scene. In one, a voice-over from Max intones: "I left Bigger feeling that everything, including justice, was still unsettled. And today I feel even less certain of innocence and guilt, crime and punishment, and the nature of man." In the other, the camera cuts back to Bigger's cell, where he collapses facedown on his bed, "while a Robeson-esque bass sings a spiritual."[157] Both of these voices fail to expand on or continue the challenge to racial politics that Wright elaborated throughout his novel. The spiritual only returns Bigger to the rural, folk identity Wright railed against in his review of Hurston. It leaves behind a voice explicitly coded as black to mourn for his impending death, recalling Silas from "Long Black Song."

On the other hand, Max's voice-over breaks with noir's established conventions in order to formally reinforce the frame and capture literalized in Bigger's jail cell. Chenal, who adapted James M. Cain's *The Postman Always Rings Twice* as *Le dernier tournant* (The Final Turn; 1939), must have been familiar with Tay Garnett's adaptation in 1946, Billie Wilder's version of Cain's *Double Indemnity* (1944), Otto Preminger's *Laura* (1944), or the host of other noir films that made it conventional to conclude with "the voice of the dead."[158] Following these generic standards, the voice-over conclusion should end with Bigger's voice. Instead, the film gives Max, the lawyer, the last word, breaking with generic expectations to assure that the fugitive's voice cannot travel any farther than the borders of the screen. Kaja Silverman observes that the male voice-over in classical Hollywood cinema usually inhabits a "transcendental auditory position," while female voices almost always remain rooted on-screen and within the diegesis.[159] In *Sangre Negra*, the voice of the white lawyer assumes this position of control, whereas Bigger remains silent. Rather than elaborate a new fugitivity of the voice through experiments in dubbing, the film adaptation traps that voice within an essentialized folk identity or circumscribes it within the voice of the law.[160]

Filming in Argentina, Chenal and Wright built the groundwork for a transnational cinema that would link local and international aesthetics and politics to show that there was an outside, a nonaligned space behind the color curtain that could critique the racist policies and actions within the United States without becoming a mere proxy for Cold War divisions. And yet, with both the film's own shortcomings and its disastrous censorship in the United States, *Sangre Negra* also reveals the limits to such spaces and the limited range of free speech. In a testament to the U.S. censors' power, the film's uncut version never made it to Europe. Chenal, infuriating Wright, obtained an injunction against the censored version from showing in France, which meant no version would show. The Sono Films Studio responsible for the production apparently lost the original, and even Wright never secured a copy. Only forty years later did the Argentine collector and film historian Fernando Martín Peña locate a copy in Argentina—which disappeared—and then another in Montevideo, which he purchased and restored alongside the Argentine anthropologist and Smithsonian Institute employee Edgardo Krebs, who had located another copy.[161] When the uncut and restored film was finally shown to the U.S. public in 2012, Krebs wrote, "The improbable network of people that knew firsthand about the film is as peripheral as the trajectory of the film itself, and an appropriate coda to its fugitive status."[162] Strangely, the film's very censorship produced a new transnational community formed by that optimistic idea of fugitivity Wright labored to create.

PART II

Occupying the Airwaves

CHAPTER 4

Tears in the Ether

The Rise of the Radionovela

What remains is a theme for a future essayist—a History of radio—that will have lasted for thirty years. But thirty years during which the radio will not have become a true art.

—ALEJO CARPENTIER, "El ocaso de la radio"

On October 10, 1922, the fifty-four-year anniversary of the Cuban Ten Years' War for independence from Spain, Cuban president Alfredo de Zayas delivered Cuban radio's official inaugural broadcast. Speaking over station PWX, the network of the Cuban Telephone Company, a subsidiary of the U.S.-owned International Telephone and Telegraph Company (ITT), Zayas addressed himself, in English and Spanish, to listeners in the United States: "From the city of Havana, capital of the Cuban Republic, I have the great honor of directing my voice to the people of the United States of America through the marvelous invention inaugurated by the Cuban Telephone Company."[1] Acknowledging Cuba's proximity "to your shores, inside the area of operations [*radio de acción*] of your powerful commercial influence," he closed his speech as a single individual speaking on behalf of one nation to another: "With my words, pronounced in the name of the Cuban people, I send my true expression of respect and admiration to the United States and your national institutions; my sincere friendship to your people and my consideration for your Government. Three 'vivas' to the glory of the United States! Three 'vivas' to the absolute independence of Cuba!"[2]

In the previous chapters I have explained how radio participated in the reorganization of listeners' felt sense of social space; however, those examples drew largely from national or domestic networks within the United States. Zayas's broadcast functions differently, simultaneously addressing two national publics and indexing the vexed question of

Cuba's national borders. This forked address, which speaks in English from Havana to listeners inside and outside of his nation's borders, delineates a certain spatial sphere that these audiences share, a "*radio de acción*" formed from material (military, economic, and cultural) influence. Conscious of the U.S. naval base at Guantánamo and the special trade tariffs on sugar exports to the United States, Zayas points to how the apparent disembodiment of long-distance broadcast remains dependent on specific embodied relations.[3] His decision to address his words to ears beyond his nation, thereby converting any Cuban listeners into eavesdroppers, speaks to how the radio in Cuba was already bound by the *radio de acción*, the area of operations of its neighbor to the north.[4] The broadcast's ability to surpass national borders, in other words, also points to the unstable sovereignty of Cuba's borders. Opening up the transnational dimension behind and beneath what we have already seen in U.S. radio networks, the signal's reach shows us how seemingly peripheral locations underwrote apparent political centers to restructure the politics of the Good Neighbor.

This chapter takes seriously the symbolic and material importance of Zayas's inaugural state broadcast because his bifurcated speech anticipates the political and cultural struggles between Cuba and the United States carried out over the radio throughout the middle third of the twentieth century. As I will detail in this chapter and the next, radio broadcasting's relatively easy transgression of nearby national boundaries made it a focal point in battles over national sovereignty and cultural control in both countries. As Cuba, a U.S. protectorate from 1898 until 1902, shifted from a neocolonial nation to a revolutionary state to a Soviet-supported satellite, radio broadcasting remained a key tool in managing the country's fluctuating foreign relations. Early in Cuba's history radio provided a commercial conduit for U.S. businesses to advertise their goods, even to test their marketing campaigns before bringing them to U.S. audiences. Later, rebel relay stations in Florida and elsewhere would help disseminate the Cuban guerrillas' broadcasts from the mountains in eastern Cuba, sending the signals back to Cuban audiences with more hertzian force. And it was not incidental that in the wake of the Cuban Revolution in 1959, the propaganda wars carried out by both sides were known as "radio wars." Zayas's speech signals the complex and evolving transnational politics that politicians, activists, corporations, and writers enacted through the radio.

Given the intimate political and economic ties between the two countries until Cuba's alignment with the Soviet Union in the early 1960s, the story of Cuban radio also deepens the claims I have made about the connection between radio and property in the previous chapters. While Dos Passos's *U.S.A.* trilogy pointed to the ways in which the creation of national radio networks in the United States depended on the incorporation of broadcasting stations throughout the hemisphere, in what follows I will demonstrate how the interwoven radio networks in Cuba and the United States helped make possible the domestic (that is, national) changes to housing policy I discussed in connection to the work of Chandler, McCullers, and Wright in chapters 2 and 3. More directly, the chapter explains how the legislative synergy between housing and radio laid the groundwork for the creation of a new narrative genre—the *radionovela*—meant to sell U.S. goods to Cuban consumers. The melodramatic serial *radionovela—the* genre of domestic listenership in Cuba—helped generate the income for the United States to support the housing policies put in place during the New Deal. Radio's commercial extension of the Good Neighbor Policy, in other words, made Cuba a good neighbor by turning the country into a consumer of U.S. goods meant to prop up the domestic housing of the United States. How the genre both enabled and resisted this role through the vocal politics of a racialized domestic servant-become-matriarch in the world's most popular *radionovela* shows yet another alternative to the racial politics of the voice Richard Wright took on throughout his career. The cultural and military uses of radio in Cuba will reconceive the transnational from the other side of the radio dial and examine how alternative models of property and equality arose as new users created new functions in radio under revolutionary political conditions.

In linking the United States' area of operations to "powerful commercial influence," Zayas was in some ways attempting to improve the future power dynamic between Cuba and the United States. Just two years after the latest military invasion by the United States in 1920, the Cuban government sought to alter the legacy of the U.S. military on the island through increased and what they hoped would be mutually beneficial trade. Zayas could recall that the United States had claimed official control over the island only two decades earlier, from August 1898 to January 1899, after intervening in Cuba's fight for independence against Spain (what U.S. historians continue to refer to as the Spanish-American War). Seizing the opportunity to usurp Spain's colonial control in the Caribbean by temporarily supporting Cuban independence, a U.S. military government in the wake of the conflict remained in charge of Cuban affairs and forced the newly elected Cuban constitutional convention to sign the Platt Amendment. As Article III of the notorious document stated, "the Government of Cuba consents that the United States may exercise the right to intervene for the preservation of Cuban independence, the maintenance of a government adequate for the protection of life, property and individual liberty."[5] Only after this amendment became law, and the uncontested election of Cuban president Estrada Palma, who was living in the United States at the time, did the U.S. military government vacate the island on May 20, 1902. Thus, while radio historians in Cuba have denounced Zayas's kowtowing speech, we should also recognize that the president's address in 1922 employed the radio as an opportunity to speak beyond borders, which in some ways invoked the earlier borders of the fall of 1898.[6] To live under the commercial influence of the United States seemed to Zayas a compromise that might at least rid the island of a U.S. military presence.

Twelve years after the Cuban president's speech, three New Deal legislative changes related to domestic broadcasting and foreign policy pushed Cuba and the United States toward Zayas's ambitions. Today we can recognize the United States' 1934 Treaty of Relations, Reciprocal Trade Agreements and Federal Communications Act as the documents that supplanted U.S. military might in Cuba in favor of U.S. commercial influence and Cuban economic dependence. What might have seemed like benign national documents at the time now represent the "soft power" of neocolonialism that would dominate relations between the United States and Latin America throughout the twentieth century. This transformation was what Franklin Delano Roosevelt had in mind when, after four U.S. military interventions in Cuba, he spoke to the Democratic convention in Chicago of a policy of the good neighbor. Roosevelt had borrowed the term and much of its meaning from his predecessor, Herbert Hoover, who, as secretary of commerce from 1921 to 1928, "was responsible for redirecting U.S. policy away from what the nation has always done poorly—military occupation—and toward what it does exceptionally well—business."[7] The policy was tested quickly, when, in 1933, the Cuban people, partially instigated by a clandestine Cuban radio station, carried out a bloodless military coup against the dictator Gerardo Machado. Roosevelt resisted calls by his special envoy to Cuba, Sumner Welles, to send the military to "police" the island and thus demonstrated his loyalty to the doctrine of the Good Neighbor Policy's respect for Cuban sovereignty in a way unmatched by Hoover.[8] By the following year, May 29, 1934, Roosevelt formalized the Good Neighbor Policy with respect to Cuba through the Treaty of Relations, rejecting the Platt Amendment's doctrine of military intervention on the island but retaining the rights to the naval station at Guantánamo.[9]

Intent on trading military strength for economic power, Roosevelt's government quickly supplemented the Treaty of Relations with the Reciprocal Trade Agreements. The first of these agreements was proposed to Latin American nations by U.S. Secretary of State Cordell Hull at the Pan-American conference in Montevideo in December 1933 and summarily signed by Roosevelt on August 24, 1934. The treaty, which has been called "the basis of New Deal foreign economic policy,"[10] quickly supplanted American military intervention with Cuban economic dependence. Economic historians agree that the trade agreement "cement[ed] the fledgling Cuban economy to the United States" by reducing Cuba's protections on its infant industries in exchange for dramatic cuts in the U.S. tariff on Cuban sugar, tobacco, and other tropical agriculture.[11] Reducing Cuba's opportunity to diversify and industrialize its economy, the agreements made Cuba dependent on sugar sales to the United States to get money to purchase agricultural and industrial commodities from the United States.[12] Given this dependent model, it should be no surprise that the historian Robert Dallek contends "the reciprocal trade program chiefly served American rather than world economic interests."[13] The historian Lars Schoultz adds, "When trade reciprocity was combined with other New Deal initiatives, especially the 1934 creation of the Export-Import Bank, the net economic result of the Depression years and the ensuing European devastation was to increase further the dominant U.S. role in Latin American markets."[14] Twelve years after Zayas's radio address, the United States was only too happy to trade with Cuba and expand its sphere of commercial influence over the island.

What historians have failed to include in this story of changing hemispheric relations is the importance of a seemingly unrelated change in domestic broadcasting that helped secure the success of the United States' neocolonial control over Cuba. Adding to the Treaty of Relations and the Reciprocal Trade Agreements in 1934, the U.S. Congress passed the Communications Act, which I discussed in chapter 1, replacing the Federal Radio Commission with the Federal Communications Commission (FCC). The Communications Act guaranteed the dominance of commercial rather than public radio and effectively handed control of the radio to private corporations.[15] The synergy between these three New Deal policies meant that U.S. military intervention in Cuba had been replaced by Cuba's economic dependency on the United States, which translated into corporate cultural influence through the commercial airwaves. Cuba, which by 1931 was second only to the United States in its number of licensed radio stations, became a major market for advertisers eager to sell the goods the Reciprocal Agreements were already promoting.[16] In other words, radio became the key means for the new U.S. policy of "soft power" hegemony throughout the Americas.[17]

With the help of this new economic and cultural relationship, the island's most popular stations—CMQ, founded in 1933, and Radio Habana Cuba (RHC), which eventually became RHC–Cadena Azul—grew in influence during the 1930s. However, the stations expanded explosively only in the early 1940s, when the U.S. government encouraged their broadcasters to find Latin American affiliates. CMQ became affiliated with NBC, and RHC with CBS. These overt links to U.S. broadcasting only reinforced the ties to U.S. corporations that some Cuban corporations had already enfolded within their funding structure. The Mestre brothers, led by the youngest brother, Goar, took over CMQ through the money they had made distributing American products from Bestov Foods, and eventually General Foods, American Home Products, and Norwich Pharmaceutical.[18] Goar, the principal inspiration behind the new CMQ, had received his business degree from Yale in 1936 and only decided to purchase the station after he flew to New York in the summer of 1942 to meet with the Mexican radio tycoon Emilio Azcárraga. Mestre knew Azcárraga because the former had hired Mexican talent to appear on his clients' radio spots in Cuba. When they met in New York, Azcárraga explained his close ties to NBC and suggested Mestre purchase an existing Cuban station.[19] Cuban radio, in other words, emerged at the intersection of multinational business affiliations, linking the commercial-cultural ambitions of the Good Neighbor Policy in a much different image than that offered by Orson Welles's *Hello Americans*. The Cuban Mestre brothers represented the policy's hold over the island. By 1943 the brothers held a controlling interest in CMQ, which became the nation's largest and most influential radio and television station. CMQ would dominate Cuban entertainment for almost two decades until Fidel Castro's revolutionary government would replace it with a "rebel radio" in September 1960. By that time Goar Mestre, his commercial empire confronted with state control, would flee to Buenos Aires and never return.

THE *RADIONOVELA*

The U.S. commercial influence in Cuba found its most powerful conduit in radio advertising aired during Cuban *radionovelas*. These advertisements, primarily from U.S. cleaning supply manufacturers who bought airtime to sell their household goods to the largely female and domestic audience that listened to the melodramatic afternoon programs, eventually gave their name to the genre of the "soap opera."[20] Thus we should consider the *radionovela* one of the major cultural products of the new relations produced by the 1934 policy decisions in Washington and Cuba.

The very name of this new genre—*radio-novela*—announces the combination of old and new media, print and sonic techniques that brought new listening audiences into the field of literature. Different from *radioteatro*, which preceded it on the airwaves, the *radionovela* drew on two techniques from the print novel: the narrator and serial form. Félix B. Caignet, the genre's most celebrated innovator, whose work I take as paradigmatic of the genre in this chapter, introduced the narrator into Latin American radio in 1934 with his detective series *Chan Li-Po*. Based on the popular Hollywood adaptations of Earl Derr Biggers's Charlie Chan novels, Caignet's *radionovela* followed those films' exoticization of the eponymous detective with implications for transnational racial community and exclusion that I will address in greater detail in this chapter. At the level of technique, the radio narrator allowed Caignet to establish new levels of discursive space made audible to listeners by changes in speakers' vocal tone and pitch, or alterations in register—syntax and vocabulary. A new voice of authority and organization, the narrator could also add flexibility to the *radionovela*'s topography and chronotope by setting listeners in a new time or space.[21]

And as with today's soap operas or the novels of Dickens or Sue, *radionovelas* followed a serial format, ending each segment and the close of an episode with a cliffhanger to retain audience attention. Of course, the commercial function of serial programming was meant to keep listeners pegged to their radios during the advertising while they waited for the show to restart. The serial format also usually meant the production of vast scripts whose eventual page count would match or exceed their nineteenth-century progenitors. The popular demand for some shows led their writers far beyond the frame of their initial ambitions, and as I will explain, this extension had implications for questions of national consolidation and the confluence of what Lauren Berlant has named "intimate publics" based on affective relations with other, older notions of imagined community.[22]

Berlant's concept, which she aligns with what she calls feminine genres and melodrama, has special relevance for the *radionovela*, whose primary mode was melodramatic and whose audience, at least in the minds of the corporations and their advertisers, was primarily female.[23] In chapter 1 I drew on Loviglio's notion of "radio's intimate public" to discuss FDR's low-volume delivery and the acoustic engineering required to produce the atmosphere of the fireside chat. In my argument, Roosevelt called on this sonic form to support the ideology of the New Deal platform intended to make a disparate listenership feel a neighborly relation to each other through their assumed closeness to the friendly paternalism of "Frank" in the White House. The melodramatic mode of the Cuban *radionovela*'s intimate public has affiliations with that of FDR's "fireside chats," especially in its reliance on domestic and familial community to create an affective sense of social belonging.[24] However, the *radionovela*, with its position in between Cuba and the United States—as a Cuban genre selling U.S. goods—raises interesting questions regarding the intimate public Berlant argues "cultivates fantasies of vague belonging" through stories and commodities.[25] To which group does the Cuban *radionovela*'s listener belong? To a Cuban listenership whose structure of feeling is conditioned by U.S. advertisers, and thus a nominally U.S. nationalism? To a transnational grouping of popular sentiment? Or to a Cuban nationalism based on its own particular languages of emotion against the imported affects from U.S. distributors?

Precisely the confusion raised by the *radionovela*'s possible communities of belonging returns us to the neocolonial problem voiced in Zayas's address: the forked tongue of that broadcast comes back as a bifurcated social and affective imaginary within which listeners seek a place. Radio did not create this situation, but it helped extend and organize it in newly intimate ways for broad new publics. While it is not my aim in this chapter to analyze the history of melodramatic and sentimental literatures' effect on Cuban publics, it should be noted that already in the nineteenth century sentimental and melodramatic literature circulated between the United States and Cuba, shaping the culture of sentiment and political action.[26] In ways that anticipate the intermixture of melodrama and politics I will discuss in the close of this chapter, David Luis-Brown argues in his study of sentimental literature and hemispheric citizenship that much of the political rhetoric of José Martí, and therefore the very rhetoric of revolutionary Cuban nationalism, was shaped through Martí's reading of sentimental literature and melodrama from the United States.[27]

Martí exemplifies a reader borrowing from another nation's melodramatic tales' creation of a sense of social and national belonging, and then redirecting those texts to form both nationalist aspirations (Cuba was still a Spanish colony) and transnational feelings of social belonging. Yet, whereas the nineteenth- and early twentieth-century print publics included only literate readers, the *radionovela* accessed a much larger and more diverse audience whose cultural models of sentiment did not necessarily depend on print culture. Rather, the *radionovela* defined much of its own newly emerging canon of stories, and the central place of U.S. commodities and advertising in these apparently Cuban narrative worlds did not leave room or distance for the kind of recontextualization practiced by Martí.

In the rest of this chapter I will specify these problems through an exploration of how the most celebrated *radionovela* in radio history—Félix B. Caignet's *El derecho de nacer* (The Right to Be Born)—reinforced the conventions of melodramatic familial belonging and contested the image of that family's makeup. In order to demonstrate how the *radionovela* emerges from both national and transnational narrative traditions and cultures, I will show how it borrows from the fading genres of Cuban *negrismo* and declamation to organize a new sense of the familial, national, and transnational domestic: both the domestic worker and the sense of Pan-American domesticity called for in the Good Neighbor Policy. Finally, I will conclude with how this radio melodrama affected the Cuban revolutionaries who would utterly transform the New Deal notion of what it means to be a good neighbor.

EL DERECHO DE NACER: MELODRAMA, RACE, AND THE TRANSNATIONAL FAMILY

In July 1948 Cuban citizens took to the streets in protest because they wanted to hear a voice on the radio. Congressmen called temporary recesses on legislative debate and huddled around office radios waiting for the voice to speak. Movie theaters interrupted their scheduled films, turned the screens black, and played radio station CMQ through the theaters' speakers. The voice the people wanted to hear was not a young Fidel Castro, recently returned from the "Bogotazo" riots in Colombia. Nor was it Castro's mentor, Eduardo "Eddy" Chibás, who would commit suicide during one of his live radio broadcasts in 1951 with the hope that his death would launch the nation into revolution. Nor was the voice that of one of the would-be revolutionaries who would overtake Havana's Radio Reloj stations in 1957 and announce that then president and dictator Fulgencio Batista had been killed. (He had not: the rebels' attack on the presidential palace failed, and they were summarily executed.) Those radio voices would arrive in due time, but in July 1948, Havana, and most of Cuba, halted every afternoon to hear if Don Rafael del Junco, a plantation owner in the country's eastern city of Santiago, would recover from the mysterious illness that had left him mute, and admit that he was indeed the grandfather of the Havana doctor Alberto Limonta.

Of course, neither Don Rafael del Junco, nor Alberto Limonta, nor Alberto's adoptive Afro-Cuban mother, María Dolores Limonta, affectionately known by listeners as Mamá Dolores, are real people. They are instead the main characters in *El derecho de nacer* (The Right to Be Born), the first serial melodrama written for Cuban radio by the screenwriter, journalist, lyricist, songwriter, and poet Félix B. Caignet. *El derecho de nacer*'s cultural impact has been vast. In daily life, the show contributed a new idiom to Cuban speech. The phrase "Don Rafael del Junco ya habló" still appears in conversation, periodicals, and blogs as an idiom to describe someone who has broken their silence about a portentous topic. More broadly, the narrative that closed with don Rafael's revelation of paternity not only provided future programs with the basic structure and discourse of feeling that has defined the *radio* and *telenovela*, but it also left its mark on a generation of Latin America's most recognized authors.

As the most celebrated writers of the Latin American Boom eulogized their careers in memoirs and essays in the early twenty-first century, they returned to Caignet's work as *the* key text in twentieth-century Latin American literature. In his final nonfiction work, *La gran novela latinoamericana* (2011), Carlos Fuentes writes, "Félix B. Caignet's *El derecho de nacer* continues on as the most faithful mirror of a certain sensitive [*sensible*] and immediate Latin American reality."[28] Mario Vargas Llosa, commenting on the *radionovela*'s importance in his early training as a writer, says the genre "had such a tremendous audience that it was said that a passerby walking down any street in Lima could hear the chapters of Félix B. Caignet's 'El Derecho de Nacer.' There wasn't a single house that wasn't listening to it. That little picturesque and effervescent world gave me the idea for one of my novels: *Aunt Julia and the Scriptwriter*." Still more emphatically, Gabriel García Márquez recalls in his memoir *Vivir para contarla* (Living to Tell the Tale; 2002) how Caignet's work rescued the former's nascent writing career: "The only thing that gave me any hope were the mixed-up relationships from *El derecho de nacer*, Don Félix B. Caignet's radio novel, whose popular impact revived my old interest in sentimental literature . . . Inspired by the success of *El derecho de nacer*, which I followed with increasing attention throughout the month, I realized we were in the presence of a popular phenomenon that writers could not ignore."[29] More than a guilty pleasure or a kitsch object that helps distinguish the aesthetic autonomy of their own work, Fuentes, Vargas Llosa, and García Márquez extol Caignet's work for its emotional mimeticism, atmospheric influence, and calming power: García Márquez went so far as to compose his own adaptation of *El derecho de nacer* broadcast in Bogotá in 1954. And Caignet's influence did not stop with the Boom. Passing allusions to *El derecho de nacer* also appear in print novels like Jésus Díaz's magisterial *Las iniciales de la tierra* (The Initials of the Earth; 1987), Eduardo Galeano's *El siglo del viento* (The Century of Wind; 1986), and Zoé Valdés's *La nada cotidiana* (Yocandra in the Paradise of Nada; 1995). At the very least, Caignet has served novelists as a kind of cultural meme, a shorthand reference for a historical moment, genre, and culture of sentiment. These writers' insistent returns to the *radionovela* call for a more complex engagement with melodramatic radio and its importance for understanding the twentieth-century engagement between media and narrative form.

El derecho de nacer's plot, modeled on a revision of the Oedipus myth and Hollywood golden-age melodramas, follows the conventions of other national romances that celebrate familial belonging and biological reproduction. When Don Rafael's daughter María Elena is left pregnant and alone by her lover, Don Rafael insists she have an abortion, which she rejects. But when the child is born, Don Rafael brings him to a cabin on his coffee plantation and pays an Afro-Cuban woman, María Dolores Limonta, to dispose of the child. Of course, she never does. Instead, Mamá Dolores leaves for Havana, where she raises the child she names Alberto Limonta. Alberto eventually becomes a doctor, and thirty years after his birth, he ends up saving Don Rafael, who has moved to Havana in the intervening years and is brought into Alberto's emergency room after being struck by a falling tree during a hurricane. However, just when the old man is going to reveal his identity, thanking the doctor for his service and admitting to him that he is his grandfather, Don Rafael suddenly becomes incapacitated with a temporary embolism. Only after almost three more months of episodes does Don Rafael finally recover his ability to speak and reveal his secret. The program concludes by returning to its frame, with Alberto in his Havana medical office recounting his life story—the story of *El derecho de nacer*—in order to successfully convince a patient to forgo having an abortion.

This anti-abortion morality tale extols the virtue of the Afro-Cuban domestic servant against the criminal actions of the white plantation owner to ultimately assure the reproduction of the Cuban nation. The *radionovela* thus reproduces the conventions of nineteenth-century hemispheric sentimental literature's "romantic racialism," which Luis-Brown, following Linda Williams, identifies as "a more benign version of racialist thought than scientific racial determinism": where "blacks and women are inherently superior rather than inherently inferior in that they allegedly possess a greater degree of Christian morality, reverence, and deeply felt emotions than white men."[30] Williams and Berlant's work on the adaptations of *Uncle Tom's Cabin* reinforces Luis-Brown's focus on a hemispheric circulation of nineteenth-century melodrama, which, in turn, extends their studies of *Uncle Tom's Cabin*'s adaptations and afterlives into twentieth-century novels and films across the Americas.[31] *El derecho de nacer*, in other words, is an utterly conventional story. But that is precisely why it matters. A study of how the *radionovela* uses these and additional conventions to structure the figure and voice of Mamá Dolores—whose name appropriately translates as "Mother Pain"—reveals how Caignet's work taps into the long history of transnational melodrama and more local Cuban cultures of racial performance to make audible two voices: the "muzzled" *bozal* voice of racialized domesticity that both supports and challenges the idea of the "good neighbor," and the "muffled" voice of plantation power heard more clearly in the advertisements in between the show's episodes. In other words, *El derecho del nacer* can tune us in to the politics and artistry of the radio voice—in particular, the temporarily silenced voice of Don Rafael and the dialect of Mamá Dolores's voice—to try to hear what makes both voices possible, and how they, in turn, reorganize the legacy of the Good Neighbor Policy.[32]

MASKS OF SOUND: VOICING "BLACKNESS" IN PRE-REVOLUTIONARY CUBA

In his definitive study of Caignet, *El más humano de los autores* (The Most Human of the Authors; 2009), the Cuban critic Reynaldo González writes, "*El derecho de nacer* is *the* work in which the stylistic and ethical ideas of radio melodrama reach their summit with authentic and didactic [*aleccionadora*] communication."[33] Key to the drama's stylistic conception is the voice of Mamá Dolores, what González elsewhere identifies as "a caricature of the *bozal* pronunciation, made even more stereotypical for its radiophonic characterization."[34] What González's text hints at here is the way in which *El derecho de nacer* marks the high point of Cuban radio melodrama through a voice that also registers the rapid decline and fading sound of a literary world that had largely defined popular, avant-garde, and bourgeois culture in 1920s and 1930s Cuba: the "negrismo" movement and the art of declamation. In Mamá Dolores's voice Caignet transforms these castoff modes of a previous generation of cultural practice into the primary materials for the most popular genre of storytelling in the twentieth century.

When González refers to Mamá Dolores's voice as "a caricature of the bozal pronunciation," he invokes the prehistory of "poesía negra," and the nineteenth-century form Jill Lane has named "discursive blackface." Lane's history outlines the invention of the literary dialect of "bozal," whose literal meaning—"muzzle"—was intended to reference the speech of "unacculturated African slaves"[35] Tracing its invention to the Cuban writer José Crespo y Borbón's character Creto Gangá in 1839, she follows the popularization of blackface in the Cuban "teatro bufo" through the "negrito," one of three social-racial types alongside "la mulata" and "el gallego." In her nuanced study of this tradition, Lane argues that blackface functioned as both "a racist desire to control the representation of black people and blackness in the new Cuban imaginary," as well as "a key site for the development and articulation of a recognizably national affect and style throughout the anticolonial period."[36] Underscoring "the difference between what we call performance and performativity itself—between the really made up and the made up real,"[37] Lane argues that the linguistic distortion of Iberian Spanish through Crespo's "bozal," and the "darkening" of an idealized whiteness through blackface performance helped produce a new *mestizaje* Cuban identity. Such an identity, she argues, finds its culmination in the paradoxical "mestizo autóctono," or "indigenously mixed-race subject," called for by José Martí in "Nuestra América." And it was this new subject that Cubans would use to define themselves and their emerging nation as they fought for independence against the Spanish Crown and the looming neocolonial presence of the United States.[38]

This prehistory begins to explain how Cuban audiences in 1948 could differentiate between the meaning of the "muzzled" speech of Mamá Dolores and the muffled speech of Don Rafael, and how sound came to carry markers of race, class, or gender. However, a more historically proximate connection also made Mamá Dolores's voice possible to write, to perform, and to hear. The twentieth-century literary movement of "negrismo" or *poesía negra, poesía mulata, poesía afrocubana*, and *poesía afroantillana*, borrowed from and altered the nineteenth-century tradition of Cuban blackface but was still written predominantly by white and mixed-race Cuban men who combined elements of the Afro-Cuban popular in music, speech, and religion.[39] Different from the *négritude* movement of Aimé Césaire and Leopold Senghor, negrista poets like Caignet, Emilio Ballagas, and their more celebrated contemporaries Nicolás Guillén, and the Puerto Rican poet Luis Palés Matos imitated Afro-Cuban speech and customs in broad strokes, often writing through the image of blackness rather than from the personal, complex psychology of an individual perspective within the black diaspora.[40] At their most sympathetic, they turned to Afro-Cuban culture to define Cubanness in the face of the U.S. neocolonial influence and, especially in the case of Guillén, to forge a sense of diasporic African culture. Perhaps the most successful of these literary relations was Guillén's friendship with Langston Hughes, and the latter's translation of Guillén's poetry into an African American poetic idiom that helped construct a way to speak across these diasporic traditions.[41] And, much as Michael North has argued that imitations of African American dialect by T. S. Eliot, Ezra Pound, and other U.S. writers helped fuel modernist formal innovations in poetry, scholars of Cuban poetry have shown how *poesía negra* similarly drove poetic experiments in *la vanguardia* through the pages of *Revista de Avance*, *Cuba Contemporánea*, and other journals.[42]

Caignet's poetry from this time played a key part in establishing and popularizing certain speech forms as recognizably Afro-Cuban, regardless of their distance from any ethnographic reality. Although his collection *A golpe de maracas: Poemas negros en papel mulato* (The Beat of the Maracas: Black Poems on Mulatto Paper) was not published until 1950, most of the book's poems were written and performed in the early 1930s. In their mix of sympathy and caricature we find a poetic style akin to other *negrista* poets like Guillén. However, Caignet evinces a particular sensitivity to how this invented stereotype of living speech can become a voice to speak up against racist power.

Caignet's poem "La conga prohibida" (The Outlawed Conga) stands out for its political stance against Cuban government censorship of Afro-Cuban drumming in the 1920s and 1930s. That poem, written in the *bozal* speech form, opens with a stanza that inverts the primitivism that characterizes some "poesía negra" and was often used as a justification for white condemnation of Afro-Cuban culture: "¡Qué balbaridá, Dio mío! / A donde bamo'a paral . . . ? / A lo negro no han dejao / lo blanco sin calnabal. / La conga ettá prohibía, / dice un bando milital . . . / Quien no tenga traje'seda / Se queda sin calnabal" (What barbarity! My God! / When is it going to stop . . . ? / The white has taken away Carnival from the black. / The conga is forbidden, / says a military ban . . . / Whoever doesn't have a silk suit / Can't go to Carnival).[43] As Thomas F. Anderson remarks, the poem's opening phrase "¡Qué balbaridá!" casually redeploys the racist accusation that African music is somehow "barbaric," and uses the popular expression "¡qué barbaridad!" to condemn the white government's "barbarism" in prohibiting the conga and Carnival.[44] Furthermore, as the last lines demonstrate, accompanied by playful musical internal rhymes on "seda" and "se queda," class politics drives this racial misalignment: only those who can afford silk suits have the power to prohibit carnival. In delivering a political message in caricatured speech, "La conga prohibida" reveals the double movement of negrismo as it celebrates, supports, and patronizes Afro-Cuban culture.

Although analyzing such poems as printed objects still grants us access to much of their meaning, we should recognize that the belated publication of Caignet's writing meant that these poems primarily found audiences in the 1930s through oral performance. If the negrismo movement and its links to "teatro bufo" provide one of the cultural histories that helped produce and make audible Mamá Dolores's speech by the time *El derecho de nacer* premiered on Cuba's radio CMQ in 1948, the other literary and performance trend that helped Caignet write for the voice was the art of declamation. Declamation emerged in Europe in the nineteenth century as an aesthetic discipline linking elocutionary training in normative pronunciation with a belief in the human voice as the ideal vehicle for carrying and producing affect. Theorized and practiced in Germany and France with the rise of acoustic science and the desire for a universal language inspired by the writings of Herder and Rousseau, declamation's supporters argued that the oral recitation of poetry brought out the inherent musical tones in speech: a putatively potent means of communication beyond the limitations of the written word and specific national languages.[45]

In Latin America, the Russian-born, Jewish-Argentinian *declamadora* Birta Singerman popularized declamation in the 1920s and 1930s in performances, concerts, and films like *Nada más que una mujer* (1934). After earning a scholarship to study literature and "the arts of reading and declamation"—proving that declamation was still a valued literary discipline—Singerman rose to such fame that she and her fans dubbed her "la libertadora de la poesía."[46] While her detractors claimed her recitations made "un mausoleo de la palabra" (a mausoleum of the word),[47] Singerman influenced a generation of writers and performers, from the Argentine poet Alfonsina Storni to the Afro-Cuban *declamadora* Eusebia Cosme. It was Cosme who brought Caignet's poems to popularity first in Cuba and then in the United States, where she moved in 1938. There she recited Caignet's poems, along with those of Nicolás Guillén, at venues across the city, including Carnegie Hall, and later, with Langston Hughes, on her CBS–Cadena de las Américas radio program, *The Eusebia Cosme Show*, in 1943.[48]

Cosme's oral performance of Caignet's written poems turned his printed words into a script or a score for physical, intellectual, and sonic interpretation. Antonio López and Jill Kuhnheim have highlighted this "interpretive" function in Cosme's performances, arguing that recitation works as a critical act that shapes meaning by tuning audiences in to the subtle means by which register, tone, pitch, and timbre index social worlds, individual psychology, and fields of affect.[49] Recitation, in this case, sounds something like another mode of critical "reading"; it is a reading or reciting practice that situates the performer somewhere between the author and the audience in the production of meaning. Cosme's years of study in rhetoric, literature, and music helped her embody meaning in sound, making changes in volume, enunciation, and rhythm constitutive of a character or emotional state.

But what is the relationship between sound and blackness in these performances, and does Cosme help us understand Caignet's "poesía negra" differently? To reformulate the title to Caignet's book of poetry, what happens when "poemas mulatos"—poems written by a mulatto Cuban man—are recited and interpreted "en papel (del) negro," in the voice of an apparent *afrocubanidad* by an Afro-Cuban woman? Strikingly, Caignet's writing in *bozal*, a speech form associated with improper, inadequate, or otherwise erroneous pronunciation as judged by an idealized Spanish norm, required a performer expert in the art of enunciation to render them emotionally powerful and legible. As Caignet himself freely admitted in an admiring letter to Cosme, "ninety percent of the myth of my verses is in your *interpretation* [interpretación] . . . My poetry exists only when you perform it!"[50] Only in hearing Cosme's "interpretation" of his poems did Caignet consider them complete and comprehend the voice he had been seeking.

From this angle, Cosme's sonic performance succeeds not because she speaks with some impossibly authentic voice. Rather, it is precisely the high artificiality, the academic training, and the aesthetic labor of this joint venture that makes the poems a popular and critical success. As Emily Maguire has commented, even as Cosme's position as "a speaking black female subject . . . visibly contest[s] the objectification this same female figure receives in much *negrista* and Afro-Antillian poetry," her readings do not speak to "one image of racial identity, whether stereotyped or harmonizing[.] [T]hey speak instead to the many acts of translation—to the *articulations* and disjunctures—inherent in representations of racial identities."[51] Cosme, Caignet, Guillén, and others helped produce the sound of a represented *afrocubanidad* in the 1920s and 1930s—not an authentic sound from within a community but an image of that community that became legible or audible as blackness to a Cuban radio audience.

This emphasis on representation, drawing, as it did, from the nineteenth-century pictorial and narrative typologies of *costumbrismo* and the "teatro bufo," always threatened to fall into mere stereotype, and in the mammy-like character of Mamá Dolores it sometimes did. Mamá Dolores, as a self-sacrificing domestic servant speaking a caricatured language, could appear to be a simple import of the racist stereotypes already popular in the United States. After all, Caignet's first radio protagonist, Chan Li-Po, was barely more than a copy of Charlie Chan. We should not forget that as much as Mamá Dolores's voice emerges from the already conflicted tradition of the "teatro bufo," it might also follow the less ambiguous minstrelsy of the United States. On the other hand, thousands of letters Caignet received from Afro-Cuban women thanking him for the character of Mamá Dolores, who was portrayed on radio and later in film by the white actress Lupe Súarez, testifies to a more complex understanding of character, performance, emotional authenticity and identity that Cosme's own work enabled.[52] Cosme herself never disavowed the role: in 1966 she played Mamá Dolores in the second cinematic adaptation of *El derecho de nacer*. Respecting Cosme's decision as more than a mere capitulation to a market that held few roles for black actors, we can approach Mamá Dolores's role as simultaneously a powerful representation of black domestic women's essential role in holding Cuban society together and an essentializing blackness that ignored the complexities of individual women in order to maintain the status quo under the military dictatorship of Fulgencio Batista.[53]

We are certainly more accustomed to understand this complex mediation of identity in the latter, more familiar way: as a consequence of an already emerging society of the spectacle—since even the voice here is meant to conjure an image—in which listeners are mere consumers, accepting the identities provided for them by a mass commercial media that shapes generic identity in order to more efficiently identify consumer demographics and marketing strategies. No doubt this, too, accounts for something of Mamá Dolores's voice. It also returns us more directly to the voice, or rather the silence of Don Rafael del Junco. Alongside Mamá Dolores's "muzzled" speech, Don Rafael del Junco's muffled, straining, and illegible voice dragged *El derecho de nacer* out for another seventy-five episodes and three months as a more direct consequence of radio's commercial ties. When the voice actor José Goula, who played Don Rafael, realized that the plot hinged on him admitting to Doctor Alberto Limonta that he was in fact the doctor's grandfather, Goula decided to go on strike until his salary was increased. While Radio CMQ's owner, Goar Mestre, ordered Caignet to fire Goula, Caignet instead developed the ingenious solution of Don Rafael's long silence, which only increased listenership and an obsessive fan base that increasingly paralyzed the country with the desperate question, "¿Ya habló Don Rafael?" In Don Rafael's muffled voice of plantation power one could hear Mestre's stifled business demands, and a minor victory in a worker's labor struggles, but also a sound that produced further profits for Mestre and his advertisers. It is the sound of art under capitalism: the dialectic of resistance and containment.

THE RADIO DISCOURSE OF THE PASSIONS: CHIBÁS, CASTRO, GUEVARA

Mamá Dolores and Don Rafael's voices, respectively "muzzled" and muffled, tell the story of Cuba's neocolonial radio world in the 1940s and 1950s, precisely the world Cuba's revolutionary actors would attack and supplant a decade later. One might assume that the radio revolution that arrived with the consolidation of Fidel Castro's guerrillas in Cuba's eastern mountain range of the Sierra Maestra in 1958 opposed CMQ's commercial and sentimental content from the start. However, if Cuban historians and scholars tend to depict prerevolutionary radio as mostly crass commercialism, the divisions between politics and melodrama were more like the studio glass at CMQ: *radionovela* celebrities could get distracted in their rehearsals while watching the leader of the political Ortodoxo Party, Eduardo "Eddy" Chibás, and his young protégé Fidel Castro Ruz deliver their political denunciations of government corruption in the adjoining studio.[54] And the studio glass was two-way. Anke Birkenmaier describes "the contamination of political discourse by the *radionovela*'s emotional discourse" and points to Chibás's on-air suicide and its legacy as the culmination of a process in which "the radio helped construct a passionate public narrative with which everyone could identify."[55] Everyone, and especially Fidel Castro, who, Birkenmaier suggests, "found the antecedent for the Revolution's passionate discourse" in Chibás's own rhetorical flourishes over the radio.[56] As I will explain in the remainder of this chapter, the history around Castro's eventual decision to use the radio for revolutionary ends required him to both borrow from and reformulate radio's previous associations with passionate discourse.

Castro's relationship to Chibás stands at the beginning of that history. Chibás, who began his radio career as a supporter of former president Ramón Grau San Martín's Auténtico Party, before establishing his own Ortodoxo Party on a platform of cleaning up the corrupt Cuban government, became Cuba's most well-known radio orator, denouncing real and imaginary corruption over the air for thirty minutes every Sunday evening from 1943 to 1951.[57] Often chastising the efficient-minded business model of CMQ's Goar Mestre for failing to pay for his program, Chibás typically ran past his allotted time, and his verbal attacks stretched the limits of what could be said on the radio. Indeed, his vehemence inspired the last democratically elected Cuban government to establish a "right to reply" law in 1950, which "permitted anyone personally attacked over radio to petition the Ministry of Communications for free air time on the offending station."[58] Compelled to supply more evidence to support his denunciations, a year later Chibás issued less than damning proof in documents accusing President Prío's minister of education, Dr. Aureliano Sánchez Arango, of embezzling funds.[59] On August 5, 1951, Chibás, his numbers down in the polls and still stung by his critics' remarks, went on air and shot himself in the stomach toward the end of his program. He thought the act would spark a revolution in the streets, as if replacing his voice with the sound of a gunshot were his only viable political response. However, in accordance with the stringent business model Goar Mestre learned from his time at Yale and NBC, the CMQ technicians, unaware of what was about to unfold, but following their boss's instructions to hold orators to their allotted time, cut Chibás off in midsentence and went to commercial just before his gun went off.[60] The death of Chibás, a victim of the marketer's bottom line and his own belief in radio's oratorical power, marks a later stage in the neocolonial radio relationship between Cuba and the United States, and a hinge to the anticolonial project to come.

Chibás's belief in radio's revolutionary potential to turn words and sounds into instant political change was not singular in Cuban history. In 1933 and 1957, rebels occupied radio stations and prematurely declared the respective governments' overthrow.[61] Neither attack proved successful, but the idea that radio alone could start a revolution remained part of the island's political imaginary. After Fidel Castro was arrested following his July 26, 1953, attack on the Moncada military barracks in Santiago de Cuba, he felt compelled to explain why he hadn't occupied a radio station instead. In his defense *La historia me absolverá* (History Will Absolve Me), delivered before the dictator Fulgencio Batista's military tribunal, Castro invokes and then departs from these antecedents. There he boldly insists: "With only ten men I could have occupied a radio station and launched the *pueblo* into the fight. It wasn't possible to doubt their spirit [*ánimo*]; I had Eduardo Chibás's last speech from the CMQ, recorded in his own words . . . and I didn't want to make use of it, despite the desperation of our situation."[62] Castro claims to have turned away from replaying Chibás's galvanizing address from two years earlier in order to defend the people. As he says, "We agreed not to take over any radio station . . . and our attitude saved the citizenry a river of blood."[63] In his explanation, the radio retains its capacity for instantaneous uprising, as it had for

Chibás in his planned "live suicide." However, precisely this power to fuse *armas y letras*, or weapons and words—the utopian belief of the twentieth-century vanguardist movements—becomes, from Castro's perspective, an unnecessarily dangerous tool.

Radio's danger, for Castro, derives from its notorious disembodiment. Castro's entire defense in this speech belabors the importance of the *body in uniform* as necessary to his military *decorum*, which seeks to assure that his fellow revolutionaries are recognized as lawful combatants.[64] In this context, his rejection of radio's *disembodied* power to call out the nonuniformed *pueblo* from their homes bolsters his adherence to international law. And yet the effect is strange: Castro seems to say that a people's revolution must not be carried out by the people.

More precisely, Castro's speech retracts from the people's spirit (*ánimo*). Without delving too far into the kind of hauntology that has come to typify accounts of radio as ghostly disembodiment, it is worth recognizing that Castro mutes Chibás's words because he fears that a voice from the past made to sound as if it lives in the present will overwhelm his own control.[65] Playing Chibás's *recorded* voice over the radio's *live* transmission threatens to enact the leaderless animation of the *pueblo* that would undo his role.[66] This belief in radio's liveness, derived from a stubborn tradition that had long interpreted even those recordings broadcast over the radio to enact the real event, produces a specific kind of listening experience that Castro fears for its power to animate the audience.[67] However, over the course of the coming years, something about this belief would change. By April 1958, encouraged by Ernesto "Che" Guevara, Castro signed on to the idea of rebel radio as essential to the guerrilla warfare that would finally undo Cuba's economic dependency on the United States. What happened between 1953 and 1958 to alter his belief in radio as an appropriate tool for revolution?

Although sources for answering this question are scarce, we can recognize that a specific change in content became central to making radio revolutionary for Castro. As the radio became a major force in the rebels' guerrilla warfare tactics, Castro, Guevara, and lead propagandist Carlos Franqui debated what content best served political ends. If Castro had rejected the idea of playing Eddy Chibás's voice, what would they broadcast instead? Franqui recalls: "Despite [Castro's] intentions, I wouldn't let him put bad poetry nor even *radionovelas* on Radio Rebelde. If he accepted my decisions, it was because my work on Radio Rebelde was effective, and he was the first beneficiary."[68] Elsewhere, adding to his criticism of Castro's taste, Franqui writes that "back in the Sierra, Fidel had wanted us to broadcast scenes from the war on Radio Rebelde, while Che and I insisted reading poetry."[69] While we should approach an exiled former propagandist's declarations with healthy suspicion, we can nevertheless understand Franqui's recollection as part of an ongoing struggle to define the meaning of radio in the Revolution's history. For Franqui, who invited Jean-Paul Sartre, Simone de Beauvoir, and other cultural luminaries to the island, Radio Rebelde's objective was to disseminate high culture to the populace, opposing melodramatic *radionovelas* and live-action battle scenes.

On the other hand, officially sanctioned memoirs of Radio Rebelde published in Cuba depict broadcasters' descriptions of the scenes of war, and even radio communications between Fidel Castro, "Che" Guevara, and Camilo Cienfuegos as a substitute for *radionovelas*. According to these memoirs, the station's "*great novel* was the epic of challenging a tyrannical oligarchy defended by a professionally armed military supported and trained by the United States."[70] For these memoirists, Radio Rebelde, station 7RR served as a cultural corrective against the commercial stations' "low level of programming," their melodrama and "neocolonialist" imposition of "Yankee" shows: "The entire country (when I say *country* I mean *pueblo*), during the nighttime hours of Radio Rebelde's broadcast, turned their back on . . . the weepy radio novels to tune in to [*pegar su oído a*] that call of fiery Martinian sermon and absolutely honest news."[71] The *radionovela* vanishes here in favor of a powerful masculine figure of politics and poetry—the passionate rhetoric of José Martí. Likewise, Martí's "fiery" language can now replace the passionate discourse inherited from the *radionovela*. In both Franqui's summation and that of the Revolution's record, the *radionovelas*, depicted as carriers for the virus of U.S. capitalism, encounter their antidote in Radio Rebelde's poetry or "honest news."[72]

While these accounts tend to ignore the diversity of programs broadcast over the country's two major stations, they foreground how a central component of the rebel attack on Batista included a radio resistance to the economic dependency amplified by the Good Neighbor Policy's three decisions of 1934. In addition to these battles over radio's content, radio became a centerpiece to the specific tactics of resistance developed by Guevara in his handbook *La guerra de guerrillas* (Guerilla Warfare; 1961). Writing about radio as if it embodied the kind of Promethean speech act we heard described by Ruttenburg in chapter 1, and rejected by Castro in his 1953 defense, Guevara insists: "The propaganda that will be the most effective in spite of everything, that which will spread most freely over the whole national area to reach the reason and the sentiments of the people, is spoken words over the radio [*la oral por radio*]. The radio is a factor of extraordinary importance. At moments when war fever is more or less palpitating in everyone in a region or a country, the inspiring, burning word increases this fever and communicates it to every one of the future combatants."[73] Radio, which does not require its users to be literate, can reach a larger audience, spreading, in Guevara's livid analogy, like wildfire or a virus, and allowing the isolated rebel army "to speak directly to the Cuban people, bypassing its urban underground and other opposition groups that had previously served as interlocutors."[74] In this semantic field of elemental oppositions, revolutionary radio counters the weepy voice of the *radionovela*'s melodrama with "fiery" sermons that burn across the country. The radio—its listeners and products understood as less complex, layered, and conflicted than I have depicted them earlier in this chapter—no longer becomes a means to speak on behalf of U.S. commercial concerns but rather bypasses those corporate models and speaks "directly to the Cuban people." In other words, Rebel Radio's action becomes a specific corrective to the *radionovela*'s popular paralysis.

The French journalist and media theorist Régis Debray, who was arrested in Bolivia in 1967 for aiding Guevara's insurgency there, would add that radio allowed the guerrillas "to play an ever larger role in the country's civilian life . . . Increasingly, everyone [in Cuba]—from Catholics to Communists—looked to the Sierra, tuned in [to Radio Rebelde] to get reliable news, to know 'what to do' and 'where the action is.' Clandestinity became public."[75] Thus, "radio produces a qualitative change in the guerrilla movement."[76] This "qualitative change" is known as Guevara's *foco* theory of revolution, in which "guerrillas comprise an elite revolutionary group (the *foco*) which stimulates opposition to the status quo among the relatively passive masses. The masses are inspired by the guerrillas' actions and would remain passive without them. Guerrillas use radio to direct the actions of the civilian opposition. Radio is an instrument by which guerrillas maintain leadership of civilian organizations even though they remain geographically remote from the cities and population."[77] In these theories, radio allows for a formal change in revolutionary armed struggle based on the location of address, its massive oral transmission of messages, and its ability to translate this speech into massive direct action.

In speaking "directly to the Cuban people," Radio Rebelde adamantly reorganizes the political use of a technical infrastructure inherited from Zayas's inaugural broadcast in English.[78]

The most powerful example of radio's importance for the Revolution came on the first official day of the revolutionary calendar. As Fulgencio Batista fled the island on January 1, 1959, a group of military personnel attempted a coup d'état, which threatened to undo the rebels' claim to the government.[79] Radio Rebelde's directors, including Carlos Franqui, went on the air immediately and asked other stations to link up with and rebroadcast 7RR's signal.[80] As Castro later recalled, "it was necessary to stop any illusion that the coup d'état could signal the triumph of the Revolution. It was necessary to alert the masses, to alert the *pueblo*. And in that moment Radio Rebelde played, I mean, carried out its last fundamental role in wartime. That was transmitting—in a network [*cadena*], practically, connected with all of the other national stations—the instructions to stop the coup d'état."[81] In a total reversal of his proclamation six years before against radio's address to the people, Fidel spoke the following message: "¡Revolución, sí; golpe militar, no!" (Revolution, yes! Military coup, no!).[82] As a result, workers across the country declared a general strike, effectively rejecting any government not led by the rebels. Castro's radio call, issued from the eastern capital, Santiago de Cuba, linked the country's two main cities through a single voice. If Castro had ultimately accepted Guevara's belief in radio's ability to speak directly to the people, he did so with his own voice, not a recording of Chibás's last words. These "qualitative changes" in the guerrilla movement, from the *radionovela* to Chibás's jeremiads against the government, to Castro's turn against the radio, to his final praise, through Guevara, of the radio's power "to alert the *pueblo*," proves the previously vague claim that Castro's "passionate discourse" extends sentimentalism into political action.

The consequences of this new use of radio would dramatically affect the very concepts of the domestic—as worker, as nation, as private/public barrier—and property. Radio Rebelde, as a guerrilla-directed infrastructure built to counter the messages of commercial capital, became the model for the Revolution's nationalization of private industry and private property, as well as the basis for its overt politicization of cultural life. In the subsequent years, as the revolutionary government took over control of Mestre's Radiocentro complex, expropriated privately owned land through the agrarian reform laws (May 1959–63) and redistributed real estate through the Urban Reform Law (October

1960) to establish “the concept of housing as a public service,” the monumental power of Radio Rebelde’s model for revolutionary notions of property and domesticity became clear.[83] In chapter 5, I will describe how radio’s role in helping to sustain that project in the ensuing decades further challenged the national property lines and the differing conceptions of property that divided Cuba and the United States.

CHAPTER 5

Radio's Revolutions

Commenting in June 1960 on his first visit to Havana, the civil rights activist Robert F. Williams, who self-identified as a "black internationalist," recalled that a white reporter from the United States told him, "I had no business talking about the race problem over here and that I should do my talking at home."[1] Williams, by this point, had been talking at home for years. As president of his local chapter of the NAACP in Monroe, North Carolina, he had spoken out vehemently against the Ku Klux Klan, and since 1956 he had published a broadside he named the *Crusader*, dedicated to denouncing racism in the United States. Williams was most infamous for advocating and practicing armed resistance to racial oppression. As a supplement to the nonviolent tactics officially supported by the NAACP and Martin Luther King Jr., Williams told the national press in 1959 that African Americans must meet "violence with violence," a position he formalized in his 1962 book *Negroes with Guns*. That book, which opens with the question, "Why do I write to you from exile?," was published in Cuba, where Williams fled in order to escape first a white mob in Monroe, and then the FBI, which was pursuing him on exaggerated charges of kidnapping a white couple.[2] Thus, two years after the reporter had told Williams to do his talking at home, he found he could no longer speak from home. In exile in Cuba from 1962 to 1966, and then in Mao Tse-Tung's China from 1966 to 1969, Williams finally returned to the United States at the end of the 1960s to advise the State Department on normalizing relations with China. Coming full circle, he now taught people at home how to talk "over there."[3]

In between these moments, however, Williams found a way to speak from two places at once. In a radial revision of the form of forked address produced in President Zayas's official inaugural Cuban broadcast I discussed in chapter 4, Williams and his wife, Mabel, broadcast an English-language radio program from Havana to the United States in order to agitate against racial oppression. They called it *Radio Free Dixie*. For many listeners throughout the United States it became the soundtrack to the struggle for civil rights.

I will return to the Williamses later in this chapter, but I open with them here because *Radio Free Dixie* and Robert F. Williams's practical theorization of what Michel Foucault would call "the enunciative function"—what it means to speak from one place or another—embodies many of the problems I have been tracing throughout this book.[4] From Dos Passos's search to represent "the speech of the people" as a "radio network," to Richard Wright's critique of the racial unevenness in "the people's voice," to Félix B. Caignet's racialized construction of prerevolutionary Cuba's most popular voice, to Fidel Castro and Ernesto "Che" Guevara's anti-imperial populist "rebel radio," I have sought to demonstrate how radio broadcasting technology became the chosen medium for writers and political activists to theorize, represent, and construct a political voice through which the people can speak. I connected this desire for political community to the challenge of real property: that is, both the challenges real property imposes on hopes for political community and the challenge to real property these writers found in radio. Finally, I detailed how writers borrowed from radio to enact specific formal changes to realist literature—shifting its focus from "things" to spaces, transforming characters from expressive subjects to conduits—to reimagine the limitations of real and personal property.

I have interpreted this work as part of a transnational, improvised, and open-ended struggle for different actors to build a "neighborhood" of the Americas that would exist as an alternative to the Pan-American, U.S.-imposed hemispheric idea of the "good neighbor." While the book's first part focused on the creative domestic response to this idea in the United States, I turned to Cuba in chapter 4 to examine how the very idea of domestic sovereignty in the 1930s United States depends on the exploitation of non-national, and neocolonial relations. Through Caignet's *radionovela*, a product of those relations and yet also a popular critique of the expansion of racial exploitation that Wright's work diagnoses in the United States, I showed how melodrama—realism's other—provided a new framework for thinking through the "housing problems" posed in the opening chapters. Additionally, the porousness between this melodramatic mode and the apparent "epic realism" of the Radio Rebelde broadcasts complicated any easy divisions one might expect to distinguish the two genres.[5]

In the following pages I continue this investigation into voice, property, and narrative form through the various "radio wars" carried out between Cuba, the United States, and their proxies in the wake of the Cuban Revolution. These wars fought for control of the neighborhood of the Americas—how that community's relations would be defined, who would be included, and where its limits would be drawn. As the revolutionary government fought against the neocolonial commercial relationship already present in the inaugural broadcast forty years earlier, they used radio to tell their story. The expropriation of Goar Mestre's Radio CMQ Radiocentro building in 1960 marked the end of commercial radio on the island for at least another fifty years, as Radio Rebelde became the model for the revolutionary government's national expropriation of private property.[6] Given the way radio had helped solidify U.S. control over Cuban markets and, on the other hand, how the medium became the means by which the Revolution came to power, it is perhaps not surprising that both countries launched "radio wars" that mobilized an audible arsenal in battles over private property and national sovereignty.

Indeed, radio's challenge to the territorial limits of national sovereignty marks its most original threat to property. Speech and speakers outlawed in one place could now reach listeners as if they were next door. Where Wright could only hear the radio as the voice of white power within the United States in the late 1930s, two decades later Williams understood it as a means to broadcast the voice of black power from outside the country's borders. In doing so, Williams's broadcasts found a new method to threaten a whole history of personhood, property and racial subjugation. Here was the power of speaking from one place to another.

This question of address structures the following pages. After examining the discourse of the radio wars and the Williamses' broadcasts, I conclude with an experimental radio play by the Cuban-born writer Severo Sarduy. My reading of his *Los matadores de hormigas* (The Ant Killers; 1978) shows how the site and reach of broadcast speech, as well as the combination of Caignet's melodramatic *radionovela* and Radio Rebelde's "epic realism," create a new audible form that undermines even the charismatic voices of rebel leaders. Taking advantage of the spatial possibilities of stereo technology, Sarduy's play reworks Williams's concern with the place of address to decolonize radio discourse and its imperial and racist history. Pushing against the limits of realist discourse and print in ways that will sharpen realism's possibilities in the closing chapters, Sarduy's play helps us to critically reconstruct listening as an anticolonial force.

RADIO WARS

In the wake of Radio Rebelde's military success, which helped secure the rebels' victory and an end to commercial U.S. radio ownership on the island, an information war began over the airwaves between the two countries.[7] With the covert help of the CIA, in 1960, Thomas Dudley Cabot, former president of United Fruit Company and ex-director of the State Department's Office of International Security Affairs, began using United Fruit's former relay station on Swan Island, off the coast of Honduras, for propaganda against Castro's revolutionary government.[8] The station, dubbed Radio Swan, broadcast programs taped in Massachusetts by Pepita Rivera, a Cuban exile billed as "Havana Rose," and rebroadcast signals from stations in Miami, including the instructions to Cuban dissidents during the Bay of Pigs invasion on April 17, 1961.[9] With the invasion's failure, the main U.S. propaganda station, Voice of America (VOA), increased its budget by $3 million, expanding the Latin American service from six to twenty-two hours of daily Spanish and Portuguese broadcasts in order to counter, in President Kennedy's words, "broadcasts from Havana . . . encouraging new revolution in the hemisphere."[10] Thus, in the wake of the Revolution, the United States used shortwave and medium wave (AM) broadcasts to establish simultaneous commercial and governmental broadcasts to Cuba. Although U.S. corporations no longer held property on the island, the U.S. government fought to help them regain control of Cuban airwaves from afar.

Whatever the intended effect of these broadcasts, critics in Cuba and elsewhere have long noted their failings. In the Cuban newspaper *Revolución*, an August 27, 1963, article subtitled "Bandera de pirátas" (The Pirates' Flag) rails against the United States' Voice of America broadcasts. The writer bemoans the North American broadcasters' "pretension of paying homage to what they hate most. Those of the 'voice' read verses from different Cuban poets about their national flag. In the station's mouth they sounded like insults. But the worst thing . . . was that for the background music for those poems they chose nothing less than pieces from the song, 'Nena,' from the movie, 'El Último Cuplé,' played on guitar. Neither the movie, nor the song, nor anything else was Cuban." As in other cases, the location of the address alters the words' meaning. Cuban poetry spoken through the Voice of America sounds like an insult to listeners in Cuba. Part of what alters the meaning is the soundtrack, which does not seem to realize it is playing a Spanish song to Cuban nationals. If, on the one hand, the poetry is insulting for its Cubanness spoken in a foreign mouth, on the other, the music is insulting because its speakers fail to realize it is not Cuban. Listeners with a sense of colonial history would no doubt hear *El último cuplé*'s Spanish provenance as a marker of the broadcast's imperial overtones.

The cultural and national asymmetry that emerges in this instance of failed propaganda further complicates how words shift in meaning as they tote their cultural baggage with them across the ether. Citing the analytic philosopher C. L. Stevenson's concept of "persuasive definitions," according to which "a term changes its descriptive meaning without substantially changing its emotive meaning—the emotion that the term expresses," Philo Washburn observes that "'freedom' on the VOA generally referred to personal liberties such as freedom of expression, assembly, and religion, while on [Radio Havana Cuba], 'freedom' generally referred to collective rights such as freedom from poverty and hunger and access to guaranteed housing, health care, and education."[11] What Washburn does not discuss is how the reception of a given message by a listener outside the intended ambit of the broadcast might inflect these terms with her own cultural formation. While the VOA and RHC broadcasters might share an "emotive meaning" with many listeners and frame their terms toward specific semantic fields, the above example from *Revolución* makes clear that listening also transforms—reframes or redirects—the words at the site of reception. The places of enunciation and reception matter in a way that radio's attenuated *radio de acción* make particularly audible.

Semantic questions about the place of listening carry specific political weight. The proliferation of techniques and stations used on all sides to "win" the radio war often ran up against, and sometimes ignored, national and international law. While the VOA's charter forbids it from broadcasting within the United States in order to prevent the government from "propagandizing" within the country, the Kennedy administration's decision to use domestic AM stations to broadcast to Cuba has been seen by some to violate the law, since domestic listeners could tune in to programs intended for Cuba, too.[12]

Indeed, as radio broadcasts leapt from Massachusetts to Miami to Honduras to Cuba, or from Cuba's Sierra Maestra to Caracas to Miami to Havana, national uprisings and binational wars were awash with hemispheric noise. The transnational interference made the metaphor of the "neighborhood" feel real. Cuban station CMBL tried to block Dominican dictator Trujillo's Radio Caribe's rebroadcasts from the U.S.-based Cuban Freedom Committee; Radio Budapest inadvertently interfered with Cuba's Radio Aeropuerto broadcasts to the United States; and Radio Havana Cuba's signal, established in early 1961, pestered Nicaraguan dictator Anastasio Somoza, who complained "that he could hear accusations against him 'in my own home as if [Radio Havana Cuba] were installed in Managua.'"[13] When, in the summer of 1963, Cuba began the hemisphere's first electronic signal jamming with a device provided by the Soviet Union, it blocked VOA transmitters in the Florida Keys, Radio Swan, WGBS Miami, WMIE Miami, WKWF in Key West, and stations in Florida, New Orleans, and the Dominican Republic that were broadcasting Illinois Democrat Roman Pucinski's aforementioned Cuban Freedom Committee programs.[14]

Conquering a signal with noise, or the noise with a clear signal came to index political power. As early as May 19, 1959, the Cuban newspaper *Revolución* depicted Fidel Castro's speech on agrarian reform as a sonic battle won over busy and contested airwaves:

> 6:28 Violeta Casal identifies the station: "Radio Rebelde here, and its Network of Liberty from the heart of the Sierra Maestra." 6:29 The reading of the Agrarian Law begins. Jorge Enrique and Violeta alternate: "Regarding . . ." 6:31 Brief pause in order to change the fuel for the portable radio transmitter. 6:36 Renewed contact with La Plata. The reading resumes. 6:45 Interference. A speaker, in English. Possibly some Florida radio station . . . 8:45 Fidel begins: "Certainly, the most common emotions invade one in these instants." 9:00 Interference. Electronic noise. Mechanical roars. Noises. Sirens. But Fidel's voice conquers. He can be heard above the interference.[15]

Staging the interference and distortion of the radio signal as a struggle against foreign influence ("A speaker, in English. Possibly some Florida radio station"), the speech's content falls away, replaced by the simple feat of transmission: "Fidel's voice conquers." The speech's purpose—to declare the law of agrarian reform that would strip U.S. businesses of their property in Cuba and deliver the land to the Cuban people—is heard in the nonsemantic properties of the broadcast: the noise-to-signal ratio of Fidel Castro's voice.

RADIO FREE DIXIE

Among the broadcasts involved in the radio war, Robert F. Williams's *Radio Free Dixie* program remains the most explicit revision of the island's neocolonial broadcasting history. More powerfully and confrontationally than the other transnational broadcasts I have discussed, Williams's program focuses on the place of his enunciation—Havana, Cuba—and the meaning that arises from this context for listeners in the United States. An African American man advocating armed revolt against "racist white America," Williams found in radio broadcasting a potent response to the ongoing limitations to imagine "the people" without forgetting or erasing the racism such a project had so often entailed. Its first broadcast only a year after the Kennedy administration's revision of the Good Neighbor Policy under the name of "The Alliance for Progress," Williams's *Radio Free Dixie* speaks—in English—from Havana to the United States to reframe the political and ethical questions of the neighbor in a transnational field that simultaneously questions the status of the "foreign" in the policy of the good neighbor.[16]

Keeping in mind the reporter's assertion from the beginning of this chapter—that Williams "had no business talking about the race problem over here and that [Williams] should do [his] talking at home"—we can ask once again what it means to speak from one place or another. Why does Williams's place of enunciation anger his listener? What does it mean to speak *toward* home as opposed to speaking *from* home? How does the geopolitical context of an utterance alter the form and content of that utterance? Where does his address fit in the history of the policy of the good neighbor?[17] In asking these questions, I also hope to come closer to understanding something more about the specificity of radio: Williams's chosen medium to allow his voice in Cuba to emanate from speakers in living rooms, cars, and on park benches *inside* the United States.

In temporarily bridging the gulf between Havana and Monroe, North Carolina, between Cuba and the southern United States, *Radio Free Dixie* reversed the current of the Good Neighbor's commercial influence, exporting music, literature, and culture *to* the United States. And yet, part of the program's originality in the radio wars derives from Williams's decision to broadcast in English and his insistence on a program with musical content exclusively imported from the United States. The only English-language program on Havana's Radio Progreso at the time, *Radio Free Dixie* did not attempt to export Afro-Cuban culture to the United States or to explain to listeners how the more varied and flexible racial categories in Cuba departed from the simple binary distinction between "black" and "white" operative in the United States.[18] Instead, explicitly directing his program to the audience of "Black America," Williams sought to export U.S. culture back to the United States. Whereas forty years earlier Zayas addressed himself in English to U.S. listeners recognizing Cuba's place in the United States' commercial sphere of influence [*radio de acción*], Williams spoke in English to inspire radio action, to propagandize for an African American revolution in the United States, and to provide an oral and musical soundtrack to the ongoing civil rights movement.

Such a project poses particular problems for theories of transnational cultural exchange intent to depict hybridity in race or language as the apex of progressive politics. *Radio Free Dixie*'s broadcasts do not resist racial binaries, and they flatly ignore the potential "translation zone" comprising the Caribbean and the southern United States.[19] In some ways, Williams's rather opportunistic use of Radio Progreso even recalls the earlier colonial and neocolonial period in Cuban history, wherein the United States included Cuba as part of its territory.[20] Invoking that period of Teddy Roosevelt's "Rough Riders," Amy Kaplan writes that "the links between disenfranchisement in occupied Cuba and the Jim Crow South point to imperialism as the *exporter of the domestic color line* and recontextualize racism at home [in the United States] as part of a global imperial strategy of rule."[21] In a strange sense, Williams takes this previous imperial model and, due to the Revolution's own successes, inverts it. Precisely the geographic proximity between Cuba and the United States, and the political entanglement of the two countries, generates the anticolonial power behind the station's slogan: "You are tuned to *Radio Free Dixie*, from Havana,

Cuba, where integration is an accomplished fact."[22] Through the radio wavelength, Williams stretches Dixie to Havana, and the wireless imaginatively snaps the color line, breaking its inevitability. He restores the dialectical relationship wherein Cuba's place in the U.S. sphere of influence places the United States under Cuba's sphere of influence. For listeners in Monroe, North Carolina; Birmingham, Alabama; or elsewhere in the South, *Radio Free Dixie* was not a utopian fantasy: it came from a real island and spoke through an embodied voice too close to home to be ignored.

Broadcasting from his anticolonial outpost, Williams borrowed from the legacy of radio along the U.S.–Mexican border, playing music without paying heed to sponsors or censors.[23] In *The Crusader*'s first advertisement for the program, Williams wrote: "The object of 'Radio Free Dixie' is to create a better understanding of the Afro-American problem in the USA, and to expose the true nature of U.S. Racism. The program will consist of Jazz, Afro-American Folklore, news, interviews and commentary. This will be the first completely free radio voice that black people have had to air their case against brutal racial oppression."[24] While Williams and his wife, Mabel, read news about racial violence in the United States and editorials from the *Crusader*, or the black nationalist theologian the Reverend Albert B. Cleage Jr., they used music, specifically the "new jazz" of Max Roach and Ornette Coleman, as well as soul albums from Curtis Mayfield, Nina Simone, and Sam Cooke, "to create 'a new psychological concept of propaganda' by combining 'the type of music people could feel, that would motivate them.' He envisioned 'something similar to what is used in the churches—[in] the "sanctified church," there is a certain emotion that people reach.'"[25] Advertising in *The Crusader* for readers to send him "jazz, Dixieland, folk-music and recordings of the current protest movement in the South," Williams received records from friends such as Amiri Baraka, Richard Gibson, and William Worthy.[26] As a result, *Radio Free Dixie* could broadcast the voices and sounds of a new musical and political vanguard to audiences who might not otherwise have access to the movement.[27] With a keen understanding of radio's unique capacity to balance words and music, political commentary about African American life throughout the United States and the musical revolution led by African American musicians, Williams helped provide the soundtrack for the civil rights movement.

As the "new music" changed, so too did *Radio Free Dixie*'s own style of address. In a January 21, 1966, broadcast, Mabel Williams praised "hep" jazz and rock musicians. Repeatedly comparing the music of "socially conscious rock 'n' rollers" to "their forefathers' spirituals," she exalted the musicians' "crying and moaning" as "symbolic of the frustrated black souls who labor and suffer in racist white America. [These musicians]" she went on, "are becoming the epic force of the Afro-American revolt . . . These new cats are no slackers. We are tuned in to their wavelength and we dig them plenty."[28] Mabel Williams, as she demonstrates in this broadcast, is attuned to the nonsemantic modes by which an alternative vocabulary and specific musical selections call out to different constituencies within Black America. Juxtaposing political speech and music, the Williamses joined fellow African American DJs for whom "soul music became a race-coded soundtrack for the assault on Jim Crow" as "black appeal radio became an important means of spreading the civil rights message."[29] At the same time, Robert Williams also broadened the audience for this music through his 50,000-watt transmitter; at Radio Progeso, Williams reached a diverse audience as far away as Los Angeles and Saskatchewan, Canada.[30] He knew from his own experience as a listener that radio would be the best medium to expand his message's audience. As he told one listener from Illinois:

> When I lived in North Carolina, I used to hear the Cuban station, RADIO PROGRESO (long wave), but the broadcasts were in Spanish, which I could not understand. When I was forced into exile in Cuba, I remembered RADIO PROGRESO and thus conceived of the idea of RADIO FREE DIXIE. Having fought for justice for my people as an official of the NAACP for six years, I didn't want the Ku Klux Klan to have the satisfaction of having silenced my voice. Formerly, I could reach only a few people while living in the "Free World" with my Newsletter, THE CRUSADER, and public speaking. From here, I am reaching the whole world, so the racists have done their cause and their country's cause a great disservice.[31]

Listeners across the country and, indeed, the world could speak to the program's success in keeping Williams's voice present and powerful in the United States.[32]

The diversity of listeners and listening modes testifies to the tremendous audience his broadcasts reached. Federal Bureau of Investigation employees interviewed listeners or, as listeners themselves, merely transcribed Williams's speeches without paying attention to the political potency of his voice's timbre or the songs he played.[33] Some listeners wrote Williams merely to report on the strength of the program's signal in Dallas or Bethlehem, Pennsylvania.[34] Others wrote to praise the music or ask the identity of a singer. (Williams often wrote back with details about an album and where to order it in the United States.)[35] Still others wrote to thank Williams for broadcasting a message otherwise unavailable from stations in the United States. A listener in Kalamazoo, Michigan, wrote to tell Williams he was "glad to know that the truth is really being broadcast," another in Miami, Florida, exclaimed, "thank goodness someone is saying what most people are afraid to say in this country," and another in Louisville, Kentucky, said, "I'm pretty far north to receive standard broadcast that far, but though the signal was weak and fading, *the message was not*. I think it's a fine program and I know that you will gain much support with it—simply by telling the truth, which is not done in Racist America."[36] Echoing Williams's intention with the program, yet another listener in New York wrote to say, "The minute it started coming in loud and clear, something told me the station must be in Cuba; must be, because U.S. stations do not care or dare to broadcast the new music our musicians are really playing today and our peoples [*sic*] hearts are beating to, because the new music is for freedom."[37] Even detractors testified to the broadcast's power. "I am sorry to say we have such a formidable enemy, the radio," one disgruntled listener in Florida wrote to Williams. "Your 50,000 watts reaches out and spreads your venomous lies all along the Atlantic coast line [*sic*]."[38] With the ability to circumvent the censors, Williams effectively altered the soundscape in the United States, changing what could and could not be said, as he framed his musical content with political speeches so that each reverberated in the other. And as independent radio stations in the United States began to rebroadcast his programs, or listeners taped his shows and passed them on to friends, his audience even expanded beyond the signal's original reach.[39] To borrow a term from the sound scholar Jennifer Stoever, we can say that Williams's broadcasts dramatically changed the parameters of "the sonic color-line."

However, it was precisely his program's music and its message that propelled Williams to leave Havana. Early in *Radio Free Dixie*'s development, the head of the CMCA English-language radio station initially opposed Williams's jazz selections as "decadent imperialist noise."[40] Although Fidel Castro approved of the program, Williams came under fire again for publishing an article in the Cuban magazine *Bohemia* that distinguished jazz from swing. "Jazz is the music of the Black people of America," he wrote, "and it's an insult to call it decadent."[41] Williams's support for jazz, and his refusal to advocate for socialism in his broadcasts—Williams never enlisted in the Communist Party and included the statement, "printed as a private publication in Cuba" in every issue of *The Crusader*—ultimately distanced him from the increasingly dogmatic Communist line in Cuba.[42] His biographers and critics Cohen, Tyson, and Thomas all agree that Williams left Havana for Mao's China because he felt alienated by Communist ideology in Cuba, and its insistence that he subordinate the African American struggle against "racist white America" to class revolution. Strangely, he found more freedom in Chinese communism. By 1966, he departed Cuba and never returned.

THE ANT KILLERS: THE EXPERIMENTAL AESTHETICS OF ANTICOLONIAL RADIO

Los que debían de leerme, que son los cubanos, no me conocen ni me pueden leer.

(The Cubans, who should be reading me, don't know me, and can't read me.)

—SEVERO SARDUY, "Para una biografía pulverizada en el número de *Quimera*"

A year before Williams left Cuba, already frustrated with the CPUSA's antagonism toward *Radio Free Dixie* and the case for African American civil rights in the United States, he began to harbor suspicions that Che Guevara had been forced out of Cuba due to his own struggle with Communist Party leaders. Whatever his motivation, Guevara had traveled to the Congo (Zaire) under a false passport, and from April 24, 1965, until November 1965, he attempted to aid in the guerrilla struggle there. However, Williams's frustration and Che's clandestine mission were interrelated in other ways. As Piero Gleijeses puts it, in their struggle against the United States, "[t]he Cubans . . . were not suicidal . . . They gave moral support to radical African American groups, but were careful not to provide them any material assistance, and certainly not military training. The Cubans tried to avoid the lion's jaw. They responded to the U.S. challenge, instead, in the Third World. In the years immediately after the revolution, Cuba had focused on aiding rebels in Latin America. By 1965, however, its attention had shifted to Africa." Although Guevara's attempt to establish a guerrilla column in the Congo largely failed, Cuban fighters and doctors who had worked alongside him were the only foreigners who participated in the successful independence movement of the PAIGC (African Party for the Independence of Guinea and Cape Verde) in their fight against the Portuguese colonials in Guinea Bissau beginning in May 1966, and ending in 1974.[43] By July 1975, the Cubans had moved on to Angola to help the MPLA (People's Movement for the Liberation of Angola) in their struggle against the Portuguese colonialists and the opposing rebel groups, the U.S.-funded UNITA (National Union for the Total Independence of Angola) and FNLA (National Front for the Liberation of Angola). African colonies and emerging nations, in particular those guerrillas fighting for independence from Portugal, became the focus of Cuban military aid and the site for exporting revolution.

Combining this history and the interconnections between the *radionovela* and rebel radio discussed in chapter 4, we can hear the Cuban-born author Severo Sarduy's radio play *Los matadores de hormigas* (The Ant Killers; 1978) as a narrative experiment in anticolonial sound.[44] Sarduy's play, originally broadcast in French translation over Radio France in 1981, is about the use of radio during the end of the colonial struggle in Angola, and that war's effects on the Portuguese Empire. It is also a play that uses sound to reorient how we listen to the radio, how we define it as a medium, and how we think of its political functions.

The play's complicated structure includes six acts, or what Sarduy names "sequences," organized through a series of events meant to repeat and reverberate across the play. In a key feature to which I will return, the play's voice actors, whom the production notes only identify as Man One through Man Four, and Woman One and Woman Two, are called on to assume different characters in the play's various settings. To sketch the play briefly, sequence 1 satirizes a group of "revolutionary tourists" from France who seek out "free love" in a Portugal itself under siege by local guerrillas.[45] Across the next two sequences, these groups interact, as the guerrillas, described as neoprimitive "wild boys," dance next to their radios and eventually occupy the Salazar Bridge on the outskirts of Lisbon. When we shift to Angola in sequences 4 and 5, we hear four other Frenchmen as they take photographs of wildlife, poverty, and rebel fighters for European publications seeking lurid and glamorized depictions of death and violence. After the photo shoot, in sequence 5 the group overhears two Portuguese soldiers describe abandoning their military outpost in the face of the independence uprising in Angola. As the play draws to a close, the different stories begin to collide. The tourists who crashed their car in sequence 1, like the photographers who crashed their car in sequence 4, meet the soldiers, recently returned from Angola in sequence 5, who drive them to a rental car service. In the play's final sequence, the tourists find themselves stopped by the Portuguese rebels on the Salazar Bridge entering Lisbon. In the argument that unfolds in the rest of this chapter, we will come to understand that this conclusion to a play about decolonization reveals the empire occupied by the periphery, and the radio's central role in this anticolonial project.

Indeed, the play's ending marks the culmination of its narrative logic and architecture. Like a broadcasting radio tower or a stone dropped into a pond, the play's events reverberate and radiate across time and space in concentric circles. The radios within the play's diegesis help enact this narrative trick; they serve as relay stations between the play's different sequences. For instance, if we hear a song hummed in one section, we find it broadcast in the next. An event, like a car crash, occurs in sequence 1 and becomes part of a news broadcast in sequence 3. The simple repetition begs the listener to link the two events, and thus the play's narrative structure succeeds in interpolating listeners into the radio's colonial role: forging these connections between a sequence in Lisbon and another in Angola imitates the ways in which the radio sustains the empire's contact with its colonies.[46]

Surprisingly, the play's anticolonial project works through the simple reversal of this radiating logic. Following Radio Rebelde's example of guerrilla radio, as well as Frantz Fanon's revolutionary theory of radio listening as an anticolonial practice, Sarduy sets out to create an experimental guerrilla aesthetics for radio drama.[47] In corner stores, in cars, at military barracks, on barren paths, the play's diegetic radios blare out with music and news of Angola's independence movements, announcing the invasion of the empire's soundscape with the sounds of the periphery. Moreover, in the extradiegetic space of the play, Sarduy attempts to enact a further sonic decolonization, in which "the basic [sonic] elements—the music from Portugal's former colonies—continually accept other sounds into their interior, which progressively push them towards the edges of the page, and of our hearing."[48] For instance, Sarduy's production notes indicate that the play's diegetic action in the first sequence should be interrupted three times. The first time we hear the music of Guinea Bissau, which his notes identify as "[1]," and that of Mozambique, marked as "[2]." Later in the sequence, after the tourists crash their car in the diegetic space, we hear the following in the extradiegetic space: the music of Guinea Bissau [1], the sounds of a car's breaks [3], and the music of Mozambique [2]. Finally, at the end of the sequence, we hear the music of Guinea Bissau [1], the screeching sounds of a car's breaks [3], the music of Angola [4], and the music of Mozambique [2]. These sonic collages connect the play's diegetic action to the music of Portugal's current and former colonies, once again pointing to the radio's colonial force in managing the far-off territories of empire, but they now also establish a possible network or constellation of anti-imperial sounds returned to deterritorialize Portuguese culture. The radio, in this case, begins to open a portal to the dialectical movement between empire and colony, imperialism and anticolonial struggle: a doorway through which a far-off event, such as the car crash, can make its way into the listeners' experience.

Sarduy's attempt to aurally represent the anticolonial process both borrows and departs from other anticolonial models. For instance, in Fanon's analysis of radio in the Algerian revolution, he argues that rebel radio becomes audible to the population when the rebels offer programming audibly different from the colonial programs' emphasis on the empire's music and culture. Fanon, in turn, models his writing practice on these radio strategies, when he imitates the rebel radio signal and "decolonizes" French by employing Arabic words or French terms with Arabic etymologies.[49] Sarduy's sonic collages similarly transform the extradiegetic soundscape, as the postcolonies dominate our hearing. Yet, strangely, Sarduy does not place Portuguese music at the center of his "basic sonorous elements" and marginalize it to the "edges of our hearing." Instead, as the play continues, the sounds of the tourists and their imported songs of revolution occupy the center of our hearing. Perhaps speaking to his French broadcast's national audience through locally legible sounds of resistance, at the end of sequence 3 the music of Portugal's postcolonies gives way to the sound of a Portuguese street protest and the French singer Michel Delpech's "Ça ira." The latter, a hit pop song whose lyrics invoke the symbolic leader of French revolutionary sentiment, ironically seems to locate Europe, and especially France, at the center of a postcolonial soundscape.[50]

Of course, we can also read and, more importantly, listen against the grain of Sarduy's self-analysis and hear the music on the edges actually containing those sounds that come between them. In this case, the empire falls into the fold of the former colonies. One selection in the second series, however, complicates even this hearing. The production notes first list Billie Holiday's album *Lady in Satin* when Man One in sequence 2 says that the "ant killers," young Portuguese rebels camped out in the scrubland outside of Lisbon, "knew *Lady in Satin* by heart and sang it with exaggerated 'ch.'"[51] In the French broadcast, we hear a different album play almost inaudibly in the background at this time. However, it is not until sequence 5, just as we transition from the photographers to the stories from Portuguese soldiers in Angola, that we hear Holiday's voice as one of the sonorous element interludes in the play, sandwiched between music from Brazil, Goa, and Portugal and the sound of African birds.[52] Why do we encounter the U.S. jazz vocalist in this play about decolonizing Angola? We hear the album twice more: at the end of the sequence, and at the end of the play. In those moments Portuguese music precedes it, and Lady Day is followed by the play's first sentence: "It takes place in Portugal."[53] What does it mean to hear Billie Holiday's voice in this context? Does it carry the same resonance as when Robert Williams played it for listeners in the United States in his broadcasts from Cuba? Does Holiday carry a specific political resonance, like Michel Delpech's "Ça ira" in the play's first half, or do the two become just empty signs for popular "foreign" singers when broadcast together? Perhaps in hearing Holiday in this context it becomes easier to hear how Sarduy's sonic decolonization works.

On one hand, Billie Holiday's inclusion participates in the disruption of the Euro-African political and cultural field shaped by the sonorous elements in the play's first half. By including music from Goa and Brazil in the play's second grouping of sonorous elements, Sarduy already breaks with the westward flow of slavery, reminding listeners that the Portuguese slave trade moved toward India, as well as the Americas. In this context, the familiar tendency to hear Holiday's voice on French radio as another example of the ongoing movement of the "black Atlantic" is disrupted and similarly disorients the listener expecting to hear more music from former Portuguese colonies.[54] Holiday's voice trails off here in different directions. The context asks us to hear in her voice the echo of that trail of bodies—recalling that the slave trade extended far beyond Portugal's imperial desires—and the potent protest against racial hatred most famously rendered in her version of "Strange Fruit." The personal ruin, audible in *Lady in Satin* (1958), the last of her recordings she heard released, would stand in for the historical suffering of an entire mass of peoples.[55]

And yet, even as Holiday's voice might resound here as part of Robert Williams's work on *Radio Free Dixie*, connected to a larger moment of black internationalism that we hear in the play's radio broadcasts of the Angolan independence movement, it also points to a commercial empire outside the ambit of the colonial project. This broader commercialism threatens to convert the musical selections into instances of "world music" and thereby flatten their meaning.[56] Caught in this vague category, the music's possible defamiliarization can find itself already normalized by the radio apparatus itself, which includes musical montage like this as a basic condition of its listening experience, as the turning of the dial.[57]

If this attempt to wrestle radio from its colonial use only opens the door to commercial power, the play resists arguing for the mere seizure of the means of production as part of the decolonial effort. Eschewing a wholly economic rationale for the region's devastating social conditions, the play returns us to think about radio's role in the colonial project. Not the specific relationship between Portugal and Angola, and the struggle for national sovereignty, but rather the micrological forms of interpersonal communication that Sarduy claims govern radio discourse. Commenting on "the decolonization of the body . . . [and] the liberated voice," Sarduy writes, "I'm trying . . . to destroy the radio dialogue, an archaic form of communication between two voices in which one always tries to 'colonize' the other."[58] Notably, Sarduy's "liberated voice" (*voz liberada*) distinguishes itself from Marinetti's Futurist "liberation of the word" (*parole in libertà*), by inverting the desire to absorb a multiplicity of listeners into the voice of the single speaker with the desire (*sexualidad liberada*) to dissolve the single speaker into a "galaxy of voices."[59] Once again sidestepping the critical commonplace to hear radio as fascist discourse, and challenging even the intimate, calm dialogic voice in FDR's fireside chats, the play's "galaxy of voices" equally resists the Bakhtinian utopia of the dialogical imagination. Anticolonial radio, Sarduy suggests, will not adhere to belief in the rational conversation of the public sphere since dialogue itself contains the colonial project in miniature.

In order to counter the colonial force of radio dialogue, Sarduy elevates listening into an art necessary for his writing. He elaborates his audile technique through an explanation of the connections between his own professional work on French radio and his literary production in a 1985 article, "I'm a Present-Day, Electronic Joan of Arc." There he writes: "I earn my living as a radio journalist, on French International Radio. However, not only the radio broadcasts, but rather everything that I write is suitable for broadcast, is *essentially vocal*."[60] Writing "for the voice" (to name the collection in which *The Ant Killers* was eventually published), Sarduy has to learn to listen. Comparing himself to Joan of Arc ("Like the warrior saint, I hear voices"), he describes how his writing practice derives from a listening practice, claiming: "I don't write for anything but those voices . . . Everything is already 'said' from the beginning. There is no other beginning necessary than listening."[61] This emphasis on listening and the voice might seem odd for readers familiar with the decadent visual vocabulary of Sarduy's essays in *Escrito sobre un cuerpo* (Written on/about a Body), and he acknowledges as much: "The radio writing is a consequence, the result of an initial listening [*una escucha inicial*]. But one could ask me: Why the voice, and not the image, for example?"[62] Like the other writers in this study, Sarduy asks what listening can add to storytelling as he constructs his own version of narrative acoustics.

To explain why he focuses on the voice rather than the image, he turns to what he calls the "erotics" of the voice. Recalling how he and Roland Barthes "explored the labyrinth of many voices . . . from their nocturnal echoes in some little plaza in Tangier," Sarduy glosses Barthes's idea of "the grain of the voice" as "a texture, an intonation, a bumpiness [*rugosidad*], a timbre, an accent [*un deje*]: something that unites the body with something else, that at the same time centers it, motivates it and transcends it."[63] No doubt Sarduy exoticizes as much as he eroticizes the voices he hears. This prejudicial listening threatens to turn these voices into pure sound, both phenomenologically dense and transcendentally abstract. Yet, we might also hear this description as the negotiation between the corporal and cultural, the individual and collective. To shift Sarduy's terms toward those of the linguistic anthropologist Nicholas Harkness, we can say that Sarduy describes what Harkness names "the phonosonic nexus": the meeting point that links "the literal phonosonic voice—the 'voice voice'—that emerges from vibrations and resonance in the vocal tract," and the "tropic extension of the voice."[64] The physical sensuousness in a voice accrues cultural meaning, such that the sound of a voice comes to index a whole series of social positions in the world. Voice, in this case, is no longer a wholly textual object, whose social register we discern through changes in vocabulary, repetition, and focalization. Rather, voice becomes inseparable from sound. It is precisely this space of sonic-social indexicality that Sarduy will manipulate in his attempt to decolonize the voice.

Sarduy enacts this decolonization through sound and voice by attempting to mute, or, in the dramatic language he employs, "destroy" them in his play. In the introductory note that accompanies *The Ant Killers* we find that this hinge between the body and the subject has been obliterated. Voice, especially the radio voice, in this play, will liberate the body: "*The Ant Killers* is a text about decolonization: of territories and bodies."[65] In praise of the acousmatic, voices no longer connect with subjects, "what one person says could be said, in the end, by another. The voice is not an indicator of a psychology or a personality."[66] This notion of the liberated voice is instantiated through the play's ubiquitous radios, and the narrative use of multiple actors to speak single strips of discourse, thereby harnessing broadcast's fantasy of the disembodied voice to challenge personal, national, and imperial boundaries. Sarduy appears, in this case, as an inheritor of Dos Passos's and Wright's work with the democratic possibilities of the radio voice. Like them, he uses his narrative to work through mediality—the multiple meanings, uses, and political effects of radio. Unlike those earlier literary modes, Sarduy works with sound on the radio to decolonize the radio voice.

To be yet more specific, we might say that for Sarduy the anticolonial project hinges on the voice, whereas the colonial project emerges from dialogue's insistence on the pronoun *I*. After all, he concludes his introductory note with this fractured sentence: "The Dissolution. Of the *I*."[67] How might the erasure of a personal pronoun contribute to sonic forms of decolonization? In the introduction to *Never Say I*, a study of the French modernist novel and the artful practices authors developed for speaking about same-sex sexuality, Michael Lucey quotes Oswald Ducrot: "What is remarkable about the pronoun *I* is not just that it constitutes a shorthand way of speaking about oneself. It is more that it obliges the person speaking to refer to herself with the same word that her interlocutor will use in turn for the same task. The use of *I* (and much the same could be said of *you*) thus constitutes an apprenticeship in and a constant practice of reciprocity."[68] Ducrot's view of this reciprocal process seems positive. The pronoun's remarkable indexical flexibility allows it to correspond to a multiple number of speakers, training them how to borrow the word and release it to the other. Sarduy, on the other hand, insists in the written introduction to his play that such "reciprocity" is "colonial"; hence his desire to dissolve the first-person singular.[69] While a writer like Proust, in Lucey's study, avoids the autobiographical use of "I" because it would tie him to a sexual identity and cultural position he still wants to negotiate in the more open space of fiction, Sarduy attacks the pronoun because it petrifies social positions into a colonial dialectic, where the "I" always struggles to remain in control. While Sarduy—a queer author in exile in Paris from Cuba's Revolution, steeped in French novels and culture—might be thought among the aesthetic and social inheritors of the novelists in Lucey's study, the former's insistence on the radio, not the novel, as the medium to challenge the first-person singular deserves particular attention.

When Sarduy calls for the "dissolution of the I," he responds to a practical problem for early radio writers: how to interpolate the audience into the drama. In a discussion of the discursive theory of the French radio pioneer Paul Deharme, Anke Birkenmaier explains that "the grammar of the radiophonic discourse . . . uses deictics that include the listener . . . in order to avoid creating a division between 'you' and 'I,' the listener and the radio discourse."[70] Deharme developed these ideas in radio commercials and radio plays as the publicity director of Radio Cité in Paris during the late 1920s and early 1930s. There he worked with the surrealist poet Robert Desnos, Kurt Weill, Antonin Artaud, and the Cuban novelist and radio sound engineer Alejo Carpentier. Picking up on Deharme's radiophonic theories, Carpentier addressed the obstructions between the listener and the radio broadcast voice by suggesting that speakers adopt a toneless delivery. As Carpentier puts it in his 1933 article "The Radio and Its New Possibilities" ("La radio y sus nuevas posibilidades"), "[The radio speaker] should dehumanize his voice as far as possible, adopting *a neutral tone*, a uniform one that stands out over the rest as a sort of talking machine."[71] This call for a machine-like voice echoes Deharme's point, made in his 1928 article "Proposition for a Radiophonic Art" ("Proposition d'un art radiophonique"), that the speaker should "forget the art of reading aloud . . . *Grey diction* is recommended, arguably a difficult task for the speaker, but after a while he will be a specialist at it, a kind of phonograph."[72] In both cases, Carpentier and Deharme's "speaker" takes the position of an objective narrator, removed from the action of direct speech and orchestrating events from above. However, as Deharme insists, this speech should not derive from text—"forget the art of reading aloud." Rather, the speaker should become something like an oral recorder, the phonograph that apparently bypasses textual mediation.

This fantasy of the toneless voice complements the suppression of pronominal obstacles to create a new fluidity of discourse in radio, "the blind art." Anticipating Sarduy's technique by almost a half century, Deharme writes: "The words of a given character can be spoken at times by one individual voice, at times by the speaker, as well as the character's deeds and gestures. The particular voice that, for a moment is at the character's service, will then become available to be used for other characters on another occasion. This should not pose a problem for the balanced listener; does it not also happen in our dreams that a sentence begun by one person is finished by another person, or even by us?"[73] In these guidelines, the voice becomes something like a mask, able to be passed from one *character* to another. Similar to what the sociologist Erving Goffman called an "animator" of speech, as opposed to an "author" or "principal," the one who gives voice to the utterance does not assume it as his or her speech alone.[74] The speaker, or voice actor, makes him or herself available to the characters. The speaker becomes a mask that the characters put on rather than the other way around. In many ways, this radio theory provides the link between Dos Passos's creation of the popular voice through a toneless conduit, and the work Sarduy does to develop an anticolonial politics of sound. Whereas Dos Passos's textual flatness made all characters sound as if their voices derived from the same radio network, Sarduy, as I will explain, uses the voice actors' individual timbres to link the different characters they inhabit. The ensuing connection between characters will ultimately bring those voices on the periphery into the center and decolonize Portugal from within.

Sarduy builds on Deharme's principle—"What one person says could be said, ultimately, by another"—and then applies radio's particular artistic form to the colonial situation. With its "galaxy of voices," the play aims at the "destruction of the individual as a metropolis—the conscience or the soul—with its colonies—the voice, the sex, etc."[75] The cosmological metaphor, in this case, stands against the cosmopolitan. The voices cannot be anchored to any one city, and therefore their movement across different territories marks only their dissolution as individual voices rather than their cosmopolitan circulation. This destruction occurs through an apparent contrast to Sarduy's interest in "the grain of the voice" as a carrier for land and materiality. Instead, the author of *The Ant Killers* seems to work against tonal difference. The absence of tone, in this theory, allows the same speaker to adopt different geopolitical and narratological positions. The voice of a Portuguese soldier sounds the same as a French photographer in Angola or a French tourist in Lisbon. As the production notes say, "*Two Portuguese soldiers:* the voices of M3 and M4, or any two other voices."[76] In this work written for the voice, tonelessness, the absence of marked speech, holds open the door for decolonization, allowing voices to travel back and forth between the eroding colony and dissolving empire.

As I said before, the early radio theorists' promotion of a toneless delivery is a fantasy, one we might associate with Arnheim's fascination with the great fiction of radio: the disembodied voice. It would be a familiar mistake to hear the *voz liberada* as another instantiation of the "floating signifier." However, we should recognize, with Sarduy's critic Peter Hallward, that "What survives this destruction [of the individual] is a certain *style*, a certain tone of voice or manner of expression . . . 'A novel is nothing other than variations around a character who is himself *not an entity but a tone given to his manner of existing*.'"[77] The "dissolution of the I" does not lead to the "free play" of signifiers but rather to a renewed attention to tone and what Goffman called "forms of talk," a play with signifiers limited by specific *rules of use*.[78] Instead of the mere dislocation of meaning, the play's dynamic work to acknowledge and rearrange context develops a more active, particular, and *pragmatic* understanding of meaning. While Hallward imagines this work as "an exercise in *Creative thought* in the Deleuzian sense,"[79] I think we come closer to understanding this abstract principle through listening that retunes our ears to the deployment of pronouns, sound, and deictic placement in the play.

Strangely, the seeming erasure of position by the fantasy of toneless discourse helps reveal how the residue of tone works to further Sarduy's anticolonial aesthetic. To begin with, his characters require tonal difference, as well as some sense of individual tonal marking in order for listeners to tune in to the "dissolution of the *I*." Thus, at the end of the play's opening sequence, "man four" tells us, "the car's radio stayed on,"[80] and then "man one" speaks "as a radio announcer," followed by "man two" and "man three," both of whom speak different lines "as a female radio announcer."[81] Speaking with the same tone, "man two" and "man three" become or "voice" the same professional register and gender. The assumption in the production notes that listeners and actors alike can distinguish the voice of a radio announcer from that of a schoolchild, for example, helps inform the performances. However, we only understand them as an example of the dissolving "I" if we realize that they are different speakers sharing the same voice, or men playing the role of a woman, or, in the case of "man three," a voice that will soon become a Portuguese soldier in Angola. On one hand, our ability to recognize the sonic register or tone of a given character gives the voices their narrative and social value. On the other hand, the residue of individual tone or timbre creates a second layer of narrative space, in which we begin to hear that one voice actor inhabits multiple characters, thereby connecting those characters through his or her timbral difference.

Tone and timbre, in this case, collaborate with referential language to allow speakers and listeners to create and dissolve frames of reference and the positions of address in the play. While some of the contextual work occurs in semantic phrases—"This takes place in Portugal"—the speaker's tone carries other information. As I have just described above, shifts in tone correspond to shifts in frame. Tone helps listeners establish their "footing" in relation to the narrative. When we hear "three chimes from the marimba," followed by a voice that speaks in complete but direct sentences, delivering news items in a professional vocabulary and syntax and a serious register, we recognize a newscaster speaking. The tone that indexes the professional role—"radio newscaster"—derives from this specific assembly of linguistic markers—syntax, vocabulary, register, timbre, pitch—encountered on the radio. Rather than conceal these complexities in Barthes's rich but inexact descriptor, "the grain of the voice," we might better think of them as "nonreferential indexicals," parts of speech that possess social or narrative value without any necessary semantic or denotative meaning.

For linguistic anthropologists, such nonreferential indexicality helps organize social space. A typical example would be the difference between formal and informal registers marked in romance languages as the shift between "tú" and "usted" in Spanish, or "tu" and "vous" in French. Although the reference remains the same—either word refers to the same object—the word choice alters the social atmosphere. Referring to one's boss as "tú," for instance, could index a familiarity that others in the office don't share or simply mark the casual space of a Silicon Valley start-up. Sound can work in a similarly indexical fashion.[82]

As the linguistic anthropologist Michael Silverstein observes, oral narratives can employ sound to organize the shift between narrator and character, as well as the infection of the character's point of view with that of the narrator, in what narrative theory names "free indirect discourse." Whereas Silverstein's study of a Wasco narrator observes that the speaker accomplishes such shifts through "attitudinal colorings . . . in the phonic substance of the language of narration," I have claimed above that Sarduy's voice actors use changes in tone (sonic register), along with their residual vocal tone (timbre), as a similar "infection" between characters and actor, similar to that between character and narrator in the more familiar example of free indirect discourse.[83] To put this same point slightly differently, sound functions *pragmatically*, in that it organizes different contextual frames. To borrow from Michael Lucey: "It is by way of its pragmatic aspect that a discourse participates in the creation and the perpetuation of a social universe with all of its divisions into classes. It is through the pragmatic function of a discourse that interlocutors either indicate their well-established places within their social universe, or make implicit claims upon a place that is not yet established as theirs, or else *put* somebody in their place."[84] In *The Ant Killers*, such pragmatic work, carried out through tone, helps construct a national, as well as a social context. When we hear an utterance repeated with a different tone in a different context, but with a residual timbre that links both characters, we can hear the two contexts dissolve into each other, thus taking an expected, or "presupposed," context and "entailing," or producing, a new context.[85] Moving from the microsocial interactions of interpersonal discourse to the colonial relations in the play, this pragmatic understanding of tone reveals that *how* something is said tunes listeners in to the play's experimental use of rebel radio to represent an anticolonial critique.

To understand how the narrative value of tone allows listeners to recognize the shifting and collapsing borders between empire and colony, we need to follow "man three" through the performance. In sequence 1 this voice plays a French tourist in a car crash in Lisbon, as well as a female radio broadcaster reading news that "Portugal recognizes her African territories' right to independence."[86] Absent in sequence 2, the voice reemerges as a different French tourist outside of Lisbon, watching a group of rebels, "naked and plastered with makeup, wearing branches on their head," dance around a radio, which broadcasts the closure of several Portuguese military outposts in Angola.[87] In sequence 4 the voice is heard again as an announcer, describing "the transfer of power to the black majority" in Angola,[88] and then, in sequence 5, we hear it as a Portuguese soldier, abandoning his military post. In the play's final sequence, the voice is again that of a French tourist, halted on the occupied Salazar Bridge entering Lisbon, describing boys with "their faces and feet painted black, wearing strange hats made of branches."[89] If we fail to recognize this voice as carried by the same performer, then we fail to hear how the "liberated voice" travels from Lisbon to Angola, into the body of a Portuguese soldier, a female radio announcer, and a French tourist. Furthermore, we fail to hear how the voice returns to Lisbon, much like the primitivist rebels, created by the colonial imagination, return to undo the imperial state.

Before turning to this last point, however, we should note how the "liberated voice" threatens to return to the charismatic, monolithic voice of the dictator all too familiar throughout the history of radio. For while I have focused mainly on the play's explicit theme of decolonization in Africa and Europe, only with difficulty can we tune out the resonances with the Cuban Radio Rebelde the younger Sarduy listened to as a novice writer in Camagüey and Havana during the late 1950s. If, as I mentioned toward the beginning of this section, Sarduy adapts something of Guevara's "*foco* theory" to the aesthetics of the radio drama, it should be clear that he works against any single voice possessing the radio. The ambition to return rebellion to radio by producing a play "where whatever one says could be said by another" turns against the unification of Rebel Radio in the single, national voice of Fidel Castro and returns to Guevara's theory, in which the radio voice, like rumor, could never be individual, and becomes an "inspiring, burning word" that spreads like wildfire across the population.

While we encounter this experimental guerrilla radio in the play's form, the rebels within the diegetic space of *The Ant Killers* more closely approximate a parody of rebel radio. Rather than sending messages from their concealed destinations to lead the passive urban masses, as per Guevara's *foco* theory, these "wild boys," whose camp is described as "a terrified anthill," safely conceal themselves behind the desert dunes where "they had maintained one symbolic contact with civilization: a radio, always turned on."[90] Dancing around the radio to news of the Angolan liberation, the sinking of a Spanish ship near the "Spanish Sahara," and the German tourists' car accident, the painted rebels with headdresses made of branches recall Marshall McLuhan's 1964 theory of "the radio as tribal drum."[91] As they stream across the Salazar Bridge in the imperial capital of Lisbon and then reappear at the play's end in blackface on the same bridge, we see the face of colonial primitivism come home. This is not to follow McLuhan's primitivist and racist logic, which attaches the radio to the "archaic forces" of "the resonating African within."[92] These rebels do not imitate any actual African rebels but rather the European fantasy of Africa as primitive. It is a fantasy passed through further fictions associated with radio and renewed primitivism, such as the 1971 novel *Wild Boys*, by William Burroughs. In case we missed the reference, the rebels scrawl the latter's name on one of the pillars of the bridge across from the name of Bakunin, announcing the fall of the fascist Estado Novo, founded by the former prime minister Salazar.[93] In this deterritorialized space, on the middle of a rebel-occupied bridge renamed for a Russian anarchist and the U.S. inventor of the "cut-up" technique, a French tourist waits for his car radio to talk to him, and the speakers in blackface finish each other's sentences as if they shared one mouth. The empire has turned colonization back on itself.

Strangely, this is the moment where we can now recognize why Sarduy, in a letter to his sister celebrating his radio plays, told her that he was "quickly becoming the author of the French version of *El derecho de nacer*."[94] Recalling his youth in Camagüey listening to Caignet's *radionovelas* and his adolescence in Havana tuning in to guerrilla's rebel radio, Sarduy brings the two models together here. His complex work with narrative acoustics finds resonance in the art of declamation, the sonic construction of character, and the bufo tradition that influenced Caignet. In composing his plays Sarduy learned from the aural training practiced by Eusebia Cosme and other *declamadores* to turn timbre, tone, pitch, and volume into character. However, he uses his knowledge of these techniques perversely: to try to turn radio against itself.

And it is at this moment in the play, when the lessons of Mamá Dolores's voice—its absolute artificiality and its reflection of white primitivism—resound in the rebels' chants, that another voice returns. Seemingly arriving from nowhere—although it has been played on one of the radios earlier in the work—we hear Billie Holiday's version of "When It's Sleepy Time Down South." The song, which appears on Holiday's album *Last Recording* from 1959, instead of *Lady in Satin* (1958), bears special semiotic weight. Released the same year as the Cuban Revolution, and infamous for its use of racial stereotype and its nostalgia for an idealized South after the Great Migration, the song signals Sarduy's two influences—Caignet's negrismo and the guerrillas' revolution. Again, Holiday's song is heard just as the soldiers and tourists come face to face with the Portuguese rebels, "their face and feet painted black, with strange headdresses made of branches." The very image of decolonization, their absurd artifice is what matters. Hearkening back to the politics of what Jill Lane described as "the really made up and the made up real," they represent the distorted opening that connects Angola and Portugal by means of colonization, and by means of the radio. In the image of these rebels and in the vocal masking gestures of the voices that describe them, Sarduy shows the surprising anticolonial force of the *radionovela*, a genre previous revolutionaries had thought of as their enemy.

And to remind us that this story has also always been about Cuba, the play ends with "a lump of sugar." The rebels, burdened by bedrolls and provisions, file in single line across the bridge:

> [M1] Some, hesitating and struggling, carried, together, a single bundle, [M2] they retreated at times: [M3] others, ready, came to replace those who had fallen. [M4] And the bundle continued advancing, [M1] without touching the ground, [M2] on high, like a trophy. [M3] Like an insect's corpse, [M4] or a lump of sugar.[95]

The speakers share the burden of the words here, just as the rebels share the bundle. But the language recalls earlier moments, when the men of a rebel camp, "the ant killers," formed three concentric circles of defense to protect "[M2] an already opened can of sardines and a packet of sugar."[96] Or again, when the Portuguese soldiers abandon their post in Angola, where, like "the ant killers" they organize their retreat progressively, "in three concentric circles,"[97] at the center of which they bury "[M3] an already opened can of sardines and a packet of sugar."[98] In other words, in this play about radio and decolonization, repeated strips of discourse spoken by different voices knit together different imperial and colonial contexts and then disrupt those contexts, breaking through the concentric circles of defense, just as the tourists in the final scene on the bridge realize, "without knowing it, we had crossed a forbidden circle."[99] When they see, at the center of that circle, men carrying a bundle, "like a lump of sugar," and they themselves imitate this procedure in sharing the utterance and voice, we as listeners realize that the voice, too, is "like a lump of sugar,"[100] the very product that began Cuba's colonial process, that fueled its uneven trade relationship with the United States, that inspired radio commercials for saccharine melodramas with advertisements for Coca-Cola, and continues to define Cuba's ailing monoculture today. Looking back at Cuba's rebel radio, Sarduy imagines rebels killing ants to hold onto the crop that has colonized them from the start.

PART III

Hand-to-Hand Speech

CHAPTER 6

House Taken Over

Listening, Writing, and the Politics of the Commonplace in Manuel Puig's Fiction

En *La traición* nadie ve nada, todos *dicen que ven.*

(In Puig's *Betrayed by Rita Hayworth* no one sees anything. Everyone *says they see.*)

—ALAN PAULS

Sens de mots diffère, d'abord, puis le ton: on trouve du nouveau dans le ton dont une personne dit telle ou telle chose. Nous prendrons le ton de la conversation, comme limite suprême, et où nous devons arrêter pour ne pas toucher à la science—comme arrêt des cercles vibratoires de notre pensée.

(The meaning of words differs, first, then the tone: one discovers something new in the tone in which someone says one thing or another. We take a conversation's tone as a supreme limit where we ought to stop in order to avoid reaching science—the limit of the vibrating circles of our thought.)

—STÉPHANE MALLARMÉ

In March 1960, Goar Mestre, the now exiled former head of Radio CMQ in Cuba, arrived in Buenos Aires. Born to a wealthy family in Santiago de Cuba, where his father ran a pharmaceuticals company alongside the fathers of the actor Desi Arnaz and the politician Eduardo Chibás, Mestre was educated at Yale and came to the radio business as a means to sell advertising for his family's various companies and their investments in U.S. businesses. Mestre had been to Buenos Aires before: his wife was Argentine, and in 1948 he came to the city to be elected president of AIR (Asociación Interamericana de Radiodifusión, or the Interamerican Broadcasting Association). As the leader of an association created by the United States to encourage the expansion of U.S. networks and a specifically commercial-focused broadcasting model, Mestre lobbied against state-supported radio and, eventually, television throughout Latin America.[1] Back in Cuba, he denounced the government of Argentina's president, Juan Domingo Perón, for interfering with radio broadcasts, and Mestre smeared his main Cuban commercial competitor, Amado Trinidad, as an "agent of Perón."[2] However, by the time Castro's revolutionary government took over Mestre's CMQ in 1960, Perón, overthrown in the so-called "Revolución Libertadora" in 1955, was in exile in Spain, where he would remain for another thirteen years. With a pro-business military government in power in Argentina, Mestre fled Cuba for Argentina. Upon his return to Buenos Aires in 1960, Mestre, with backing from CBS, purchased the city's channel 13. With his Good Neighbor connections to U.S. businesses and commercial networks, he once again built a media empire. It would not last. Perón would return to power in 1973, and, as in Cuba, Mestre's channel would once again be expropriated by state power. Yet, by the time Perón's second wife and widow, President Isabel Perón, would take control of the channel in 1974, Mestre's influence over the medium would transform him into "el rey de la televisión" (the king of television).[3]

I open with Mestre's story here to point out how the media networks linked to transnational politics and corporations interwove the cultural and political lives of Argentina, Cuba, and the United States from 1932 through the 1970s. As the biographies of Ernesto "Che" Guevara, Virgilio Piñera, Rodolfo Walsh, and others can testify, this circulation did not depend on business networks alone: political and artistic alternatives circulated as well. Movement from country to country seems more typical than exceptional during the midcentury period I have traced over the course of this study. However, these connections between the countries had been forged on radio even earlier. As I mentioned at the outset of this book, the first transnational radio broadcast in the Americas to reach a mass audience, and the moment that made radio a truly mass medium in Argentina, was the 1923 fight between the United States heavyweight champion Jack Dempsey and the Argentine boxer Luis Ángel Firpo.[4]

Rather than bring listeners together in the kind of utopian "logosphere" Gaston Bachelard proposed in "Radio et Rêverie," this transnational hookup instantly conveys violent nationalist ambitions with significant literary repercussions.[5] When Firpo knocked Dempsey out of the ring and onto the typewriter, delivering, as it were, a knockout punch to the medium that radio's live broadcast voice was replacing, the scene, reported in Julio Cortázar's 1967 "El Noble Arte," reveals a discourse network that brings together North and South in a conflict that sets writing, violence, and radio at center stage.[6] For not only does radio seem to emerge as writing's rival here, but writing seems desperate to make violent contact with reality, and to alter it through action. When Cortázar, speaking in Cuba, told audiences that any successful short story "wins by technical knockout," he made the analogy clear. The 1967 return to the radio broadcast from 1923 made the latter into an emblem for precisely the kind of *littérature engagée* that would define a hemispheric turn to imagine writing as direct political intervention.[7]

That is one origin story of radio and writing in Argentina. However, we should also note that while the 1923 broadcast marked the "birth of radio" for Cortázar, historians of technology provide a different birthday for radio in Argentina. Officially, the first radio broadcast (*radiodifusión*) in Argentina took place on August 27, 1920, two months before the inaugural broadcast in Pittsburgh, Pennsylvania that so many histories of the medium claim as a first.[8] Argentina's paper of record, *La Nación*, reported the event on August 30:

> In the Coliseo they're performing a curious and extremely interesting experiment . . . In a great space around Buenos Aires, one can attend on these nights a truly marvelous spectacle [*un espectáculo realmente maravilloso*]: a group of people, with an attentive and transfixed attitude, listen without seeing [*escuchan sin ver*] the performance of the lyrical spectacles from the Coliseo. The audience [*el auditorio*], in the living room or in the dining room, encircles a small apparatus that *resounds with better clarity than the phonograph and without the noise* [rumor] *of the needle on the disc*: the receiver's membrane desperately vibrates in its magnetic box and an amplifier expands the environment of the grave chords of "Parsifal" or the burning melodies of "Iris."[9]

As with many other inaugural moments in media history, the wonder expressed in the now ordinary experience of listening to radio is remarkable to read. And this first Argentine broadcast, with its "desperately vibrating membrane" that seems to refer equally to the device and the listener's eardrums—both "receivers" taut with attention—makes audible a different national story. The fact that radio broadcast in Argentina was born on that winter night matters to cultural and media historians because it troubles the standard narratives that cast technology as radiating outward from Europe or the United States to the global South, and far-flung colonies. And yet, in the broadcaster's choice of content—Wagner's *Parsifal*—the broadcast also tends to reinforce those narratives of cultural dominance. The amateur radio aficionados who decided to leave their workshops and share these sounds with a larger audience did not choose to broadcast a conversation from the street, a reading of gaucho poetry, a local politician, or a vanguard manifesto. If Cortázar's story marks radio's beginning as a popular, massive, transnational event that connects writing to political action, the *Parsifal* broadcast expresses its nationalist claim to primacy and technological innovation through a European paradigm of high culture. Instead of listeners streaming into the streets, the 1920 broadcasters began Argentine radio history, and one could argue, popular broadcasting everywhere, with the German composer most closely associated with European radio criticism, including Theodor Adorno's insistence that we hear radio as fascism.

The difference between the Firpo-Dempsey fight and the *Parsifal* broadcast also represents the struggle between two cultural paradigms for broadcasting. This battle between commercial broadcasting and public broadcasting quickly came to settle on the politics of language in Argentina. As I will argue in this chapter, the keenest critic in this linguistic and sonic debate was the Argentine novelist Manuel Puig. Puig's work invites us to pay attention to the historical context of linguistic reform and radio legislation in order to understand how his subversive use of radio in print challenged governmental doctrine and sought to transform the value of belittled and prohibited forms of speech. Through commonplace expressions, the careful imitation of radio listeners' speech, and a keen sense of melodrama's political power, Puig transforms the narratological space of the novel to forge a communal sense of property. From the late 1960s to the early 1980s, his best-selling novels rework the cultural products of the Good Neighbor Policy alongside the populist Peronist politics of Argentina's melodramatic radio culture in order to question the very possibility of the neighborhood of the Americas.

THE VOICE OF THE NATION?

If radio did not begin the debates about the "proper" use of language in Argentina, it brought oral speech into the legislative realm and provided a platform for political nationalists to push for linguistic norms in a country with a still fluctuating ethnic and national identity. Summarizing some of the country's multilingual history, Beatriz Sarlo argues that for statesmen like Juan Batista Alberdi, the nineteenth-century search for a national language and culture in Argentina often promoted the "'good' heterogeneity of European immigration" against the "bad heterogeneity" of the "Hispanic-creole and Indian" populations.[10] By the early twentieth century, the terms of the debate had shifted, as intellectuals and writers subordinated these good and bad polylingualisms to the master binary of high and low culture.[11] These class distinctions within a language also existed across languages, as the cultural elite determined that certain languages, such as French, English, or German, were legitimate and would not disturb the constitution of Argentine writing, while other languages, such as Yiddish, Italian dialects, and Russian, threatened to deform the nascent Argentine tradition.[12]

For Sarlo, Jorge Luis Borges is the key transitional figure who seeks to create a national literature from these debates through two operations: first, he retrospectively casts the nineteenth century as a literary utopia in which writers were able to "capture a national semantics in the tone and connotation of the oral voice." Second, in his story collection *Historia universal de la infamia* (Universal History of Infamy), he borrows narratives from the European and largely Anglophone tradition he considers "universal" in order to fashion a national literature. Thus, Borges establishes the "good" mixture between national literature and foreign languages, while sidestepping the charge of a provincial nationalism.[13]

Borges argued in essays and story collections for an Argentine language that could only be inherited but not adopted, and yet, if inherited, should make use of specific foreign languages (English, French, German) in order for it to reach the status of these "universal" models. His fellow writers and contemporary legislators, meanwhile, focused their energies on a newly massive disseminator of oral language and culture: the radio. Matthew Karush argues that Argentine politicians in the 1930s and 1940s tried forcefully to shape Argentine radio into an "alternative modernism"—like a mass culture version of Borges's writing—that fused national traditions (tango, gaucho culture) with cosmopolitan modernity, producing class-conscious melodramatic tales in which working-class heroes struggled against wealthy oligarchs.[14] Politicians and writers had to take radio seriously at this time because radio had the numbers. Argentina had not only staked a claim to the world's first broadcast; as early as 1929 the country had more than five hundred thousand radios, more than twice the number in Italy, Russia, and Denmark. By 1934 the country accounted for 66 percent of all of the radios in Latin America.[15] Radio, more than film, photography, or print, defined mass, popular culture in the country.

As the listening public expanded, radio became the centerpiece for linguistic reform. In 1933, two years before Borges published his "Universal History," and one year before the United States established its own Broadcasting Law, Argentina put in place its first Radio Communications Rule in order to assure that broadcasting programs were "highly artistic."[16] One year later, the government's "Instructions for Radio Stations" clarified the term "artistic": programs that used "slang terms [*modismos*] that deform the language" were prohibited, as was any use of "humorously low tones that support mimicking other languages, misunderstandings, angry exclamations, harsh or discordant voices [*voces destempladas*], etc."[17] These Argentine linguistic regulations were intended to conform to the guidelines set out in the first "International Radio Conference" in Washington, D.C., in 1927.[18] Paralleling nationalist vernacular reforms from the beginning of print culture through the nineteenth century, the radio reforms shaped regional Argentine speech. One casualty was the *lunfardo* dialect, excluded from the airwaves in the effort to match ideal speech forms first regulated in Washington. Yet, whereas nationalist print reforms had structured an already institutionalized educational standard, the radio reforms sought to regulate the sensorial habitus of everyday speech and prevent its possible ascendancy in national broadcasting. Radio's reach—far beyond that of print—could discipline a larger community of listeners and the potential irritation their voices might bring to the nationalist project.

As divergent wartime alliances in the 1930s and 1940s weakened Argentina's relationship with the United States—its main source of radios—the Argentine government further nationalized the radio, instituting more language regulations meant to curb the commercial and cultural influence of other countries. Unlike the U.S. attack on Argentina's film industry I discussed at the close of chapter 3, Argentine broadcasters weren't as restricted by a lack of physical materials. With radio perhaps strengthened by the material constraints of Argentine film, the nationalization process reached its legislative climax in the radio reform coincident with Juan Perón's rise to power. The May 1946 *Radio Broadcasting Manual* fixed a taxonomy of radio genres ("Works of Imagination," "Oratory," "News and Information," "Miscellany and Publicity"), insisted that stations play at least one "radionovela" connected to "Argentine tradition or history," and expanded the rules of 1935, regulating actors' tone. It mandated that "The tone of voice, including sighs or the emotional relief of figures of intonation, will be appropriate for each situation, but will always avoid the falsettos of low comedy, excessively high-pitched timbre, feminized distortions, etc."[19] The desire, as Matallana argues, to "educate the listener's cultural ear"[20] was part of a nationalist program the Manual highlights by informing radio-station owners and orators that "the control of the national language will be one of the fundamental demands that every speaker must satisfy, as much as this concerns the propriety and correctness with which one ought to articulate phonemes, as it does the correct intonation of each word, phrase, sentence, etc."[21] Through these dictates, the tone and timbre of voices, the particular diction of a speaker, or the verbal inventiveness of a comedian became freighted with political and legal import. Once again, the sound of a voice, not just its semantic content, signaled complex national, regional, and political alliances.

Such linguistic clashes were not lost on attentive listeners. Among them was Manuel Puig, an early fan of the radio comedian Niní Marshall, whose comedic sketches, a mixture of sociology and humor, played up and down the tonal spectrum that the state sought to delimit; her studied shifts in pitch and timbre called up a host of different social, racial, and gender positions and inspired critics to dub her a radio "Chaplin in skirts" and an "American Cervantes."[22] As Christine Ehrick demonstrates, Marshall's voice arose amid a "gendered soundscape" in Argentine radio that included radio librettist Nené Cascallar, Uruguayan socialist feminist Paulina Luisi, and, of course, Eva "Evita" Duarte de Perón. Marshall's shifts in sonic register ran the gamut from empathy to irony, from producing a feminist, working-class voice that stirringly advocated social mobility, to a condescending tone that mocked her characters' inability to hear their status as outsiders. The radio laws that censored Marshall—eventually unable to work in Argentina, she fled to Mexico—sought to mute precisely this tension in the ongoing negotiation of what constituted the national, popular voice. Although Marshall was eventually banned from the radio, her impersonations, which Puig continued to perform later in life for friends and family, provided him with a model for how to imagine social types through a shifting spectrum of vocabulary, accent, and tone.[23]

More than twenty-five years later, Puig expressed his conviction that radio might "educate the listener's ear," albeit in ways opposed to the state's desires. Moreover, he relates such listening to a particular practice of writing. In an interview following the publication of his 1969 novel *Boquitas pintadas* (Heartbreak Tango), Puig comments:

> When I wrote this novel, I was very interested *in working with the language of characters, because the way they spoke tells more about them than anything the author could explain.* Most of the characters in the book are first-generation Argentines, of Italian or Spanish parents, most of them peasants who hadn't been able to give their children any cultural heritage . . . This meant that *the children of these people had no models of conduct at home, and, least of all, no models of speech.* They therefore had to invent a language of their own, using the culture they had at hand.[24]

Puig goes on to specify the "culture at hand" as radio culture—"radionovelas, the lyrics of tangos and boleros, comedy shows"—whose oral programs create a model for immigrant speech. This speech, in turn, provides a model for his writing. Against the grain of the government's desires, that is, Puig identifies radio's pedagogical role in *unofficial* national culture and uses this kind of countercultural pedagogy as a model for his own process of writing. Attentive to constructing characters through "the way they spoke," through those pragmatic markers the government sought to control in *radionovelas*, Puig realizes a formal principle: to imitate the speech patterns of immigrants amounts to representing the insinuation of radio speech into the newly forming language. The author's role is not to explain but rather to listen and record.

Puig's double imitation, what we can begin to think of as his *mimesis of mediation*, runs against not just the government's language politics but also those linguistic ideas Borges had worked to establish throughout the first half of the century. For Borges, who was intent on constructing a coherent Argentine identity in language, recent immigrants' attempts at Argentine speech forms risked distorting a language that was not "truly theirs," exaggerating the language's loud linguistic signals without realizing that "nationality resides in the nuances of *tone*."[25] As Borges had written in "El escritor argentino y la tradición" that what proves the authenticity of the Koran as an Arabic book is its absence of camels—that is, of the obvious and ubiquitous—he likewise heard the authenticating sounds of Argentine speech as invisible or inaudible to an outsider, strictly impossible to imitate. What makes an Argentine speaker for Borges, as Sarlo argues, is "an interior spontaneity and property that are not acquired but rather transmitted like an inheritance across time."[26] Sarlo's term for this, the "detail effect," recalls Barthes's famous "reality effect" and matches the construction of reality to the construction of a voice. The idea is helpful here less because of what it tells us about Borges and more because of what it reveals about how Puig modifies these realist impulses built on inheritance. Equally attentive to tone, Puig's writing attempts to record precisely those imitations Borges condemns, as Puig seeks out a language defined by its lack of property—a language that is not "truly theirs" in any case—whose realism derives, once again, from his imitation of their imitation of radiophonic speech.[27]

THE UNSTABLE PROPERTIES OF "LITERARY LISTENING"

While earlier critics have often overlooked the ways Puig's work with language intervenes in specific sociolinguistic debates fundamental to the rise of radio in Argentine national politics, some have recognized the connection between what Alberto Giordano calls Puig's "literary listening" and his writing program. Most emphatically, the novelist and critic Alan Pauls, writing about Puig's first novel, *La traición de Rita Hayworth* (Betrayed by Rita Hayworth; 1968), explains how "Each voice is in itself a mosaic of rumors, a conflagration of echoes. The voice, in *La traición*, doesn't make simple circuits of emission: it always establishes mediations, screens, citations. Each voice takes up, refers to, deforms, or reproduces the voices of the others . . . [T]o speak is an exasperated exercise of *oratorio obliqua*."[28] Like Andy Warhol, who published a tape-recorder novel the same year Puig, living in New York City, published *Betrayed by Rita Hayworth*, Puig renders the individual voice into a transmission circuit for other voices. The voice itself becomes part of a media circuit, or discourse network, influenced by the novel's ambient media (cinema, radio, newspaper) and equivalent with them. More specifically, the voice's reproductive character, its passage on through imitation, transforms speaking into an act of listening that theorizes the media with which it interacts.[29]

The nonproprietary literary listening and its links to a larger media network analyzed by critics like Pauls, Giordano, and others was a theory

Puig himself wrote into the very origins of his career as a novelist. In an often recounted story, Puig was in the midst of writing a screenplay in 1962 when what he intended to be a brief voice-over based on his aunt's gossip turned into a thirty-page monologue.[30] The resulting writing, known as "Pájaros de la cabeza," formed the basis for *La traición de Rita Hayworth* and placed the acousmatic voice and the role of the listener at the center of his novelistic project. Turning from the screen to the page, Puig inscribed overhearing into the very origins of his writing project.

While this moment has become a cliché inscribed into the mythos of Puig's literary biography, his readers have generally been quick to pass by his interest in the sound of a voice to focus instead on his well-known love for cinema. The emphasis on film's importance in Puig's writing, however, has missed how his own cinematic training turned him toward the ear rather than the eye. Indeed, his auditory style emerges from the combination of the popular radio culture described above and his studies as a screenwriter at the neorealist Centro Sperimentale de Cinematografia in Rome in 1957. Working in Rome alongside fellow students who would become some of the most significant filmmakers in Latin American history—Nestor Almendros, Tomás Gutiérrez Alea, Julio García Espinosa, and Fernando Birri—Puig took the lessons of the Centro in a different direction. He criticized the neorealists for their misrecognition of their medium and their sensory misperception. An overemphasis on an objective visuality, according to Puig, led to a failure in narrative: "In some ways that ideological purity, that absence of a personal point of view, that lack of narrative devices made the focus real, but purely photographic. That photographic reality was not dramatic. One remained on the surface of the problems, and without drama the audience lost interest."[31] Wedded to a static "photographic" perception of reality, neorealism fails to capture reality's movement as did the melodramas Puig adored on screen and on the radio. While Puig might have followed neorealism's static age into the pop aesthetics of boredom in films like Andy Warhol's *Sleep* (1963) or *Empire* (1964), he instead infused his listening style with sentimentalism and suspense. Under the influence of Carson McCullers and Raymond Chandler, as well as a massive personal film archive, Puig would seek to add narrative dynamism to the neorealist method.[32] Heralding listening as a dynamic art in his writing, Puig would borrow from neorealism to transform reception into narrative.

For despite his objections, Puig also repurposed the neorealist strategy for his own work. He simply replaced the camera with the listening ear. Although he dismisses "photographic reality," his method borrows heavily from the techniques outlined in the neorealist critic André Bazin's "The Ontology of the Photographic Image" (1945). In that essay Bazin insists that the photographic lens—the *objectif* in French—provides an objective, antipersonal registration of material impressions that eludes the imposition of the photographer, author, or painter's personal point of view. Surprisingly, Bazin's description of photography comes strikingly close to Puig's own celebration of the listener's nonproprietary relationship to a story. Indeed, his later interest in tape recording seems to pursue this neorealist impulse to reduce authorial interference. Puig, however, rejects the "objectivity" of recording by insisting that the method, if not the theory, of neorealism registers the medium's specific interpretation of the real. What I called Puig's "mimesis of mediation" exhibits a second-order realism, and a sense of communication as always densely mediated. This position opposes Bazin's belief that further mediation and newer technology will bring one closer to the phenomenological reality of the material world. In Puig's work, the listener always listens to others, who, in turn, borrow their words from things they have heard elsewhere through other media. Like Dos Passos's J. Ward Moorehouse, Wright's Bigger Thomas, or Sarduy's tourists, no one can claim their voice for themselves in Puig's writing.

To speak as a listener, to appropriate other voices, to borrow from others' words, brings with it a specific politics of listening. While such listening derives from the discourse network of ambient media, the structure for this listening is gossip. As a narrative art that combines listening and telling, gossip opposes a proprietary relationship to the voice and language and instead turns characters into spokespeople (*portavoces*), carriers (*portadores*) infected by what William Burroughs called "the word virus."[33] In Puig's work, as with what Gertrude Stein called "ladies' voices," such gossipy listening-as-telling or telling-as-listening relates to his broader politics of style, the citational method that reproduces other generic discourses (*radionovelas,* noir fiction, science fiction).[34] As Pauls argues, Puig "conspires at all levels against the notion of literary style as an idiolect (the private use of language), and postulates a style as the generic and social formation of enunciations."[35] This pitched battle against a private use of language carries over into the generic modes manipulated throughout Puig's career. Thus, he transforms the lack of personal style into style. While his reuse of genres and stylistic resistance to private language made him reprehensible to an early generation of writers—the Uruguayan novelist

Juan Carlos Onetti dismissed him by saying, "After reading two of Mr. Puig's books, I know how his characters talk, I just don't know how he writes"—it is precisely Puig's resistance to authorial style where we find the politics of Puig's approach. As I said before, we might think of such writing as reception driven, as a writing style that listens and thus produces a nonproprietary relationship to language and identity.

LEARNING TO LISTEN TO THE "COMMONPLACE" IN *BOQUITAS PINTADAS*

After completing *Betrayed by Rita Hayworth* in between shifts as a ticket agent for Air France at John F. Kennedy Airport in New York City, Puig returned to Argentina at the end of the 1960s to finish writing *Boquitas pintadas: Un folletín* (Heartbreak Tango: A Serial Novel; 1969). The latter novel, an attempt to write in the serial form popular in magazines intended for female readers, closed the two-book reflection on his childhood and the rise of Peronism, and intervened, in its own subtle way, into the debates about radio, language, and national culture in Argentina.[36] Puig turned to marginalized culture coded as feminine for the materials to locate a common place that served as an alternative to the narrow nationalism of Perón's radio reforms or the elitist cosmopolitanism of Borges. Like its predecessor, *Boquitas pintadas* examines 1930s life in a province of Buenos Aires and sets its primary action between 1938 and 1947, with occasional forays into the novel's contemporary 1968. Without mentioning the broadcasting manual of 1946, which was instituted the year before the novel begins, and without discussing the government's nationalist censorship of speech forms, the narrative's repeated recourse to *lunfardo*, the "voseo," tango and bolero lyrics, U.S. radio commercials, and extranational *radionovelas* presents an archive of the radio culture government censors sought to mute.[37] If, as Pauls and other critics have concluded, *Rita* creates a new novelistic style through the incorporation and transformation of the formal language of 1930s Hollywood melodramas, *Boquitas* mixes this engagement with the cultural products of the Good Neighbor Policy and the national politics of Argentine radio. However, unlike Borges's attempt to insert an Argentine language into the "universal history" of European or North American traditions, Puig's novel develops a new form by mimicking the regionalisms and slang popularized over the radio in a language neither Borges nor the government recognized as their own. Rather than worry about whether or not the immigrant's language is "theirs," Puig picks up on "commonplace" phrases and clichés that apparently belong to everyone.

One of *Boquitas pintadas*'s clearest examples of the contagious or commonplace speech Puig imitates is found in those scenes where characters listen to tango and bolero over the radio. Both musical genres carry specific social indexicals, with tango primarily identifying a working-class Argentine public, and the Cuban bolero indicating a listening public with extranational pretensions linked to Caribbean and, especially in the 1930s and 1940s, U.S. entertainment markets.[38] In the novel, Raba, a housekeeper who lives in a converted pantry in a bourgeois home, listens to tangos, as does the man who fathers her child, the construction worker Pancho. On the other hand, the more bourgeois Nené, a friend of Mabel, whose family employs and houses Raba, listens to boleros. As if to underscore the thematic importance of the division, the novel's first chapter features Nené listening to a radio program titled *Tango versus Bolero*, whose alternating plots give different versions of what we come to recognize by the end of the novel as Nené's own failed love story with the tubercular Juan Carlos: the tango version is dark and violent, while the bolero is romantic and hopeful. Additionally, Nené's letters, which she writes while listening to the program, portray the desires she turns to the radio to fulfill.[39] Nené, who Pancho brutally observes "was not a dumb Indian: she talked like a radio speaker and didn't forget to pronounce her 's's at the end of words,"[40] not only learns pronunciation from the radio but thinks through uncited lyrics from boleros. In the novel's penultimate chapter, Nené's unfiltered absorption of bolero lyrics into a stream-of-consciousness narration marks the advent of an "intimate public." Here Nené's supposedly private thoughts borrow from public forms, thereby unfolding the logic of the radio's public broadcast reaching the intimacy of private homes. Moreover, the bolero's transmission, from its rise in Cuba to its subsequent popularity in Argentina, categorizes Nené's mixed discourse as participating in not just a "public" realm but a transnational market in musical forms dependent on the radio. This kind of subjectivity catches and produces thoughts in what one might call a "gulf stream" of consciousness distinct from privatized and/or locally or nationally bounded interior monologues.

In the novel's four chapters most explicitly organized around the radio, Puig links this play with linguistic property to a more specific interrogation of real property or real estate. Radio emerges, in these scenes, as a medium that simultaneously carries the utopian possibility of shared property and also signals that project's destruction. At the center of these scenes Raba murders Pancho, the construction worker–cum-police officer who has fathered her child. In the first half of the novel, Pancho, then employed as a bricklayer in the construction of the new municipal police station, seduces Raba after a dance, and the two make love in the field around the construction site. In a fact that bears on the archaeology of sentiment and power in the following scenes, this turn to the construction site sets their amorous connection in the property neighboring Mabel's family, where Raba lives and works.

By the second half of the novel, Raba has given birth to their son, and Pancho, who has been hired as a police officer in the completed police station next door to Mabel's home, refuses to acknowledge Raba or the baby. Instead, Pancho has turned his attention to Raba's employer, Mabel. As he walks along the wall dividing Mabel's property from the police station, he flirts with her while hooking up the station's radio antenna. The two agree that he will come to her room that evening, as she looks at the phallic radio antenna and asks, "and the antenna? Did you already get it up?," while she thinks, "I'd die to kiss a real man."[41] Thus, Puig unsubtly sets up the radio as an instrument of social mobility, a hookup that permits Pancho to cross into Mabel's bourgeois home and thus cross the race and class borders the novel has firmly held in place through the romances up to this point.

This rendezvous is put in place in spite of Pancho's earlier relationship with Raba on the other side of the wall. Although Pancho doesn't know it, the next chapter reveals that Raba had, in fact, returned from Buenos Aires to Mabel's home on the same day that he put up the antenna. Raba returns with the tango songs she has listened to throughout the novel still ringing in her head. Similar to what the reader encounters in Nené's radio consciousness, as Raba waits hopefully to reconcile with Pancho, she mixes her memories of their last night together, or her imagination of that night, with tango lyrics:

> no te voy a dejar, Raba, yo te prometo que nunca te voy a dejar, soy albañil y soy bueno ". . . ellas me regalan la ilusión del alba . . ." te quiero, Raba, te quiero para siempre "en la noche triste de mi ceguedad . . ." él se aprovecha que soy ciega y trae a otra más blanca, la sirvienta del Intendente Municipal.[42]

> [i'm not going to leave you, Raba, i promise you that i'll never leave you, i'm a construction worker and i'm a good man ". . . they gave me the illusion of the dawn . . ." i love you, Raba,

i'll always love you "in the sad night of my blindness . . ." he takes advantage of my blindness and brings a whiter woman, the servant of the Municipal Supervisor.]

The interpolated tango lyrics, some of which reference the same songs used as epigraphs at the beginning of each chapter, intercut Raba's reflections on her life. However, with the invocation of another woman, the tango begins to anticipate the novel's plot—Pancho's affair with Mabel—and shape Raba's actions as she cites the tango again,

> "eran mis pupilas como dos espejos donde se miraba la felicidad [. . .]" saltan los vidrios rotos, una astilla en punta, y a la chica del taller le sale sangre: un pedazo grande de vidrio le tajeó como un cuchillo la carne, pasó entre las costillas ¡y le partió en dos el corazón! Y de un cuchillazo le corté el ala a un pollo pelado, la cabeza, las patas, le saqué el hígado y el corazón del pollo.[43]
>
> ["my pupils were like two mirrors where one saw happiness [. . .]" the broken glasses shatter, one splinters off into a point, and the girl from the workshop starts bleeding: a big piece of glass slashes her like a meat knife: it went through her ribs and split her heart in two! And in one fell swoop I cut the wing off a plucked chicken: the head, the feet; I pulled out the chicken's liver and heart.]

Here the tango's lyrics seep into Raba's memories of working in the factories of Buenos Aires, but she reinterprets an accident there as part of the tango, which now becomes a violent awakening in which Raba's "blindness" and illusory joy ("my pupils were like two mirrors where one saw happiness") is shattered by the kitchen knife she uses to prepare the chicken in her role as a domestic servant. This almost surrealist juxtaposition forces the two events to bleed into one another. The scene breaks off here. When the scene changes frames, an unidentified consciousness opens by mentioning "a mature fig," last seen in Pancho's hand as he and Mabel flirted next to the radio antenna. As the consciousness, now recognizable as Pancho, leaves a room (Mabel's) through the window, the stream-of-consciousness narrative proceeds: "la sirvienta está durmiendo . . . estoy temblando, de frío, hay un gato . . . no hay nada . . . pensé que eras un gato, que en la mano te brilla algo, ¿uñas puntudas de gato?, la cuchilla de la cocina" ["the servant is sleeping . . . i'm shaking, from the cold, there's a cat . . . it's nothing . . . i thought you were a cat. something is shining in your hand. sharp cat's claws? the kitchen knife"].[44] While the following chapter's police report makes it clear that Raba has killed Pancho (although it misidentifies her reasons and thus undermines police knowledge), the link between the image from Raba's thoughts of the tango and Pancho's last vision of the kitchen knife establishes a peculiar network of connections between radio, melodrama, and state and domestic violence. As Josefina Ludmer observes, "[Raba] kills [Pancho] with a kitchen knife; she applies the justice of the tango and of the radio serial."[45]

In this 1939 murder of a police officer and former construction worker by a domestic servant with a kitchen knife on the border between a municipal police station and a bourgeois home, in a revenge inspired by the radio's tango lyrics against an affair arranged around a conversation about a radio antenna, Puig reassembles the narratives of the New Deal and transnational cultural commerce in order to critique the collusion between state violence and domestic melodrama in Peronist Argentina. When Puig brings the radio into his novels, and even alters the novel's form to mimic the serial dramas of the *radionovela*, he retells radio culture, retrains the reader's ear to appreciate radio's intervention in the political life of Argentina from the 1930s to the 1960s. At the same time, recalling those New Deal narratives of radio and property from Chandler, McCullers, and Wright, Puig presents a bleak picture of radio's potential to create an improved neighborhood of the Americas and marks radio's property concerns within a very different political framework.

Without loudly proclaiming his political stance, Puig threads his novel into the country's history and its transnational affiliations. Taking one step further into the lower frequencies of Argentinian history and its commonplace or clichéd connections to New Deal products, we might recognize how Puig's work with radio melodrama in this novel also imitates the story of Argentina's most famous export. After all, Eva "Evita" Duarte de Perón began her radio career in 1941 as a soap opera actress with a "high-pitched and trembling . . . childish, clumsy, and unrehearsed . . . plain voice that resembled her listeners' voices."[46] Her style, at first, sought to imitate 1930s Hollywood stars like Bette Davis (*Elizabeth and Essex*, 1939), Ethel Barrymore (*Rasputin and the Empress*, 1932), and Marlene Dietrich (*The Scarlett Empress*, 1934).[47] However, just two years later, now president of the radio union Asociación Radial Argentina (Argentine Radio Association), she pitched a *radioteatro* program in which she would find a new voice by imitating famous women throughout history. As Juan José

Sebreli, Emir Rodríguez Monegal, and Christine Ehrick have argued, these roles impersonating famous women served as the training ground for Eva to become Evita.[48] (Indeed, the show's scriptwriter, Francisco Muñoz Azpiri, would later serve as her speechwriter.)[49] While some listeners found her performances "hilarious"—"What a daily pleasure, this nasal voice who played the empress with rural tango accents!"—others, especially working-class urban migrants like Duarte herself, heard themselves and their own ambition in the voice's divergence from radio's metropolitan norms[50] By the time she organized a popular uprising in favor of the future president, but then recently deposed vice president, secretary of war, and secretary of labor Juan Perón on October 17, 1945, she had herself begun to fulfill what Perón, visiting her in a radio studio during one of her performances of some of these famous women, called "the social function of radio."[51] That same year Duarte's new show, *Hacia un futuro mejor* (Toward a Better Future) propagandized on behalf of Perón, borrowing from the sound of melodrama to create "a female voice—loyal, self-sacrificing, and maternal—[that] becomes the voice of the morally pure nation": a voice, as one introducer proclaimed, that was "the voice of a woman of the people—that of the anonymous masses."[52] This new voice, lower, more powerful, and still recognizably passionate and feminine, disturbed the American neighborhood enough that Nelson Rockefeller's Office of Inter-American Affairs—the key cultural wing of the Good Neighbor Policy—broadcast a counter program, *Hacia un mundo mejor*, throughout Latin America.[53]

In turning from Hollywood melodramas in *Rita* to the radio novel in *Boquitas*, Puig mimics the cultural movement in which Hollywood's melodramas provided a vocabulary Eva Perón and her writers transformed into a radio voice to carry a national politics in support of Juan Perón's state. The products of the Good Neighbor, in other words, coalesced with Argentine commercial media and, later, state-run broadcasting in Evita's voice to contest the very ambitions of the Good Neighbor's Pan-Americanism. Once again we hear the reformation and reformulation of the new neighborhood of the Americas.

With this context in mind, we can return to the murder in *Boquitas pintadas* as an exemplary and revealing attempt to think through the power relations that inhere as the transnational enters the domestic sphere to create another version of "radio's intimate public." When Raba murders Pancho with the kitchen knife on the border of the two properties, the domestic home and the municipal police station, the narrative indexes radio's dialectical movement between the openness of its engineering—the *broad*casting of wireless, or what Marinetti's radio poetics called "parole in libertà," "free words" or "words spoken freely"—and the conflicting social desires of its content: Evita's dream of rising to power from the provinces, and the melodramatic *radioteatro*'s promise of love surpassing all social class distinctions.[54] By placing the murder on the property line and under the radio antenna, Puig signals the collision between radio's wireless promises and the material constraints of real property. At the same time, he diagrams the collusion between melodrama and state violence. The bourgeois Mabel, whose desire for the working-class Pancho mimics an early bolero in the novel, transgresses social divisions without punishment. On the other hand, the representative of the state, Pancho, receives a vigilante death penalty for a similar transgression up, rather than down, the social ladder.[55]

Taking seriously Ricardo Piglia's assessment that the murder in *Boquitas pintadas* "condenses Puig's narrative world,"[56] we can also recognize how Pancho's death represents, in plain sight, a stylistic issue that has troubled Puig's critics for decades. In placing the event in this common place, on the margin between two properties, Puig allegorizes his aesthetic politics and writes the unrecognized history of radio in Argentine literature as a history of the struggle around common places—both those expressions the government sought to ban on the radio, and the desire for housing and radically democratic equality made iconic in the famous photographs of working-class "descamisados" bathing in the fountains in front of Argentina's presidential palace in popular support of Perón on October 17, 1945. Much as McCullers's and Wright's interest in real estate served as a figure for thinking through the local, national, and international politics of the neighborhood in chapters 2 and 3, Puig places this murder at the converging point of construction sites, radio, and neighborliness, foregrounding what Daniel Link calls the central issue in Puig's work: "How do we live together?"[57] Which we might also hear as, "How do we be good neighbors?"

One potential answer to this question emerges in Puig's formal work with "commonplaces." From critics José Amícola to Graciela Speranza, those attributes I identified above as deriving from what Alberto Giordano calls "literary listening" have been attributed to a more general 1960s aesthetic of pop art.[58] On the one hand, the "pop" label distinguishes Puig's novels from another type of 1960s literature. As Graciela Goldchluk writes, "at a time in which writers fought to 'be the voice of those who didn't have a voice,' Puig decided that he would prefer to remain silent and listen with attention. With his silence he was able to quiet, momentarily, the overwhelming rhetoric established as literary culture or political commitment, and create a space where a singular writing, a writing like any other, could emerge."[59] Unlike the *littérature engagée* promoted by his 1960s contemporaries, Puig achieves this paradoxically singular and general writing through listening and an adherence to and manipulation of popular generic forms. Once again intervening in the central challenge of populism—to speak *for* the people or to serve as a conduit for the people's voice—Puig chooses the latter. In his integration of borrowed expressions, clichés, and set phrases, Puig's listening and recording style is best thought of as an aesthetic of the "commonplace."[60]

The "commonplace" approaches Warhol's pop art, admits room for Pauls's analysis of gossip, and highlights the importance of place in Puig's emphatically local, and always transnational texts. In the aesthetics of textual production, "place" refers to Puig's texts situated between mass culture and "high literature,"[61] or between kitsch and camp.[62] In terms of gender and language, "common" amounts to a "third term that transcends and makes evident the fragile difference of sexual binaries."[63] For Daniel Link, Puig's writing exists in the "*casi*," the "almost," which is neither one thing nor the other, neither masculine nor feminine: "the voice in Puig's novels is the voice of the *casi* in all its forms."[64] And Alan Pauls finds in Puig a Barthesian neutral that is "*ne-uter*: neither one, nor the other, between the two," and also "entre-medio," in-between, neither "a place of synthesis nor a scene of complementarity."[65] Somewhat against the grain of Link's argument, and in contrast with Josefina Ludmer's assertion that Puig's female characters "belong neither to the middle class nor the *middle ground*,"[66] I am arguing here that precisely what troubles these critics is Puig's "commonplace" aesthetic, which indexes a location equally difficult to define as Link's "almost," and yet one that Puig places at the center of *Boquitas pintadas*.

WRITING FOR THE COMMONPLACE, WRITING AGAINST PARODY

Such a reading sides with the author against a tradition that hails his work as a parodic critique of media's alienating relationship with its consumers. The most influential and well-known of these readings is the 1971 essay "Note to the Notes to the Notes . . . A Propos of Manuel Puig" ("Notas") by Puig's friend and fellow novelist and radio aficionado Severo Sarduy. Opening with an epigraph from Jacques Derrida, Sarduy situates himself within the *Tel Quel* group's poststructuralist practice, arguing that *Boquitas pintadas* is "a parody of the novel," "the parodic transgression, the derisive double of the serial novel."[67] Sarduy's Bakhtinian reading—he footnotes Bakhtin's notion of "parody" in particular—counters the novel's potential pop reception, as he argues, "it's not a pop book," "the chapters are not themselves common places [*lugares comunes*], but rather they draw attention to the syntax that has produced [*ha dado lugar*] a kind of fiction that is now a common place."[68] As opposed to the novelist's apparent mastery, his characters, argues Sarduy, are "intoxicated with common places [*drogados de lugar común*]—their consumption must precede every dialogue; conversation, alienated by this vertigo, has been reduced to a simple '*connection*.'"[69] As the pinnacle of Puig's apparent parody, Sarduy points to the "serial novel inside the serial novel," the chapter immediately following the report of Pancho's death, in which Mabel and Nené reunite in Buenos Aires and listen to a *radionovela* as they talk for the first time in two years.

For Sarduy, the scene of listening serves three narrative functions: (1) it pacifies the characters' *lack* by situating them in the story's fantasy; (2) as with Hamlet's "Mousetrap," it makes the reader aware of her potential "being-in-fiction"; (3) it makes the parody's model emerge in its midst, reveals the "common place . . . alienating us," and allows the novel to "question" this model, a critical function that renders *Boquitas pintadas* "subversive." Sarduy's emphasis on alienation and parody as key terms meant to elucidate the text's "subversive" potential makes his own writing appear somewhat commonplace forty years later.

Rather than follow this familiar critique of the culture industry, we can locate the narrative's political heft elsewhere, in its comedy of manners rather than its parodic play with genre and semantics. This subtle distinction between comedy and parody retunes the reader's attention to the narrative's use of radio to create a medium through which characters organize social space by listening. In turn, as Sarduy admits, the reference to a *radionovela* from the early 1940s also draws attention to the historical period: a period, as I have demonstrated above, that operates under particular linguistic codes and rules and regulations associated with the radio.[70] Lastly, Puig's turn away from parody recalls his own listening technique, in which he insists, "For me, parody means mockery, and I don't mock my characters. I share with them a number of things, among them their language and their taste."[71] He adds, "The point is that the ordinary speech of these people is already a parody. *All I do is record their imitation.*"[72] Moving beyond the problems with listening I identified in Bigger and Max's failed exchange in *Native Son*, Puig's listening technique, which records imitation, allows for a discourse that "shares" language between author and character and comes closer to Sarduy's ambitions in his own radio writings than in his critique of Puig. And yet, while it recalls Dos Passos's contagious utterances, Puig's listening technique transforms a scene that might appear to subsume meaningful dialogue into mere "consumption," and produces instead a depiction of listening as a communicative act.

Puig's nonparodic art of listening is in fact on display in the very scene Sarduy highlights as paradigmatic parody: Mabel and Nené listening to the afternoon's *radionovela*. More a game of social cat-and-mouse than simple alienation, Mabel and Nené use the *radionovela* to organize their discourse even as it guides their conversation. From the moment Mabel interrupts Nené's questions about life in the provinces to ask if they can listen to the "afternoon novel," the latter feels she has lost in the social game: "Nené remembered that her friend always knew before she which was the best movie, the best actress, the best star, the best *radionovela*. Why did she always win?"[73] Realizing that "as hostess she ought to entertain her guest," Nené is forced to turn on the radio, which places Mabel, the listener who can recount the series, in position to guide the conversation. At the same time, it allows Mabel to resist answering questions about the past—"after the show don't forget to tell me about Raba"—and displace these personal questions onto the radio story.[74] Unlike the isolated reading of a print novel, the radio novel allows the two to continue their conversation, tuning in and out of the story, drawing attention to the radio to highlight another meaning and drowning it out when the *radionovela* too closely approximates a personal detail one of the women is trying to conceal from her friend. In one instance, Mabel explains that in the previous episodes of the serial radio novel *The Injured Captain*, set at the end of the First World War, a French captain from an aristocratic family disguises himself in the uniform of a fallen German soldier in order to escape detection in German territory. Here he encounters a farmer's wife who had grown up near his castle in France and had made love with him in the nearby forests until his parents forced him to marry someone of his social class. Nené responds, "'he can only truly love one.' Mabel preferred not to respond. Nené turned on the radio, Mabel observed her and it was no longer through the veil of her hat but through the veil of appearances that she was able to see Nené's heart. There was no doubt about it: if the latter believed it impossible to love more than one man it was because she hadn't succeeded in loving her husband, since she had certainly loved Juan Carlos."[75] Even merely recounting the *radionovela*'s plot allows Mabel to determine Nené's intimate secrets. At the same time, the scene points back to Mabel's affair with Pancho, and thus Nené's comment about the story also serves as a veiled critique of Mabel's past. To recall again the jargon of linguistic anthropology, Mabel and Nené use language metapragmatically, attuned to the social function of terms rather than their referential meaning alone.

As these personal manipulations of the radio program continue throughout the episode, the conversation becomes increasingly marked by national and transnational radio and linguistic politics. Most notably, the program, set in the First World War but broadcast, according to the story's diegesis, in 1941, references the ongoing Second World War and, by association, Argentina's contentious political neutrality. Furthermore, as a program based on extra-Argentine themes, it goes against the policies outlined in the Argentine broadcasting manual. And as these "foreign" voices enter the room and play out over the radio—the same medium that brought news of the war into domestic households throughout Argentina—the listeners' language is marked by the incursion. Whereas the rest of the Spanish-speaking world uses the word "tú" as the informal second person, Argentine speakers commonly use "vos." In *Boquitas pintadas* the "tú" appears only when characters listen to boleros—the Cuban song form played over the radio—or in this scene of audition to the radio novel. Switching between the "tú" and the "vos," Nené and Mabel establish a fluid territorial language in the domestic space, which indicates their attention to the program. As their attention wanes, they switch from the "tú" to the "vos," indicating a reemergence of the national language. While such code switching might not appear as elaborate or explicitly political as the "deterritorializing" function I mentioned in chapter 5's discussion of Fanon's use of Arabized French words in his writing about Algerian revolutionary radio, Puig's seemingly inconsequential additions carefully map the small social and pragmatic power games—the comedy of manners—linked to the changing linguistic rules of radio use that went on to define the Peronist era.[76] The seemingly inconsequential talk in this domestic exchange marked off as women's culture reveals a sophisticated encounter with the proprietary and proper codes of Borges's nationalism and universalism. Through attention to commonplace language and scenes, Puig's recording of imitation, or his mimesis of mediation sets the groundwork for his final novelistic experiments with listening.

POP EPIC REALISM: RECORDING IMITATION

Puig's "commonplace" writing around the radio in *Boquitas pintadas* recalls Sartre's reading of Dos Passos. As with J. Ward Moorehouse, the characters in *Boquitas* speak as if a thought "was not formed inside [them], it came from afar."[77] In Dos Passos's case this formation of the utterance produced a "flat" style, a nonpsychological, nonexpressive indirect speech that Sartre characterizes as "slack air," representing the perspective of the chorus. Such a "populist" form countered Roosevelt's "false intimacy" by imitating the very sound technologies that dampened the "sound of space" and allowed "voices from afar" to sound like a voice in the next room. In its lack of psychological depth and the apparent wholeness it draws between the perspective of the individual and the chorus, I noted that Dos Passos's writing could be said to approach the kind of Homeric mimesis theorized by Eric Auerbach, or what Mikhail Bakhtin named "epic" mimesis. Auerbach writes that "a problematic psychological condition . . . is impossible for any of the Homeric heroes, whose destiny is clearly defined," and that "Homeric poems conceal nothing . . . Homer can be analyzed . . . but he cannot be interpreted."[78] Similarly, Bakhtin comments: "[T]he epic world of the absolute past is inaccessible to personal experience and does not permit an individual, personal point of view or evaluation . . . [I]t is impossible to experience it, analyze it . . . [T]he important thing is this formal constitutive characteristic of the epic as a genre . . . its reliance on impersonal and sacrosanct tradition, on a commonly held evaluation and point of view—which excludes any possibility of another approach."[79] Strangely, while these summations of "epic" or "mythic" mimesis resonate with Sartre's descriptions of Dos Passos's mimetic strategies, they also intersect with what I identified earlier as the pop aesthetic critics find in Puig's writing.[80] Graciela Speranza sums up Puig's style as "suggestively flat,"[81] and, in a similar vein, Sarlo writes, "Like the pop artists, Puig resolves to erase [*llevar a cero*] the personal traces of his writing and, also like the pop artists, he is opposed to expressionism."[82] Pop's opposition to "expressionism," like the epic genre's foreclosure of any "problematic psychological condition" and inaccessibility to personal experience, leads both modes to produce "a commonly held . . . point of view." The epic's collective community and pop art's collection of consumerist objects come together to define flatness as a populist narrative mode. However, what differentiates Puig's flatness from that of Dos Passos, and even Sarduy? Is there a "voice of the chorus" in Puig's novels? Why would a pop aesthetic valorize an epic mimesis meant, in Bakhtin's analysis, to stand against the novel's democratic heteroglossia? What does such a narrative mode allow Puig to say about his cultural moment that we can't already find in the period's other (social, magical, primitive) realisms?

While Dos Passos distances himself from his characters, Puig identifies himself with his. Against Dos Passos's ultimately cynical procedure, Puig's melodramatic narratives could be said to produce the "voice of the chorus" according to the pop strategy of blankly imitating the ubiquitous genres from the mass media (radio novels, Hollywood cinema, serialized novels in so-called "woman's magazines"), clichés, stereotypes, and commonplace objects. Like Warhol and Lichtenstein, Puig privileges the genres of feminine consumption at the same time that he inverts gender roles in male genres (the tango) to the point that a woman (Raba) can occupy the male's place by murdering him.[83] In this sense, *Boquitas pintadas* indexes the sociocultural tension that tends to separate the opposing terms he brings together.[84] Moreover, unlike the parody Sarduy claims for Puig, the boleros, tangos, and *radionovelas* in the latter's novels offer his characters languages and affects through which they can communicate.[85] Against the kind of parody that assumes a scornful distance from alienation, Puig attains a strangely affecting pop art. Just as surprisingly, the erasure of his personal style, and production of epic mimetic flatness, allows for a new conception of the multivoiced text, in which the author identifies with his characters to the point that he actually lets them speak for themselves. I call this mode "pop epic realism" because of the way it reorganizes these traditions to lay a new claim on the real. Against rival magical and socialist realisms, Puig fashions this new mode through the radiophonic discourse in his early novels and then perfects the strategy when he picks up a new listening technology in his final works.

DICTATION, *TESTIMONIO*, TRANSLATION

Across the novels Puig published over the next decade—especially *The Buenos Aires Affair* (1973) and *El beso de la mujer araña* (The Kiss of the Spider Woman; 1976)—he would elaborate the practices and themes from *Boquitas pintadas*. On one hand, Puig imitates the way others listen in order to develop a critical practice of writing about the ways radio (and later, other sonic media) structures popular power. On the other hand, Puig borrows from radio broadcasting, whose very engineering poses problems for different forms of property, as sound crosses national borders and permeates the walls of private homes. Puig uses radio as a model to imagine the utopia of a cultural and infrastructural common place. And yet, radio did not seem to resolve some of the narrative problems of property and social power that continued to bother him. Puig aspired to further connect the symbolic and the material conditions of writing in a communal space where his listening would flatten the hierarchy between author and character, interviewer and subject. As a result, he sought out other modes of storytelling and, ultimately, another sound technology: the tape recorder.

In Puig's context, the political impulse to "erase the traces" (Sarlo) of the author derives from a particular engagement with crises of memory and disappearance in the Argentine dictatorship of the 1970s, and the particular recording technology of tape. Much as I described Richard Wright's turn away from radio for its affiliation with white power, Puig looks to a new technology to restructure the formation of the intimate public he aims to create and represent. With *Eternal Curse on the Reader of These Pages* (*Maldición eterna a quien lea estas páginas*; 1981), Puig attempts to take up the tape recorder as his prime instrument in composing his novels. I say "attempts" because while he intended *Eternal Curse on the Reader of These Pages* to be a tape-recorder novel, his interviewee declined to be tape-recorded.[86] Instead, Puig had to take written dictation. Despite this significant difference, I will read *Eternal Curse* as a novel inspired by the techniques of the tape recorder in the remainder of this chapter to observe how Puig works with the technological imaginary of this new medium. With this new sound technology, Puig could restructure the radio network into a yet more intimate occupation of a common place.

An already culturally burdened technology, the tape recorder references both the techniques of a dictatorial state intent on building an intelligence archive, and the genre of testimonial literature meant to respond to the state's program of "disappearing" citizens. The *testimonio* offers a counterarchive of eyewitness accounts of government atrocities and personal memories to tell the stories the state wants to forget and exclude from its official history. If, as I said above, Puig's early novels provided an alternative to the engaged literature of the 1960s, these final novels work through the other dominant political form of literature in 1960s Latin America, the *testimonio*, in order to produce a new form of realism dependent on the tape recorder; it is a new realism that, as Ricardo Piglia writes, "fictionalizes the testimonial and erases its traces."[87]

The fictionalization of a genre whose intended purpose, political value, and ethical weight hinge on its ability to present the real poses questions about Puig's new and potentially parasitical form, which seeks to draw its meaning, in part, from the preexisting genre's contract with its readers. It is a provocation that would seem to threaten and even undermine *the* counterhegemonic form of the period. As Beatriz Sarlo comments regarding the testimonial's archival function: "Memory is a common good, a duty [*deber*] (as was said in the European case) and a legal, moral, and political necessity. Based on these characteristics it's difficult to establish a perspective in which one proposes to critically examine the victims' narratives."[88] At the same time, Sarlo points out that the *testimonio*, which arose on the political left at the same moment that Derrida denounced Western philosophy's logocentric belief in the voice as immediate presence, poses particular problems to a critical tradition skeptical about the self-evident value and authenticity of personal experience.[89] In combining the techniques of the testimonial with the techniques of the novelist, Puig, as in his previous fiction, charts a course that avoids the critical condescension of parody but relieves the testimonial from the burden of establishing a belief in the one-to-one relationship between the author and the content.

Fiction, in this case, produces a space that opens up the speaker's words for appropriation by the reader. As María Moreno has written after listening to recordings Puig made for a book I will discuss in greater detail in the final chapter, Puig manages to merge his writing with listening: "His major intervention is during the recording, through questions that repeatedly interrupt the turn of the story to demand that his speaker stop on some details, forcing them by systematic induction. It is as if Puig decided to extract the writing from the oral story in the moment [*en directo*]. Each question allows the phrase-by-phrase emergence of that which is still not a text."[90] We might say that the writer, in this case, becomes an editor. He compiles—in the terms of Angel Rama and John Beverley—and orders the transcript, but the composition emerges in the dialogue.[91] The writerly listening attends to the properties of the voice—"it possesses 'its own musical and pictorial qualities'"—at the same time that it seeks to undo the writer's claim over the text, as well as the speaker's claim over his or her voice. However, the tape-recorder novel goes beyond even the hundreds of pages of notes Puig took interviewing political prisoners and organized into the dialogues in *The Kiss of the Spider Woman*. There, like Richard Wright, who sought to unhook the voice from the speaker and therefore opposed Zora Neale Hurston's ethnographic dialect, Puig erased any mark of "prison jargon" from the finished novel.[92] In the later works, Puig moves still further away from print; his editing occurs through his questions, in the way he listens and records.[93]

With Puig's fictionalized testimonials in mind, we can recognize a culmination of the path I have been tracing throughout this book. While the novelists in this book's opening chapters attempted to revise realism to make room for a larger popular community, and Richard Wright, in chapter 3, sought in novel writing a method to detach the racial identification between the speaker and his voice, later writers developed new methods. In chapter 4, for example, I examined how Félix B. Caignet worked with stereotype to create a flexibly inauthentic voice, whereas in chapter 5 Robert Williams attempted to provide a radiophonic "voice-over" for political action against racism. Finally, Severo Sarduy employed the radio play as a means to "decolonize the 'I'" and write against the racist colonization of Angola. Puig, as I will show, finally encounters in the genre of the testimonial and the technology of the tape recorder, the means to write alongside his characters and provide both author and character with that "humble self-enlargement" Nancy Ruttenburg identified in the United States' eighteenth-century religious speech as a marker of "democratic personality." Puig's recordings create a new literature that allows his characters to become authors.[94] The populism in his pop epic realism turns the people's voice into the authorizing function of his texts.

If the pathways throughout the chapters have shown that the trail to what I am calling Puig's pop epic realism have relied on technological changes and transnational exchanges, we must realize that Puig's turn to tape derives not only from the influence of the testimonial literature of Argentina's dictatorial 1970s but also from Puig's own return to the United States and his desire to write in English. After all, *Eternal Curse on the Reader of These Pages*, written in English but translated into Spanish and published in 1981 before its eventual publication in English in 1982, emerges from a series of interviews Puig made with a young NYU lecturer named Mark: "I wanted to be him and, it turns out, he wanted to be me," Puig later recounted.[95] The transcripts, in this case, become the means for Puig to learn to write as he listens, and, at the same time, allow his character, his interlocutor Mark, to speak for himself. *Eternal Curse* confirms Speranza's assessment that Puig "writes a literature irreducible in its freedom from the national tradition and the consecrated foreign models."[96] Different from global English or world literature, Puig uses another's language to write in a transnational mode he both follows and creates. He embraces multiple linguistic, national, and mediatic materials to construct a commonplace discourse intended to be open to all readers.

In his English-language narrative Puig incorporates dictation's process into the novel's content. Lacking a narrator for more than two hundred pages, readers of *Eternal Curse* "overhear" a series of dialogues between two characters: Larry, a thirty-eight-year-old nurse with a Ph.D. in history, and his frail patient, Mr. Ramirez, an amnesiac whose fragmented memories slowly reveal his life as a former labor organizer jailed and tortured by the Argentine government. Puig establishes his themes of memory, experience, and linguistic homelessness on the novel's opening page, where he unites the material and formal *common place* in the novel's first sentences:

> "What is this?" "Washington Square Mr. Ramirez."—I know "square" but not "Washington." Not really.—Washington is the name of a man, the first President of the United States.—Yes, I know. Thanks so much.— . . .—Washington . . .—Forget it, it's not important, Mr. Ramirez. It's just a name and nothing more.—Was he the owner of this land?—No. They simply named it after him.—"Named?"—Yes. Why are you looking at me like that?—Named . . .—My name is Larry. Yours is Ramirez. And Washington is the name of this square, all right? The square is named Washington.—I know. What I don't know is what one is supposed to feel when one says "Washington."[97]

Ramirez's aphasia derives, in part, from his national displacement and encounter with the symbolic centerpiece of the United States in Washington Square. However, as he points out to Larry, his difficulty lies less with the semantic problem of naming and more with the experiential issue of finding the proper emotion to fit the name. Ramirez's encounter with learning to feel in a foreign language points back to the novel's composition, in which Puig learns to write a novel in English by listening.

Given my reading of *Boquitas pintadas*, we might wonder whether Ramirez's puzzlement in his encounter with Washington Square is another example of the "common place," or a rejection of this idea. Arguing for the latter point, Ilse Logie and Julia Romero claim that "the city in this novel functions as a 'no-place . . . that seems organized to forget and lose one's identity."[98] Clarifying this "no-place" in a reading of another of Puig's novels, Francine Masiello contends that "Puig responds to the danger of stasis by playing in transient spaces, favoring a no-man's-land of uncharted wilderness in the field of representation . . . [He] prefers to confuse a single story of origins that might *link location with locution*."[99] With Masiello's observation in mind, Washington Square—the symbolic public commons of the nation-state—appears more like the "no-place," or no-man's land of transnational wandering. In this case, the novel's constant movement around the square, while Ramirez becomes paranoid about his former captors in Argentina following him, and Larry complains about his own emotional displacement, his feeling of abandonment by the American dream, calls into question Washington Square's potential status as a public sphere and common place.

This apparent rejection of the common place is linked to Puig's work with tone. Rather than follow the more general ethnographic realism of the works of authors such as Zora Neale Hurston or José Hernández, Puig does not turn to the hallmark of literary fidelity—writing in dialect—in order to reestablish a connection between the speaker, the utterance, and the place of enunciation.[100] Instead, he amplifies sonic fidelity into an equally troubling issue: when Ramirez tries to acclimate himself to not just the affective association of a word's *content* but also the signifying sounds that accompany it. As Ramirez repeats at numerous times throughout the novel: "I still don't know what to make of certain tones. The expression of sincere regret . . . I wouldn't be able to recognize it."[101] And if this is a particular problem for Ramirez, it is also a general problem involved in moving spoken speech to writing, in *entextualizing* a sonic event and making tonal difference available to readers.

However, as Puig demonstrates across the course of this novel, writing from listening in a foreign language can teach us not just how to write in another language, but literally in and through another person, thereby incorporating the lesson of translation into the realist project. Thus, we might revise Masiello's earlier point by saying that in *Eternal Curse* the narrative "link[s] location with locution" precisely in order to "confuse a single story of origins," and to find a way to live in another country by writing through another's mouth, by writing as a practice of listening. We arrive at some sense of this productive confusion, this collocation, by noticing how the absence of tonal markers in the opening pages allows utterances to come from both mouths. This narrative confusion is further thematized in the novel when the exiled Ramirez inverts *testimonio*'s typical investment in the disclosure of a victim's torture. In *Eternal Curse* the victim becomes a questioner of his guardian, as he turns Larry's imagined past into a portal to shared experience. Ramirez constantly asks Larry to invent stories and describe sensations in order for the older Argentine man to be able to connect language and affect.[102] Ostensibly in order to help Ramirez avoid chest pain or his own suppressed memories, he interpolates Larry into a role-playing exercise by asking: "Talk to me like your father and teach me. I will listen as if I were your son . . . Repeat what your father says to you."[103] Eschewing the psychological overtones of his comment, he insists: "It's not that I want to find out anything about your personal life. What I want to know is what a father says to a son."[104] Later, he interrogates Larry's memory, begging: "Larry, please, repeat more of your mother's words . . . Is your memory failing? . . . Just give me a few of her words." When Larry responds: "[T]hey weren't hers, either. She didn't have thoughts of her own. Her own needs . . . didn't find language to express themselves," Ramirez insists: "I'm too tired. I can't follow you. But if you repeat her words . . . maybe . . . I'll . . . be able . . . to understand."[105] Finally, Ramirez demands: "Since you can't explain what you feel, could you at least tell me what it resembles? Something I might experience or remember in my present condition?"[106] When Larry eventually responds, in chapter after chapter, with descriptions of his experiences, narratives about his childhood, and stories he *invents* through Ramirez's guiding questions, we realize that the listening and writing practice María Moreno analyzed in listening to Puig's tapes actually enters the composition here. Puig's creative listening produces the discourse through questions.

The narrative thus mimics the scene of its production and the desire to experience through another's words. Moreover, as a South American writer interviewing a North American academic, Puig inverts the normal power relations between ethnographer and ethnographic subject. And the narrative, resisting the desire to recover or entextualize Ramirez's suppressed and traumatic past, turns him into the anthropologist, asking after Larry's emotional world. Puig thus challenges the experiential basis of *testimonio* and also opens it out as a model for writing beyond Latin America. In addition, Puig makes a translation from his own writing—shifting from dictation to fiction, from Spanish to English—in order to borrow from translation's own challenge to property. As Ricardo Piglia says:

> Language is a common possession, in language there is no such thing as private property . . . I have always been interested in the relationship between translation and property, because the translator writes a new text that belongs to him or her, and at the same time does not. The translator is placed in a strange position, in the sense that what he or she does is move to a new language a kind of experience that does and does not belong there. The translator is a strange figure, moving between plagiarism and quotation; he or she makes a strange figure as a writer.[107]

As Larry, the ex-doctoral student in history, learns more about Argentine history and thus helps Ramirez remember the terrible facts of his life in Argentina, Puig foregrounds the question of history, memory, and experiential translatability central to testimonial writing. And yet, in his desire to make experience available, in his unceasing novelistic search for forms that communicate, to find voices that are "communicable," Puig fictionalizes the testimonial, as Piglia notices, erasing its traces. Puig creates, instead, a new realism born of a transnational network of media, mass culture, literary and political history, and personal experience meant to imagine the "real" in the image of language, as a "common possession" where "there is no such thing as private property." The mimesis of mediation in Puig's compositional technique, or Ramirez's imitation of Larry's parroted versions of Freud and Marx, represents a fidelity, if we can still call it that, far removed from the sonic source.

As I said above, I think we should understand this narrative pivot, away from the particularity of a character's tone and toward a more easily confused, less defined, and more communicable discourse, as a shift from older versions of literary fidelity to a pop epic realism. Puig's notoriously styleless style, his sometimes awkwardly flat English, cobbled together through an imitation of another's discourse, represents this transnational language. For much like the epic described by Bakhtin as "the absolute past, inaccessible to personal experience and cut off from an individual, personal point of view or evaluation," Ramirez strives to imitate Larry's speech, not to recover his traumatic past but to keep it away. In Ramirez's actions we see him overdubbing his own experience with another's. This recording *and* erasing enables Ramirez to at first share and then create an experience personal neither to him nor to Larry. Such forgetting serves a political purpose, as well, in that it will allow Ramirez to resist providing secretive knowledge to his torturers if they should ever return.

One of the novel's dark jokes is that Larry, the lapsed academic and current nursing home assistant, becomes a different iteration of Ramirez's old interrogators. Notably, this change occurs with the arrival of another technology, the book, or more particularly a series of French novels, including *Dangerous Liaisons*, *The Princess of Cleve*, and *Adolphe*. The books, sent by a humanitarian agency that had recovered them from Ramirez's by now long-forgotten jail cell in Argentina, include a series of numbers attached to specific words, phrases, or paragraphs. Larry sets about interrogating Ramirez about the books and then decodes the system, which reveals a coded memoir Ramirez wrote through the words of the French novels. Soon enough, Larry is in touch with a friend at Columbia University and, with a fellowship from a university in Montreal, is ready to return to his academic life by publishing Ramirez's memoirs with an introduction by Larry himself. However, when Larry reads a long passage aloud, Ramirez responds: "I don't believe a word of it. It's all twisted, according to your whims. I can't see what you got out of that. Changing the text."[108] The novel closes soon thereafter, as Ramirez dies, and Larry, attempting to steal the memoir, is rebuffed by the human rights center and left jobless.

Admittedly, this is a strange turn in a novel I have been reading up to this point as invested in the communicability of experience. It returns us to question the novel's opening at Washington Square. In striving to create a style of the commonplace, a pop epic realism that attempts to use listening and recording to write a novel in another language, it should not surprise us that Puig opens his novel in the symbolic commons of the United States. However, now we are left to ask again why Puig turned to English to write *Eternal Curse on the Reader of These Pages*. Was it, as his longtime English translator, Suzanne Jill Levine, has suggested, that Puig was dissatisfied with the troubles of working with translators and thus wanted to avoid putting the process in someone else's hands?[109] If so, the novel registers an anxiety about sharing language rather than an openness to collaborative experience. Larry's attempted treachery at the novel's end would support this claim. And yet, perhaps despite himself, Puig composed a novel that confuses a single story of origins that might link location with locution. He uses listening and recording to sculpt a place in his fiction for other voices to resound. And he does so through a flatness of style that reduces literary fidelity in favor of an experimental realism that makes his two speakers sound like each other, as their dialogue blurs together on the page. He uses dictation to arrive at a new written language. In reducing these voices' singularity, and unhinging them from proprietary claims, his novel uses the idea of the tape recorder not to access the real but rather to fictionalize experience and expand its affective range. He replaces the search for the real, with a new, shareable realism, one situated at the commons of the United States, and asking what it means to listen and write in English.

CHAPTER 7

The Ends of Radio

Tape, Property, and Popular Voice

Puig: The fact of telling me the story, of unburdening himself, was already positive for him. What's more, I paid him per hour while he was talking.
Interviewer: The reverse of psychoanalysis.
Puig: Yes. And on top of it all he got a house, which was so symbolic of all he had lost.
—MANUEL PUIG, interview by Kathleen Wheaton (1989)

A year after Manuel Puig published *Eternal Curse on the Reader of These Pages*—a novel based on interviews that tactically redeployed the themes of trauma, memory, and political violence associated with Latin American testimonial literature—he composed his tape-recorder novel, *Sangre de amor correspondido* (Blood of Requited Love; 1982). Reappropriating the critical comment that most haunted his career—the Uruguayan novelist Juan Carlos Onetti's contention that Puig was more a recording device than a writer—Puig recorded his conversations with a Brazilian bricklayer working on the author's house in Rio de Janeiro. Puig paid the man for his time and promised to split the profits when the book Puig planned to produce from these taped conversations was published.[1] When the recording was complete, Puig had the tapes transcribed and translated from Portuguese to Spanish before editing them himself.[2] Fully adopting the position of the recorder, Puig made writing a practice of listening.

I open this final chapter with Puig's novel because it marks both a culmination and point of departure in the story I have been telling throughout this book: the interrelated development of housing (real property), sonic technology, the popular voice, political community, and narrative fiction in the Americas from 1932 to 1982. Put simply, while radio offered Puig the means to critique property relations and the imbrication of transnational culture with state and provincial politics, he felt he needed another technology—the tape recorder—in order to finish collapsing the distance between author and character and further remediate sound in print, in order to create a more communal space in the actual practice of writing. The tape recorder allowed Puig to turn listening into writing in a way that radio did not. As I will explain, tape transformed the characterological "conduit" of the vox populi—Moorehouse in *U.S.A.*, Singer in *The Heart Is a Lonely Hunter*, Jake in *Lawd Today!* or Bigger in *Native Son*, Mamá Dolores in *El derecho de nacer*, the panoply of voices in *Los matadores de hormigas*, and Raba or Mabel in *Boquitas pintadas* or Ramirez in *Eternal Curse on the Reader of These Pages*—into a collaborative position occupied by author and character.

In *Blood of Requited Love* Puig shares in the voice of the speaker, inserting his questions into the narrative as if they were the speaker's narcissistic self-interrogation. But what occurs after the interviews and outside the published work also bears on the story. Puig produces a printed novel from this collaboration that results in enough profits for the interviewee to buy a house. Transforming shared speech into a space intended to be communal, and insisting on the continuity between symbolic and material space that would reconfigure the novel's traditional narratological and printed limits, Puig tells the *Paris Review*, "On top of it all he got a house, which was so symbolic of all he had lost." The dividing line between the symbolic (fiction) and the real (in this case, real estate) is blurred here, as one feeds into the other and back again. More than the blueprint Richard Wright provided for writing in response to the housing crisis of the Great Depression, Puig's novel shares a voice to find a literal home for his real character.

The tensions beneath this utopian claim, however, are already evident in Puig's frustration with the unnamed stonemason's anger about the novel. Reminiscent of the struggle between Larry and Ramirez at the end of *Eternal Curse*, Puig's Brazilian interlocutor fought for ownership of his story, relenting only when Puig reminded him of the signed contract that granted Puig—the only author named on the novel's cover—property rights over the book.[3] Thus, the neat completion of a symbolic lack with the wholeness of real property falls to ruin over a dispute about privacy, intellectual property, and profit. The novel's larger tale must include the regret of a man who sold his life story for literary publication.

The story around Puig's 1982 novel—the tale of its entextualization and the property disputes that ensue—stands at the culmination of a twenty-year period of a genre I call the tape-recorder novel. While I do not have space here to detail that genre's rise and fall—that is a theme for another book—it is important to recognize Puig's turn from radio to another technology as part of a generational shift in writing and media. From 1962 to 1982 the tape-recorder novel emerged as a new realist genre adapting sound technologies as a new medium with which to compose against literature. Drawing from the avant-garde desire to fuse art and life, the ethnographic investment in documentary authenticity, and the militant belief in evidentiary denunciation, the tape-recorder novel challenged the authority in authorship and replaced individually authored texts with collaborative works. Like Puig's book, their productions dramatize the process of property and privatization involved in the very institution of novel writing. Puig's engagement with this genre marks the limits of writing in the radio age. However, it also establishes the continuities in the ongoing attempt to give narrative form to the voice of the people in order to create equitable property relations in the Americas.

THE RETURN OF THE REAL

In chapter 6 I argued that Puig's *Eternal Curse*, which would have been a tape-recorder novel had its interlocutor not forced Puig to take dictation instead, accessed a pop epic realism. This aesthetics translates the flatness of pop art's visual display into a tonal homogeneity capable of resisting the individuating ticks of an idiolect. What emerges is a "contagious utterance" that challenges the links between location and locution on the grounds of a metonymic public commons. *Maldición eterna a quien lea estas páginas / Eternal Curse on the Reader of These Pages* fused the "distracted listening" that one can find in a pop work like Andy Warhol's *a, A Novel*, with a critique of individual subjectivity and identity politics associated with Latin American *testimonio*.[4] Puig thus upsets the ethnographic order and the investigative politics of the denunciations that emerged alongside these traditions. Despite the fact that it was dictated and not recorded, *Eternal Curse* seems the culmination of the tape-recorder novel as trans-American form: a novel that gathers and critiques all of the previous iterations in order to critique the property of the voice connected to the property of national territory against the hope for a popular commons.

So why did Puig compose a tape-recorder novel the year after the publication of *Maldición eterna*, and does it add anything to our understanding of writing as an act of listening, or the representational politics of popular voice? Unlike its companion work, *Sangre de amor correspondido* holds onto the uneven power relationship between interviewer and interviewee, author / ethnographer and ethnographic subject. A man already in Puig's employ, a man he paid to build his house, the Brazilian stonemason whose story he records is a laborer Puig exploits for his own material gain. While Puig might be said to lend this story the literary celebrity that makes it a profitable novel, the asymmetry remains obvious: Puig sustains his position as employer, as boss, profiting from the man's story to pay the man for his work building Puig's house. While Puig, in the epigraph that begins this chapter, claims the novel's profits allowed the stonemason to buy his own house and thus resolve the symbolic lack—the desire for domestic power—that drives his story, what Puig conceals are the real property consequences of the capital transaction within the literary institution. Ultimately, the stonemason buys his own house and pays himself for building Puig's house by selling his life story. Meanwhile, Puig, to quote the Argentine artist Eduardo Costa's enthusiastic endorsement of tape-recorder literature, "holds a microphone and pushes a few buttons."[5]

One might dismiss the story of real property outside the text as inessential to the novel's symbolic world. However, it is Puig himself who opens the door to make this connection when he says the house resolved the stonemason's symbolic lack. Reading the other way, we can understand how the real property undermines the symbolic work the novel achieves. As María Moreno shows in the reading of the novel I described in chapter 6, Puig's "major intervention is during the recording, through questions that repeatedly interrupt the turn of the story to demand that his speaker stop on some details, forcing them by systematic induction. It is as if Puig decided to extract the writing from the oral story in the moment [en directo]."[6] Puig's artful audition creates a story through another's voice. Based on the reaction of his critics, moreover, he succeeds in the testimonial or documentary novel's ambition to turn oneself into another.[7] Jean Franco's review of the book in the *New York Times* perversely testifies to this success, as she observes: "Mr. Puig is adept at parodying people's awkward attempts to find a language for their emotions. 'Blood of Requited Love,' however, often seems to come dangerously close to the banality the author wants to parody." Puig is able to sound like his character "Josemar" because he has used his stonemason's words. The novelist attempts to move past parody to the "banality" of popular speech by purchasing the words of another.

Indeed, Puig both buys these words and parodies them, and this difference from his other works' mimesis of mediation strangely produces a divide between Puig and his character. And this parody enters the novel despite the fact that he has recorded the words directly. While Puig once again borrows another voice to compose in a foreign language—Portuguese, in this case—his artful questions now work to establish a parodic distance they formerly eschewed. Josemar's narration is interrupted by questions that seem to come from his conscience or the voice of Maria da Jose—a woman he might have raped—by which means Puig undoes not only Josemar's misogynist's fantasies but ultimately transforms him into a victim seeking help from his mother. The mocking tone in these questions, which Franco calls "the reality principle," makes the narration parodic. While the novel might challenge misogyny, it upholds the property relations and power dynamic that exist in the book's very mode of production. The book becomes a living denunciation that exploits the stonemason for his story and condemns his perspective on the world. Against the complicated but hopeful symbolic demand for the transnational commons in *Maldición eterna,* Puig's *Sangre de amor correspondido* upholds the testimonial model only to wield the power of capital, as the Argentine-born Puig constructs his house in Brazil by exploiting a local stonemason's story. In a cynical reversal of the ethics of the good neighbor, the institution of literature erects a house on the ruins of another's story.

CODA: NO TRANSMITTER: RADIO AND TAPE AND THE REBEL ART OF LISTENING

While Puig's work with radio and tape ended in 1982, a decade later the Argentinian author Ricardo Piglia would return to that work in order to reimagine the radio network through mixtape collectors, thereby reworking narrative codes once again to imagine storytelling as an act of listening. In the wake of the fifty-year history of experiments in narrative form and struggles over popular voice, Piglia's novel provides a look back at the advantages and limitations of the radio as a medium in politics and literature. I close with Piglia's work because its reflection on that past also turns us toward the possible futures of the novel as it begins to reorganize the relation between radio, property, and popular voice for another stage of new media.

At the center of Piglia's 1992 novel *The Absent City* (La ciudad ausente) is a storytelling machine built from an unfinished technology: a radio without a transmitter. As the device's inventor explains: "I had begun to combine certain formulas . . . and I applied them to a radio receptor. Back then I hadn't yet been able to make a transmitter, only a recorder. My closet was filled with tapes, recorded voices, song lyrics. I couldn't transmit. I could only capture the waves and the memories from the ether."[8] The engineer's description begs the simple question so central to this book: What is radio? As with my introduction, in Piglia's novel this ontological question, which emerges from a series of mechanical questions—Do all radios have recorders built inside them? Does a radio need a transmitter?—leads to a series of cultural questions: What is the difference between receiving and transmitting? How does radio connect one listener and another? What kinds of political community does radio make possible?

Addressing the forms that radio might take goes some way toward imagining the social worlds it might produce. The engineer's insistence on the radio as a receptor without a transmitter, for instance, points to the complicated legislative, technological, and cultural histories that lie behind the common understanding of a radio as an aural medium intended for listening to the sonic transmissions of others. The expectation that a radio would, by necessity, include a transmitter points back to early wireless radio, an age defined by amateur radio enthusiasts receiving and transmitting signals before the consolidation of broadcast networks in the early 1920s in the Americas and Europe led to the proliferation of radio receiving sets—those devices we have come to commonly refer to as radios. Beyond that moment, as governments regulated bandwidth and parceled out frequencies, largely criminalizing amateur radio, the practice of radio transmission and reception continued in clandestine and rebel or pirate radio networks throughout the world, helping to define musical subcultures, and, as the critical writings of Frantz Fanon, Ernesto "Che" Guevara, and others explain, to galvanize and shape anticolonial political communities.[9] As we saw in chapters 4 and 5, these radio communities formed sonic counterpublics—their broadcasts public, by technological necessity, and geared toward countering the messages from stations of a colonial government, commercial corporation, or state-run news service. By insisting that he made only a radio receiver and not yet a transmitter, Piglia's inventor situates his device within this lineage and reminds readers how a given medium's political function depends in part on how cultural users shape its technological possibilities, and how some technological forms—such as a radio with a transmitter—index the histories and practices concealed or forgotten when we understand a medium like radio as solely a device intended for tuning in to other voices.[10]

While the invocation of the transmitter includes the inventor's technology in a rebel radio genealogy, he then turns away from transmission to grant this notion of political power to the receptive function he insists on instead. It is this receptor, and the idea of reception itself, that will become the novel's central preoccupation, as a community of political rebels forms around the storytelling machine the inventor will build from his radio receiver. The invocation of transmission, and its subsequent rejection, thus begins to point toward *The Absent City*'s construction of an alternative idea of the political power of a community of rebel radio *listening*. While the idea for such a community responds to and derives from Argentina's particular national history, a history I will return to again in greater detail below, Piglia's more general critique of radio's political possibilities rests on and against a belief that activist politics requires a speaker who transforms the world through his or her speech. As we have seen throughout this book, such a belief pervades both left- and right-wing political ideology. In order to sharpen our sense of how Piglia's novel proposes a nuanced alternative to the political positions he usually endorses, we might take Bertolt Brecht's influential 1932 essay "The Radio as a Communication Apparatus" as a counterthesis that remains especially representative of the critical attitudes surrounding radio and media more generally.[11]

Bemoaning a moment wherein radio has already forgotten its amateur transmitters, Brecht famously argues that radio fails as a social medium because it isolates the listener instead of "bring[ing] him into a network." For Brecht the medium houses a simple solution: "The radio must be transformed from a distribution apparatus into a communication apparatus." Or, in market terms, it "should step out of the supply business and organize its listeners as suppliers." Only when radio is able "to receive as well as to transmit" and "let the listener speak as well as hear" will the medium become "the finest possible communication apparatus in public life, a vast system of channels."[12] Seize the means of production, Brecht writes, and radio listeners will liberate themselves.

Piglia's novel, drawing on the novelistic tradition I have unfolded throughout this book, moves directly against the current of Brecht's rallying cry. While Brecht's call to transform the passive listener into an active speaker has become a staple of academic and popular media theory, reminding us how old our debates about new media tend to be, Piglia's listeners and engineers depart from that theory to value reception over transmission.[13] With this departure, *The Absent City* anticipates the renewed attention to listening that has accompanied the surge of investigations into the history and theory of recording and broadcasting media developed from around 1876 onward.

As with the novels I have discussed up to this point, Piglia's novel is a work that represents the development of listening practices in relation to sonic technology—especially radio and tape—and then converts these modes of listening into a writing practice, as well as an effective resistance to governmental censorship, surveillance, and oppression. As with Puig's writing, the listening practices represented in Piglia's novel help develop a writing practice that would serve as an alternative to the testimonial genre sometimes thought of as the most useful form of political resistance to tyrannical power, but that often seems to merely invert the authority of the monologic radio voice.[14] Piglia's new narrative mode, a narration that listens and a listening that narrates, would take seriously Roland Barthes's argument that "freedom of listening is as important as freedom of speech" and that "it is not possible to imagine a free society, if we agree in advance to preserve within it the old modes of listening."[15] However, as with the movement I have traced from Dos Passos to Puig, in *The Absent City* even those "old modes of listening" Barthes and other critics might have dismissed for their interest in dislocation, nostalgia, and escapism provide the weak but necessary freedom to form a clandestine listening and recording community whose political value derives from the willingness to let other things speak.

In Piglia's novel the movement toward this surprising political position begins with the story of a journalist, Junior, investigating the source of a tipster's telephone voice.[16] Negotiating his way through this hard-boiled dystopian thriller set in a thinly veiled Buenos Aires of 1992's near future, Junior discovers that this voice belongs, if that is the right word, to Elena, the name of a consciousness presumed to be the late wife of the Argentine author Macedonio Fernández, programmed into a machine housed in the city's central museum. Due to a mechanical glitch, or what cybernetics and systems theory might call autopoetic adaptation, the same machine has gone from producing strangely modified versions of classic nineteenth-century tales, such as Edgar Allen Poe's "William Wilson," to issuing troubling political testimonies. These insurgent tales then circulate on cassette tapes through the city's clandestine underground to the consternation of the police state. The stories, which Junior listens to on his Walkman throughout the book, come to the reader as fragmentary chapters that comprise the bulk of the novel's discourse. To read Piglia's novel, then, is to study the discursive art of listening and overhearing.[17]

Critics have not missed the role of listening in *The Absent City*. Gareth Williams has remarked that the novel opens with "an ear" and with the "foundational role of radio in the formation of popular culture and politics in Argentina."[18] More precisely, the novel opens with two ears, and two distinct scenes of listening to two different forms of media. The first describes Junior's father, an English immigrant to Argentina's southern hinterlands, come to work as the station chief at the end of the railway line: "He spent his sleepless nights [*las noches en blanco*] in Patagonia listening to the short wave transmissions from the BBC in London. He wanted to erase the traces [*borrar los rastros*] of his personal life and live like a lunatic in an unknown world, hooked into the voices that came from his country."[19] Although couched within a scene of lonely listening, nostalgic for empire, this two-sentence overture evokes the themes of exile, nationhood, aural transmission, and disappearance that obliquely anticipate the narrative's reflection on Argentina's dictatorial period and the Dirty War that left more than thirty thousand Argentine citizens "disappeared."[20] Whereas Junior's father might merely want to forget his individual pains, the radio's complicity in "erasing the traces" of a personal life will later resound in the novel as disturbingly coincident with the Argentine national military's project to make personal lives vanish throughout the country.

The passage's connection between the personal, the national, and the imperial—a network of relations already threaded into the social conditions of broadcast technology's engineering—also links this apparently local scene to a more general historical and conceptual affiliation with radio. While I have largely portrayed an alternative to the most familiar histories and theories of radio, the association raised here between radio and a certain traceless archive of genocidal acts recalls an entire tradition of predominantly European-focused scholarship that exclusively connects radio with a dictatorial fascist voice.[21] In an exemplary commentary on late 1930s French radio plays and the radio theory in Europe during the same period, Denis Hollier writes: "The characteristic specific to radio is that it is live . . . Radiophony demands live speech, voices flowing freely from the source, without leaving behind any archivable precipitates. In this sense, radiophonic utopia is a glorification of the drive that Derrida called 'anarchival' and Michel Deguy qualified as 'lethal.'"[22] Radio's live character makes it lethal, in this schema, because the radio voice fails to leave behind "archivable precipitates," indices, or inscriptions—the material traces that Friedrich Kittler and others identify as the ontological features of early sound recording technology like the gramophone.[23] Without recourse to such inscription, "radiophonic utopia" would seem to render obsolete the material support of these technologies, as well as the printed book; hence Hollier's title, "The Death of Paper: A Radio Play." Indeed, the radio plays and theory that aired in France and Germany in the 1930s were also contemplated alongside broadcasts of Nazi book burnings, a political event symbolically linking the death of paper to human extermination.

Radio's status as a "live" and "traceless" medium would mark it as ontologically distinct from those other media—gramophone, film, typewriter, etc.—called on to define modernity's discourse network. However, as I noted in this book's introduction, as well as chapter 1's discussion of Dos Passos and chapter 5's analysis of Sarduy's radio play, it should be noted that radio's "live" status derives more from its *simultaneous* communication with a mass audience than its apparently "anarchival" quality. Radio does leave traces—only an overly literal rendering of Derrida's argument would exclude the social and psychological impact registered by listeners of even early radio broadcast—and, as media historians have shown, almost coincident with radio's rise critics began debating the relative success or failure of a broadcast's ability to transmit a sense of place through the sonic indices or traces of a room, street, or voice across the ether.[24]

While historically and technologically inaccurate, interpreting radio broadcast's simultaneity as an ephemeral, noninscriptive event complicit with European fascism and genocide has dominated the medium's reception in critical theory, and throughout this book I have argued for an alternate genealogy of radio in politics and aesthetics. On one hand, Piglia's choice of words situates his character within this tradition: radio's apparent tracelessness connects the scene to the events and legacy of World War II, a legacy Junior's English immigrant father brings closer to the novel's plot through his work for the British railway that was nationalized by Juan Perón in 1946.[25] On the other hand, the slight difference from Hollier's description of early radio theory is also worth noting. Unlike in those formulations, the radio in *The Absent City* does not simply fail to leave a trace. Instead, Junior's father listens in order to *actively* "erase the traces" of his personal life. It is precisely this tension between *being disappeared by another* and the potentially empowering possibilities of *choosing* to erase one's own individual subjectivity in favor of being "hooked in to other voices"—in a much different sense than that proposed by Brecht—that will resound as the core political concern in the novel's treatment of mediated listening.

Just one page after this brief theory of radio listening the reader hears a story about the father of Junior's fellow journalist, Renzi:

> "It reminds me," Renzi said, "of my old man's sleepless nights [*las noches en blanco*] listening to the tapes from Perón that a member of the Movement brought to him under cover . . . We were around the table, in the kitchen, at midnight, engrossed just like Junior's dad, but believing in that voice that came out of nowhere [*venía de la nada*] and always sounded a little slow and distorted . . . It should have occurred to Perón to speak by short wave radio . . . from Spain, in nighttime transmissions, with the discharges and the interference, because that way his words would have arrived in the moment that he spoke them. Or no? Because we listened to the tapes when the things were already different [*los hechos eran otros*], and everything seemed delayed and out of place [*fuera de lugar*]."[26]

In this second account, radio retains its link to "live" broadcast, understood in this case as the utopian possibility of a simultaneity that would successfully collapse the distance of exile and allow Perón's words to fall into pace with the event. In contrast to the radio, however, the tape in this scene neither erases the traces of life nor sits nicely in the groove of time. The machinery affects the phenomenal experience of listening, rendering Perón's speech "slow and distorted." Likewise, the voice's misalignment with the present makes "everything seem delayed and *out of place*." The tape's temporal dislocation, furthermore, unhinges Perón's voice from a localized space: "the voice came out of nowhere." In this domestic scene, the voice's acousmatic homelessness indexes the message's social and technological mode of production: it speaks of the exile of Perón, the famed nationalist radio orator, heard through the displaced character of the recorded voice. An alternative to the way the BBC tunes Junior's father in to the empire from the periphery, the cassette tape, reliant on the movement's clandestine network of hand-to-hand delivery, constructs an estranged second city inside the Argentine capital, a city listening to a distorted, delayed voice always out of place.

This historical precedent set by the Peronist rebel movement establishes a model for the tape network that will cohere around the cyborg's stories later in the novel. Renzi makes the connection explicit as he reflects on the fact that Perón's tapes remind him of the stories produced by the cyborg in the museum, with which he finds fault for their characteristic asynchrony. "It would be better if the story was live," he says. "The narrator should always be present."[27] Although he quickly qualifies this aesthetic judgment—"Of course, I also like the idea of those stories that seem to be outside of time and start again whenever you want"—the connection between Perón's speech and the machine's stories underscores the message of political resistance that inheres within the tape technology's temporality. The machine's stories, neither guerrilla directives nor topical folk songs, and absent any immediate or obvious connection to the present, derive their political force from their lack of fit, their disruption of the state's seamless control over the national message. Their imperfect temporality, at once of the present and "outside of time," makes them hybrid objects, containers of the "noncontemporary within the contemporaneous,"[28] or representatives of "the untimely present" whose pull between multiple, contested temporalities throws the present into question.[29]

Lastly, Renzi's insistence that the tape was "distorted" and that he would have preferred to hear Perón's radio voice through "discharges and interference" places this scene within a specific genealogy of mediated listening. Distinct from the massive address of "the voice of the Führer," the emphasis here on distorted domestic listening for political revolutionary ends recalls the anticolonialist sounds in Frantz Fanon's writing, which I discussed in chapter 5's exploration of Sarduy's attempt to decolonize the voice.[30] Readers familiar with Fanon's "Ici la voix de l'Algérie . . ." will recognize in these reflections on Perón's distorted voice something of Fanon's theory of revolutionary listening. As John Mowitt and Brian Edwards have convincingly argued regarding Fanon's text, distortion, in the Algerian context, signaled the French colonial authority's attempts to jam revolutionary broadcasts.[31] Attentive listening to the distorted radio message became an expression of solidarity with the revolution and helped constitute a collective, anticolonial community. Such revolutionary and anticolonial listening recalls the work with sound and radio carried out by Radio Rebelde in Cuba, and Severo Sarduy's extension and critique of that movement in his radio play *Los matadores de hormigas*. Piglia's implicit invocation of those traditions situates *La ciudad ausente* as a further reimagination of revolutionary listening under new conditions of global capital and technology.

Renzi's emphasis on the distorted sounds throughout this passage similarly mark Perón's words as part of a rebel, clandestine network. In other words, the distortion signals to readers familiar with Argentine politics a nuanced but powerful divide in the complicated legacy of Peronism. Serving an explicitly ideological function, the distortion separates Perón's utterance from his first two terms as the populist president of Argentina, when his voice could be heard clearly over the national radio or at mass rallies, and, more important still, it distinguishes his voice from his third term, after his return from exile in 1973, when his politics swung explicitly to the right, laying the groundwork for some of the policies that would lead to the Dirty War.[32] Listening back through history, Perón's distorted voice resounds as the most ideologically compatible for a novel that wants to connect it to a revolutionary political underground.

In choosing the tape over the radio in presenting Perón's voice and, later, the voices disseminated through the cyborg Elena, Piglia urges us to consider the differences between the rebel radio voice and the fascist voice, therefore complicating the tendency in the European continental tradition that exclusively links the radio voice with fascism. More emphatically, Piglia takes up the cassette tape as a hybrid technology that exists only through the remediation of the radio and the phonograph and attempts to use the newer medium of tape to rethink radio's political history. The turn to tape restructures the idea of the radio network, moving it away from the single broadcasting station and toward an underground community of rebel listeners making hand-to-hand deliveries. Strengthening the connection, Junior, strolling through an exhibit at the museum where the cyborg is housed, notices that among the objects on display and discussed in the machine's first taped story, "Steven Stevensen," are "Stevensen's phonograph, with a tape recorder and a radio."[33] Recalling the inventor's "radio without a transmitter," in this media ecology the tape takes as its content both the phonographic archival desire detailed by Lisa Gitelman, Friedrich Kittler, Jonathan Sterne, and others, as well as the radiophonic association with absent traces, in order to create a new discourse network based on recording, erasing, and copying, wherein we will uncover a theory and practice of mediated listening that functions as political action.[34] Furthermore, tape technology's easy manipulation of recorded material—part of the apparatus for terror and intimidation in the interrogations of the Dirty

War—provides a model for Piglia's novelistic response, as his immanent theory of listening mines the dangers of aural media—passivity, erasure, isolation, and manipulation—to construct a nuanced counternarrative able to account for the usefulness of specific types of forgetting and the importance of objecthood in the formation of political communities.

MINOR FORGETTING

The Absent City's most ambitious representation of how listening and technology can engage and refashion bodies and consciousness, as well as personal and (multi)national history arrives halfway through the novel. Purported to be the machine's last story, the chapter "The White Nodes" ("Los nudos blancos") recalls already in its title the sleeplessness, "*las noches en blanco*," that framed the reception of the radio and the tape in the scenes of Junior and Renzi's fathers' listening. The connection does not seem accidental, as this story elaborates how a certain Elena, who shares her name with the consciousness housed in the machine, has checked herself into a psychiatric ward as an undercover investigator. While there, Elena seeks an "eternal state of insomnia" in order to prevent an engineer at the asylum,[35] Dr. Arana, from reading her thoughts and uncovering the whereabouts of Mac, the leader of the popular resistance she supports, and the coinventor of the machine, Macedonio, the husband who programmed his late wife Elena's consciousness into the cyborg.[36] Against the search for "names and addresses" by Arana (whose name sounds suspiciously close to the Spanish word for spider, *araña*, the emblematic arachnid of networks), Elena's only defense is to forget the facts: "She had to forget, she couldn't compromise the plans. She destroyed the [memory of her] meeting on the platform at Retiro Station, [her memory of] the bums toasting stale bread over a small fire . . . *She had the ability to erase her thoughts*, like someone forgetting a word they were about to say. They would not be able to make her talk about what she didn't know."[37] Much like Junior's father, who spent his sleepless nights listening to the BBC in order to erase the traces of his life, Elena stays awake to erase her thoughts. Yet, her forgetting politicizes what was nostalgic longing in Junior's father's case. Unlike the broader tendency to simply imagine the radio's lack of a trace to collude with fascist power, Elena's *minor forgetting* helps her escape the police state whose ultimate goal would be the total erasure Junior realizes the government has planned: "That's why there's so much control, Junior thought, they're trying to erase what is being recorded in the street."[38] A minor erasure, in Elena's case, allows these recordings to continue to proliferate. As she is perhaps both a rebel operative and the cyborg responsible for producing these tapes, her ability to erase parts of her memory, to forget facts in order to protect clandestine information, reveals her consciousness to function like tape technology, a medium capable of recording, erasing, and dubbing. Incorporating the lessons of the machine into the structure of consciousness allows her to further the political struggle against total erasure.

THE FREE MARKET OF MEMORIES

The analogy between consciousness and tape technology helps Piglia further restructure the idea of the radio network to more radical political ends. Rather than the familiar mass audience tuned in to a single station like the BBC, the novel turns to tape technology to imagine how dubbing, erasing, and recording might reconfigure the relationships between listeners. Distinct from either the Frankfurt school's passive culture industry victims, Fanon's anticolonial audience, or Jean-Paul Sartre's "practico-inert ensemble" of radio listeners who imaginatively construct their collective connection through a shared alienation from and inability to seize the means of production, Piglia's listeners form a collectivity based on a shared listening that effectively edits their individual identities into patchworks of stories, analogous to the storytelling mixtapes they pass through the underground.[39] This new notion of networked identity borrows from pirate radio as well as the black market to create a new form of narrative expression located in the novel's representative for this complex kind of ensemble (listening) identity: Elena.[40]

Piglia's attempt to narrate Elena's experience pushes his writing toward precisely the kind of listening narration I mentioned above. Ultimately unable to stay awake in Arana's laboratory, Elena slips in and out of what may or may not be hallucinations, as the reader and doctor follow her train of thought, and the narrative oscillates between the asylum and an escape scene in which Elena evades agents of the state in the city streets. At this point the distance between these two worlds, the city street and the asylum, breaks down, as Dr. Arana directly repeats one of Elena's thoughts, which leads her to wonder, "Maybe she was thinking out loud [*pensando en voz alta*]."[41] No sooner has this thought crossed her mind or left her mouth than she announces: "This is a place free of memories [*un sitio libre de recuerdos*] . . . Everyone pretends to be and is somebody else. The spies are trained to disown their own identities and use somebody else's memory."[42] The deictic confusion of place, the struggle in her spoken speech to regain her footing between private and public, between the merely thought and the spoken expression, these ideas inhere in Elena's proposal that identity and memory function like exchangeable goods.[43] This reading retains the ambiguity in the original Spanish, "*un sitio libre de recuerdos*," which I've translated literally as "a place free of memories" (a place without memories), but which we might also hear as "a free place [made] of memories," or rather, "a free market of memories."[44]

The language of the market is more than an analogy, as the memory exchanges Elena describes pertain to the resistance movement's clandestine cassette-tape network: a network formed in opposition to the state's control of information and in league with the black market of electronic commodities attached to the city's infiltration by global capital. The connection between the black market and the dissemination of the machine's stories through this rebel underground finds its link in the machine's museum guard, Tank Fuyita. Described elsewhere in the novel as a former "smuggler of cheap watches on the free market [*el mercado libre*]," Fuyita is also an inhabitant of the "Korean sector in the cellars of the Mercado del Plata" or Money Market,[45] the latter of which is situated between the streets Sarmiento and Lieutenant General Juan Domingo Perón: the two most powerful presidential figures in Argentine history. Thus, Fuyita embodies the globalization of national history through capital, as well as a new immigrant image of the Argentine revolutionary hero.

In this dialectical model Fuyita represents precisely the globalized technocracy the novel's protagonists fight against, as well as the revolutionary possibility that a "marketplace of identities" holds open for the rebels. On one hand, the novel threatens to lose itself in the cyberpunk genre's techno-orientalist paranoia, as the state awaits Japanese technicians, the only ones apparently skilled enough to disable the machine's storytelling.[46] And yet, on the other hand, Fuyita's role as the cyborg's museum guard, as well as the agent who sneaks Junior the transcript that is the chapter "The White Nodes," points to the radical role of the black market. The cyborg quotes Fuyita as saying that "the end of contraband . . . is the end of Argentine history,"[47] but this apparent "end" is also the means to proliferate the tapes' reproduction, as cassette copies continue to travel through the shadow markets, the goal or end of this contraband being the ongoing narration of an Argentine history the state wants its citizens to forget.[48]

Recalling Elena's "free site of memories," the underground commodity exchange is also a marketplace of identities wherein the exchange of subjectivities operates on the model of the clandestine tape network. On one level, we can read Elena as the character who struggles to resist Dr. Arana's interrogation, part of a rebel cell, actively erasing her thoughts as her story moves back and forth between the psychiatric hospital and the streets, tunnels, and backrooms of the city. On another level, what we experience as readers of "The White Nodes" is Elena the cyborg's consciousness as cassette tape, a consciousness passed from hand to hand, from scene to scene, as rebels erase strips of the recorded stories—stories that are, in fact, her memory—and edit and splice them with others.[49] *She narrates, in other words, the act of being heard.* If a cassette mixtape could write the story of its existence as it moved through a community of listeners, overdubbed with new songs or stories, erased, and rerecorded, that story would read like "The White Nodes."

Moreover, this narrative practice mimics the machine's storytelling mode, in which the machine tells only the stories of other people: aphasic children, shell-shocked soldiers, nuns possessed by Satan; people whose voices come from their bodies but no longer seem to belong to them.[50] Even in her closing "monologue," an homage and revoicing of Joyce's *Ulysses*, she becomes a circuit for other voices, as she declares: "I am Amalia, if they rush me I'll say I'm Molly . . . I am her and also the others, I was the others, I am Hipólita . . . I am Temple Drake . . . Those stories and others I've already told. It doesn't matter who speaks."[51] Thus, "Los nudos blancos" (The White Nodes) depicts not so much the novel's climax as its own conflicted attempt to untie itself ("nudos" means "knots" as well as "nodes") from the first-person genre of *testimonio* and, in a maneuver that recalls Junior's father, link itself to a network of anonymous voices speaking from a variety of narrative standpoints. Here we encounter the lesson Piglia learned from Puig. As Piglia would write of Puig a year after the publication of *The Absent City*, Puig "begins to use the tape recorder, and submits the transcription of a voice and a true story to a complex process of fictionalization . . . Puig fictionalizes the testimonial and erases its traces [*borra sus huellas*]."[52] Borrowing from this lesson, the tape network, and, by implication, the novel we are reading, embraces radio's link to erasure and then resists both radio's state consolidation under the sign of the single charismatic voice and the traditional political response to that voice: the first-person singular of *testimonio*.

Different from either the radio orator or the testimonial speaker, the cyborg tells stories by listening to other voices that inhabit her. We should remember again that the novel's machine was originally built from a radio receptor without a transmitter, and that the political action in the underground consists not so much in telling stories but in listening to them. Even the name of the machine's inventor, Emil Russo, a homonym, in Spanish, for Émile Rousseau, or Rousseau's *Émile*, points to the idea of reception, and the openness to inscription that undergirds the idea of the cyborg constructed on the basis of an unfinished radio.[53] Following only the first part of Brecht's call for a revolutionary radio—"to receive"—without the second half—"let the listener speak"—Russo transforms the radio into a listening machine, and "The White Nodes" stands as the written record of this new form of storytelling.

CASSETTE-TAPE COUNTERPUBLICS

The machine's storytelling both produces and is a product of the political network through which it moves. It is a network that arises, in a somewhat generic fashion, in response to the oppression of censorship. Indeed, in a surprisingly similar example to the novel's rebel community drawn from outside the fictional boundaries in which these underground rebel groups pass their voice recordings in cassette copies, the anthropologist Charles Hirschkind has documented the social practice of listening to clandestine taped sermons in Egypt. As in *The Absent City*, Hirschkind notes that the electronic underground structures itself in response to state censorship: "Through the medium of the cassette tape, acts of individual audition—properly performed—could compensate for the limitations the state had imposed on public discourse. Silence the speaker, but his words would still proliferate and acquire agency through the actions of listeners within the electronic network."[54] Like Renzi's memories of his father, these networks of listeners emerged in relation to the radio, much as the radio speeches of the Egyptian president Nasser, as well as the radio concerts of Umm

Kulthum, established the protocols of a national listening community in Egypt that transformed into a cassette-tape counterpublic beneath the silence of later state censorship.[55]

In a move that further elucidates Piglia's novelistic use of such tape networks, Hirschkind turns to Walter Benjamin's essay "The Storyteller" to establish the very different relationship to authority inherent in what he names "effective audition": "Effective audition, an act that enables the integration of the narrative into the listener's own experience, requires a subordination to the authority of the storyteller and thus, in some sense, a heeding to the story itself. This is Benjamin's point about the 'naïve relationship' between speaker and listener."[56] Listening institutes a new social structure, one that does not stop at submitting to the storyteller's authority but yields in order to understand and pass on the story. In this social practice, listening leads to storytelling, and subjects learn to become conduits for the story's transmission. In *The Absent City*, Macedonio, the machine's programmer and Elena's widower, says: "The key . . . is that one learns through narrating. To learn means that one remembers what has already happened, and therefore gains experience."[57] What distinguishes Benjamin's idealized oral communities of auratic laborers, and even the cyborg's stories, from the secondary aurality of Piglia's electronic underground is that the recorded story in Piglia's novel is *passed on without being retold*. The story circulates, collects, and mobilizes a popular community without necessarily rerouting itself through another's mouth. It both draws from the novel's social space, the isolated and individual experience Benjamin describes, and enacts the collective formation Benjamin posits on the side of orality.

My invocation of Hirschkind's study returns us to *The Absent City*'s struggle to imagine a political alternative to the charismatic and monologic male voice over the radio: a central concern for a novelist attempting to construct an alternative to the dictatorial legacies of not just Argentina but a whole series of countries throughout the Americaa. As I have described above, the posthuman but still gendered cyborg Elena represents one means of resistance. Discussing her construction, Russo claims, "we were trying to make . . . a female defense machine, against the experiences and experiments and the lies of the State."[58] If her polyphonic closing "monologue" presents us with the Scheherazade of 1,001 female voices, the mostly anonymous listeners who make up the rebel underground might be thought of like the mute girl in one of the machine's stories: as a collective "anti-Scheherazade, who heard the story of the ring each night from her father, one thousand and one times."[59] This crowd listens as part of their activism, but they do not speak. We never encounter their listening through any thick description. Unlike the scenes with Junior or Renzi's fathers, no potential theory of "close" or "sensuous" or "ethical" listening emerges from the text. These listeners exist only as implicit characters, an almost silent crowd nearly out of earshot from within the narrative. And yet, they are Elena's necessary corollary, responsible for transforming the stories she recounts into the connective tissue of a political community.

Or rather, the tapes themselves are the connective tissue, or, as Macedonio says, the "*trama común*," the common weave, or common plot, that the listeners' hands stitch together. As I have argued, in "The White Nodes" Elena's ability to circuit stories through her consciousness, to take on other identities, to cut and splice her personal narrative with those of others, helps untie the knots that might lead to any original, single, unified identity. That analysis approached the nodes/knots from Elena's vantage point. However, we might also think of these nodes and knots as the listeners in the network. They are the ones who pass Elena on in "The White Nodes." They "tell" the story by recording it and sending it to another. The tapes, and perhaps by extension, Elena, are also a meeting point, a node, "the minimal link of a *thing in common*," a connection formed through recording and recounting.[60] In finding this "thing in common" and forming their tape network, the rebel listeners do not surpass or abandon the radio. Rather, they reimagine the radio, not only through combining the techniques of radio's absent traces and the phonograph's insistent recording of those traces but through the ability to hook in to other voices. In their hand-to-hand trading and copying of tapes, they bring the radio network into the street and into physical contact, spreading stories as if spreading a contagion of rebel consciousness.

The desires in *The Absent City* for a rebel radio that would grasp the most politically utopian aspects of the history of radio in Latin America—constructing a populist network that never consolidates into the single dictatorial voice—requires Piglia to dismantle what we have come to know as radio, and reimagine its auditory and social relations as if a vast number of invisible hands had been required to knit the "wireless" world together. Of course, those hands *were* there, always holding aloft the voice and making it possible. This is why the novel remembers Perón's voice on the tape rather than his more famous oratory over the radio. Within the rejection of that radio voice is a critique of the most tyrannical moments radio history in the Americas. It is a critique that hears in radio's first crackling signal the advance of the collapse between private and public spaces and the consolidation of the authoritarian nation state. Dr. Arana stands as a symbol for both. Yet, the resistance to such state power does not merely reside in the act of seizing the means of production. For unlike so many revolutionary theories of radio, Piglia's novel presents us with the *radio without a transmitter.* The political community forged by the cassette-tape network recasts radio technology through an insistence on reception. Listening in to other voices, taking up the object position to let other things speak—these marks of effective audition listen in opposition to the subject position of the single, charismatic radio voice. Piglia's turn to the cassette tape, therefore, attempts to produce another story of the age of radio in the Americas, and, perhaps in anticipation of a file sharing future as well as a look back to the early years of clandestine radio, he seeks to acknowledge private listening experiences as acts of political audition.

Notes

PREFACE

1. Allen and Martin, "Obama Unveils 21st Century New Deal."

2. Keane, *Ireland and the Problem of Information.*

3. Ellison, *Invisible*, 581.

4. Pickard, "Before Net Neutrality"; Pickard, *America's Battle.* If some new media critics now misidentify radio as a control system or "homeostatic machine" of "dead voices" rather than the precursor to online real-time interactivity, perhaps that is because legislative fights in the United States affected the structure of radio content and how listeners used their radios (Thacker and Galloway, *The Exploit*, 126).

INTRODUCTION

1. For more on the event's significance in changing radio's popularity in Argentina, see Ricardo Gallo, *La radio*; Ulanovsky, *Días*; Matallana, *"Locos,"* 83; Sarlo, *La imaginación técnica*; and Claxton, *Perón.*

2. Salwen, *Radio*; Cortázar, "Noble arte," 124. All translations mine unless otherwise noted.

3. Fridman and Sheinin, "Wild Bulls," 57.

4. Cortázar, "Noble arte," 124.

5. Matallana, *"Locos,"* 35–36.

6. There are competing definitions of what makes a medium. Jonathan Sterne writes that "a medium is a recurring set of contingent social relations and social practices, and contingency is the key here. As the larger field of economic and cultural relations around a technology or technique extend, repeat, and mutate, they become recognizable to users as a medium. A medium is therefore the social basis that allows a set of technologies to stand out as a unifed thing with clearly defined functions" (*Audible Past*, 182). W. J. T. Mitchell, criticizing the broadness of Raymond Williams's definition of a

medium as a "material social practice," claims: "A medium just is a 'middle,' an in-between or go-between, a space or pathway or messenger that connects two things—a sender to a receiver, a writer to a reader, an artist to a beholder, or (in the case of the spiritualist medium) this world to the next. The problem arises when we try to determine the boundaries of the medium . . . If media are middles, they are ever-elastic middles that expand to include what look at first like their outer boundaries. The medium does not lie between sender and receiver; it includes and constitutes them" ("Addressing," 4).

7. Matallana, "*Locos,*" 38.

8. In many ways, the new versions of the vox populi and the corollary transformations in the constitution of literary character as a relay, conduit, or carrier that I will study across this book match Foucault's configuration of the subject and the utterance at the conclusion of *Les mots et les choses* and throughout *L'archéologie du savoir.* Kittler took up these models in his historical theorization of new "discourse networks" that brought together pedagogy, literature, and electric media to turn "so-called man" into a product or epiphenomenon made possible through the particular relations and organization of knowledge in the media system. Kittler launched his attack on Literature—which he generalizes from European literature and primarily German romanticism and later French symbolism—because Literature, in his view, participated in the proliferation of a belief system and institution of power and knowledge at once misogynist, expressive, vitalist, and mystical. Hermeneutics, the analytical mode most closely associated with this conception of literature, elevated human intelligence's self-estimation through the claim that one could read into and reveal the truth of things. Thus, Literature produced an arrogant and chauvinist humanism that ignored the human's contingent and dependent status in the world of other animals, things, and nonhuman existences. Literature, as an institution, became the alibi for the grotesque belief in human power whose potential for violence found itself dramatically amplified through new technologies in the war machine of the twentieth century. As much as his chapter on the gramophone mocks printed literature as an outmoded means of transcribing the voice surpassed by electric media's physiological inscription of the voice's vibrations onto shellac or cylinder, he represents Literature as the training ground that enabled the worst uses of electric media. Literature is coextensive with these media. The gramophone is not only important because it embodies the next step in media evolution, but because it bypasses the human's exceptional status and the symbolic system upon which Literature relied to help constitute humanist power (Kittler, *Discourse*; Kittler, *Gramophone*).

9. Katherine Bergeron has shown that we might see Mallarmé's *Un coup de dés jamais n'abolira le hasard* (A Throw of the Dice Will Never Abolish Chance) as "the record of a moment of impact, a blow that has produced on the page a shock, like sound waves rising from a drum's vibrating surface" ("A Bugle," 59). Bergeron's ingenious reading extends how print might capture "noise"; however, even this analysis retains the division between transduced and represented sound.

10. For more on "audile technique," see Sterne, *Audible Past*, 87–178.

11. Ehrick, *Gendered*; Karush, *Culture of Class*.

12. Groundbreaking work from Alejandra Bronfman, Luis Cárcamo-Huechante, Susan Douglas, Christine Ehrick, Rubén Gallo, Joy Hayes, Michele Hilmes, Jason Loviglio, and others has explored the different genres of radio history in the Caribbean, Mexico, Peru, the Southern Cone, and the United States. One remarkable study of narrative form on radio is Neil Verma's *Theater of the Mind: Imagination, Aesthetics and American Radio Drama*. (See Bronfman, "Weapons"; Cárcamo-Huechante, "Indigenous"; Douglas, *Listening In*; Rubén Gallo, *Mexican Modernity*; Hayes, *Radio Nation*; and Hilmes, "NBC" and *Radio Voices*).

13. Anderson's foundational study already calls for a further study that would appreciate the limits of his argument's focus on the newspaper and print culture for establishing concepts of national community. In a footnote, he writes: "Radio made it possible to bypass print and summon into being an aural representation of the imagined community where the printed page scarcely penetrated. Its role in the Vietnamese and Indonesian revolutions, and generally in mid-twentieth-century nationalisms, has been much underestimated and understudied" (*Imagined*, 35).

14. In other words, the "live" effect derives from simultaneous transmission, or what Jean-Paul Sartre identifies as the "seriality" of radio listening and its "indirect gatherings." Sending the same sounds to a massive audience at the same time, radio's simultaneity turned even recorded songs into live experiences, since the network of shared listening existed only as long as the broadcast airing the event. For Sartre, "the mere fact of *listening to the radio* . . . establishes a serial relation of *absence* between the different listeners" ("Collectives," 271; emphasis in original). The "absence" does not refer to the missing body in the voice but the missing bodies of the listeners, each of them simultaneously listening to the radio voice. This *shared* absence, therefore, produces an imagined collectivity based on the knowledge of a shared temporality. Radio broadcast distinguishes itself from inscriptive media—the newspaper, the printed book, the phonograph—not by failing to leave traces of the event but rather by producing the experience of a previously unfelt massive simultaneity threaded with absence. Sartre argues that the knowledge of other listeners that gives rise to what I am calling the "live effect" of seriality has been produced by mass media over time: "Thus, the impotent listener is constituted by the very voice as an *other-member* (*membre-autre*) of the indirect gathering: with the first words, a lateral relation of indefinite seriality is established between him and *the Others*. Of course, this relation originated in knowledge produced by language itself in so far as it is a means for the mass media. Everyone gets their information about French radio programmes from newspapers and the radio itself. But this knowledge (which is itself of a serial order by virtue of its origin, content and practical objective) has for a long time been transformed into a fact" (ibid., 273; emphasis in original).

15. O. Luis López, *La radio en Cuba*.

16. Ehrick, *Gendered,* 230.

17. Unlike the community's intentional coexistence, neighborhoods depend on contingent spatial relations, bound by certain territorial limits, which

tend to remain more fluid than other borders. While some might be fortunate enough to be able to choose a neighborhood to live in, one doesn't choose one's neighbors. Neighborhoods, in this sense, form accidental and dynamic collectivities.

18. Sexton, *The Monroe Doctrine*, 185–98.

19. See Pérez, *Cuba under the Platt Amendment*.

20. Although titled *Sangre Negra* in Spanish-speaking countries, the film retained its title *Native Son* in the United States.

21. FALICOV, "Hollywood's Rogue Neighbor."

22. Best, *The Fugitive's Properties*.

23. Joseph Jacquard's development of the punch card in 1805, and then Charles Babbage's 1821 Difference Engine before Henri Fourneaux's development of the player piano rolls in 1863, link this early history of sound and property to the history of cybernetics, and the development of computing. MP3s are only the most recent popular format in this trend (see Wershler-Henry, *Iron Whim*).

24. John Picker explains the corollary situation in Victorian England as the transformation from "what Romantics had conceived of as a sublime *experience* into a quantifiable and marketable *object* or *thing*, a sonic commodity, in the form of a printed work, a performance, or, ultimately, an audio recording, for that most conspicuous legacy of Victorianism, the modern middle-class consumer" (*Victorian*, 10).

25. "For in those days a major shortcoming was ascribed to my invention: the possible interception of transmissions" (qtd. in Kittler, *Gramophone*, 251).

26. Mills, "Deaf Jam"; Sterne, "Nature Builds."

27. Lucey, *Never*, 5; Berlant, *Complaint*, 3–4.

28. Cassirer, *Logic*; Panofsky, *Perspective*; Bourdieu, *Les règles de l'art*.

29. In a special issue on "qualia," Lily Hope Chumley and Nicholas Harkness observe that "the term 'qualia' is used to refer to perceptions and sensations 'in' or 'of' the mind, and the terms 'quality' and 'property' are used to refer to material properties of entities in the world" ("Introduction," 4).

30. LaBelle, *Background Noise*, 123.

31. Alvin Lucier's 1968 performance "I Am Sitting in a Room" remains the clearest artistic and theoretical statement about this combination of acoustic properties. At once a work about linguistic disability, acoustic theory, recording technology, and the voice's relation to the self, Lucier's performance begins with him reading out an explanation of his process:

> I am sitting in a room different from the one you are in now. I am recording the sound of my speaking voice and I am going to play it back into the room again and again until the resonant frequencies of the room reinforce themselves so that any semblance of my speech, with perhaps the exception of rhythm, is destroyed. What you will hear, then, are the natural resonant frequencies of the room articulated by speech. I regard this activity not so much a demonstration of a physical fact, but more as a way to smooth out irregularities my speech might have.

As the performance proceeds over forty-five minutes, and the recordings continue to reproduce this explanation, Lucier's semantic statement becomes increasingly abstracted and then totally illegible. The vocal fills with echo, feedback, and shimmering bends in pitch. Lucier's stutter, heavily disruptive when he says "rhythm" and "smooth," dissolves into the wash of sound. The stutter continues to contribute to the performance's sound, but we no longer hear it as a break in rhythm; rather it is one of many factors, including the room's resonance and the recording medium itself, that amplify the sonic qualities or acoustic properties already available in the voice. To say "Lucier's voice" seems strange at this point because copy and repetition—a certain version of mimesis—have produced a composite sound that no one properly possesses. However, we cannot forget that the voice began with semantic content. It is not sheer, sensuous sound; rather, Lucier's extreme mediation brings us back to hear the physics of the voice as constitutive of even the voice's most everyday semantic statements.

32. We might also think of Franco Moretti's claim in *Modern Epic* that "the secret (and frustrated) desire of every world text," those texts he names "modern epics," is "to heal the great fracture between avant-garde exploration and mass culture," an achievement central to all of Puig's work (107).

CHAPTER 1

1. Dos Passos, "Hookup," 406.

2. Ibid. For more on NBC's role in moving radio in the United States toward a commercial medium rather than a state-run system as with the BBC in England, see Hilmes, "NBC." For more on the general rise and consolidation of commercial radio in the United States, see Doersken, *America's Babel.*

3. If, as Dos Passos claims, the audience listened through "jaded" ears, it could be because many of them had already heard the convention's nominee, Franklin Delano Roosevelt, speak in support of Al Smith in the first nationally broadcast convention from Madison Square Garden in 1924. Thus, when he arrived in Chicago to deliver a "new deal" to U.S. citizens, Roosevelt spoke to an audience in the stadium and at home by now accustomed "to gather and listen to loudspeakers broadcasting reproduced sound" (Thompson, *Soundscape*, 233).

4. Sartre, "Prière d'insére"; Sánchez, *La guaracha.*

5. For more on the Estridentistas and radio, see Fernández, "Estridentistas"; Townsend, "Radio/Puppets"; and Gallo, "Radio." For more on Dos Passos's time in Mexico, see Rubén Gallo, "John Dos Passos." For more on Carpentier and radio, see Birkenmaier, *Carpentier.*

6. Thompson, *Soundscape*, 5.

7. The writers of the 1939 Federal Writers' Project *New York City Guide* call Rockefeller Center "an organized city" and an architectural phenomenon that "stands as distinctively for New York as the Louvre stands for Paris" (333, 334).

8. Thompson, *Soundscape*, 7.

9. Ibid., 233–34.

10. Ibid., 234; emphasis added.

11. Changes to the science of acoustics were not limited to concert halls. According to the *Oxford English Dictionary*, 1933 also marked the first use

of "acoustic phonetics," recognizing the institutionalization of an epistemic change in the relation between sound and language as well.

12. As Alan Williams makes explicit, while the suppression of acoustic space described by Thompson derives from the ongoing electrification of space, an ideological function already inheres in the apparatus itself: "[As] *microphones [are] more like ears than they are like rooms* (they function as points and not as volume [in the spatial sense]), it is never the literal, original 'sound' that is reproduced in recording, but one perspective on it, a *sample*, a *reading* of it" (Williams qtd. in Lawrence, *Echo*, 20; emphasis added).

13. Arnheim, *Radio*, 143, 142; emphasis added.

14. Thanks to Brían Hanrahan for pointing out, in his manuscript "Index and Diegesis in Weimar Broadcasting," radio's sonic improvements over the phonograph. As Roland Gelatt notes in his *Fabulous Phonograph: 1877–1977*: "The radio receiver of 1924, for all the inadequacies of its amplifier and loud-speaker, gave a quality of sound reproduction that the phonograph could not even approach" (218). Hanrahan adds: "When American record companies advertised the new electric recordings, later in the decade, they sold them by referring to this 'radio sound'" (9).

15. Even in its metropolitan location, Rockefeller Center had international connections: "Sharing the eastern exposure [of 'Radio City'], four lesser buildings serve as Fifth Avenue showcases for foreign nations: the British Empire Building, La Maison Française, the Palazzo d'Italia, and the International Building East. Slightly behind the latter two rises the forty-one-story International Building" (Federal Writers' Project, *New York City Guide*, 333–34). Diego Rivera's controversial mural *Man at the Crossroads*, commissioned by Nelson Rockefeller and destroyed because Rivera refused to remove a portrait of Lenin included in the work, helped portray a fuller sense of the different national and political forces at work in Rockefeller Center. For more on the mural and its relationship to modern technology, see Rubén Gallo, *Mexican Modernity*.

16. Rita Barnard points out that Dos Passos's "bitterly ironic biography of Keith," which she calls "one of the ugliest portraits in *U.S.A.*'s gallery of American villains," is based on Samuel Crowther's *The Romance and Rise of the American Tropics*, "an offensively partisan and jingoistic story of the role of U.S. corporations in the Caribbean" (*Great Depression*, 243).

17. Their worries about other countries monopolizing the limited number of radio frequencies would vanish in the mid-1920s with the development of the shortwave (Winkler, *Nexus*, 265; see also 99).

18. Winkler, *Nexus*, 90.

19. Ibid., 91.

20. Hilmes, "NBC," 14.

21. The Latin American connections to U.S. radio networks were also present at NBC's rival, the Columbia Broadcasting System (CBS). CBS's first president, William S. Paley, was the son of a former reader at a cigar factory in Chicago. Following a Cuban tradition, Samuel Paley would read newspapers, novels, and anything else to keep workers entertained while they rolled cigars. The elder Paley went on to found his own cigar manufacturing company, La

Palina, which became the name of one of CBS's first sponsored broadcasts (see Barnouw, *Tower*; and "About," *La Palina*, Web, n.d.). For more on the Cuban tradition of the "lector de tabaco," see Kuhnheim, *Beyond*, 6–7; and Mormino and Pozzetta, "Reader."

22. One of the major proponents of the theoretical possibilities of disembodiment in radio and other acoustic technology has been Allen Weiss (see Weiss, *Breathless*). Weiss and other critics in this vein derive much of the authority for the argument from the radio artist and theorist Douglas Kahn (see esp. Kahn and Whitehead, *Wireless*). Film theory has treated the "off screen" voice and the body with more sophistication (see Lawrence, *Echo*; and Silverman, *Acoustic Mirror*).

23. Dos Passos, *All Countries*, 413.

24. Agha, "Voice, Footing," 39; emphasis in original.

25. Ibid., 39; emphasis in original.

26. Qtd. in Peters, *Speaking*, 213. Hoover's complaint is comparable to Walter Benjamin's comments, in his essay "The Work of Art in the Age of Mechanical Reproduction" (1935) in *Illuminations*, about the actor whose body is carried out to an invisible audience as he stands before the apparatus of the film camera.

27. See Mindell, "Deaf Jam," for more on the relationship between radio and the rise of engineering feedback in computing.

28. Dos Passos, "Hookup" 410.

29. Chion, *Audio-Vision*, 71.

30. Silverman argues that audiovisual synchronization "anchors sounds to an immediately visible source, and . . . focuses attention on the human voice and its discursive capabilities" in order to "suture the viewer/listener into what [Stephen] Heath calls the 'safe place of the story,' and so to conceal the site of cinematic production" (*Acoustic Mirror*, 45). By disjoining the two moments Dos Passos calls attention to the audiovisual fissures not yet accounted for by the New Acoustics.

Mladen Dolar, who discusses the acousmatic in *A Voice and Nothing More* (2006), claims "every emission of the voice is by its very essence *ventriloquism*. Ventriloquism pertains to voice as such, to its inherently acousmatic character: the voice comes from inside the body, the belly, the stomach—from something incompatible with and irreducible to the activity of the mouth. The fact that we see the aperture does not demystify the voice; on the contrary, it enhances the enigma" (70).

31. Connor, "Modern Auditory," 221–22; emphasis in original. Steven Connor observes that "the classic example" of the *acousmêtre* "is the figure of the Invisible Man in James Whale's film of 1933" (222), thus reinforcing the interrelationship between narrative, radio, and film and what Thompson identifies as the consolidation of the New Acoustics in 1932.

32. Nancy Ruttenburg identifies a similar epistemological challenge in the *acousmêtre*, which she translates as "'an acousmatic personality': a disembodied presence made sensible to others through the faculty of hearing, but not sight, a 'kind of talking, acting shadow' who, in this case, interferes directly in the lives of those around him by manipulating their epistemological

assumptions concerning 'the reassuring fixation of the voice in the residence of the body'" (*Democratic*, 216). In the representation of Hoover, the text's narrator both participates in and observes the epistemological manipulation that appears to derive from the acoustic engineering more than Hoover himself.

For more on the familiar gothic examples, see Picker, *Victorian*. For more on the uncanny, the gothic, and architecture, see Bernstein. Frances Dyson, listening back to Edison and forward to new media theory in her book *Sounding New Media*, notes Edison's claim that the phonograph's stylus can "'effect a restoration or reproduction of vocal or other sound waves, *without loss of any property essential to producing on the ear the same sensation* as if coming from the original source'" (qtd. in Dyson, *Sounding*, 47). Edison's statement gave technological authority to a belief that the recorded voice could make dead voices live by means of an exact reproduction that seemed to almost erase mediation. This dream of a medium that erased its own mediation anticipates the New Acousticians, among them Harry F. Olson, who, Dyson writes, "defined the ideal of sound reproduction as 'the elimination of distortion and the reproduction at the listener's ear of sound waves identical in structure to those given out at the source'" (47). Literary and cultural critics and theorists have tended to embrace the metaphysical, uncanny character of the radio and phonograph voice, while ignoring or relegating to the background information about the engineering at work.

33. Connor, *Dumbstruck*, 227; emphasis added.

34. In this, Connor's analysis aligns with the media history supplied by Kittler in *Discourse Networks*.

35. Ruttenburg, *Democratic*, 24.

36. The voice of democratic personality eschews attachment to particular bodies in order to achieve political empowerment for bodies usually excluded from the political process: "Unlike race or gender, however, democratic personality is conspicuous for having no prior material referent in the body. (This is not to deny the body a role in the formation of democratic personality—far from it, as we shall see. It does deny the body a delimiting function and thus denotes the radically inclusive nature of democratic personality as a category of being potentially open to everybody)" (ibid., 10).

37. Ibid., 16.

38. As Ruttenburg puts it, "[I]ndependence of the body lent [democratic personality] a flexibility that exceeded the text's insofar as it required no artifact (such as the book) or expertise (such as reading) and disseminated itself without requiring a network of supportive institutions" (ibid., 186). The social and historical background Ruttenburg provides alters the status of the theological speech enacted in Salem and during the Great Awakenings from passive "possession" to a more complicated notion of collective speech and political action. Less enthusiastic about the gender binary emphasized by Connor, Ruttenburg reads the same text he takes as indicative of the shift from female possession to male ventriloquism, and views the two forms as intimately connected. Whereas Connor's reading of Charles Brockden Browne's *Wieland* (1798) barely mentions the sister, Clara Wieland, Ruttenburg seizes on Clara's role as a narrator who retains her innocence despite her *authorship*

by repeatedly asserting her ethically superior status as a victim, a hapless *character* in Carwin's insidious narrative (ibid., 186; emphasis added). In so doing, Clara attempts to vindicate herself from the possible charge of making use of the ventriloquial power of authorship by redefining it as a legitimate form of humble self-enlargement, a form of benign ventriloquism, an innocence of power in which the author's ontological difference from his or her character might be denied and their identity asserted (ibid., 186). When Clara forces the criminal Carwin "to ventriloquize himself" and then listens with feigned awe alongside her brother when the disembodied voice speaks, she assumes the position of both author and character and therefore reimagines the former in the guise of the latter.

What we might add to Ruttenburg's argument is that through the character of Clara Wieland, ventriloquism and humble self-enlargement resist the association with spatial control Connor associates with throwing voices; they are conceived instead as a means to embrace listening as a creative act. Attention to acoustics, the sound of space, helps Clara unmask or "de-acousmatize" Carwin and becomes the means for Clara to access a tradition of popular speech, "the voice of democratic personality," which allows her to assume a position neither wholly contained by nor excluded from the narrative action. Standing in between the position of the character and of the author, the disembodied and the gendered body, Clara becomes an imaginative response to the acoustic problems of power posed by radio and film theory. Key to such listening is that it occurs within the discursive space of the scene. Thus, it is a listening not aligned with surveillance, but a participatory act within the frame (see Silverman, *Acoustic Mirror*, 88–90, 96–97).

39. Dos Passos, "Hookup," 406, 407.

40. Ibid., 414; emphasis added.

41. In the words of the radio theorist Edward Miller in *Emergency Broadcasting and 1930s American Radio*, FDR enunciates his "broadcasted speeches almost as if they were 'narrowcasted' missives, seemingly particularized for the private home, giving the listener the illusion that he was being spoken to in a point-to-point transmission, *rather than overhearing him in a radio address*" (89; emphasis added). Miller adds that "Roosevelt's use of the plural pronoun 'you' and the possessive pronoun 'your' was rare in commercial broadcasting as a way to address the audience . . . Dramas and comedies presented their narratives as if the listener were *overhearing* the programming" (88). The president's use of the second person "strengthened the idea that he entered the home via his transmitted voice" (89). However, in *Radio's Intimate Public*, Loviglio observes that noncommercial broadcasters such as "Roxy" Rothafel, Will Rogers, and the notorious Father Coughlin already provided a context for FDR's chats, in that the latter "drew on radio's already established convention of the intimate mode of address" (6).

42. Dos Passos, "Hookup," 414; emphasis added.

43. Loviglio, *Intimate Public*, xiv–xix. For a different use of the term "intimate public," see Berlant, *Complaint*.

44. Loviglio, *Intimate Public*, 4, 7, 5; emphasis added. Dos Passos's attention to the significance of tone in shaping this new social space finds support in

the critical taxonomy applied to Roosevelt's speeches: "In addition to elaborate preparations in the writing of [the Fireside Chats], great care was taken in the timing and aesthetics of the broadcasts. Scholars of speech have dissected Roosevelt's impressive radio performances from a myriad of perspectives, generally concurring with Robert T. Oliver that he had 'the best modulated radio voice in public life.' Studies of his pitch (tenor), intonation ('vibrant with enthusiasm'), pronunciation ('eastern' mixed with elements of 'general American'), cadence ('measured and deliberate'), and speaking rate (a comparatively slow one hundred words per minute), and volume (great dynamics) all bear out the general scholarly, journalistic, and popular consensus that Roosevelt was a master public speaker. In a study of his wartime Fireside Chats, Waldo Braden points to Roosevelt's 'short declarative sentences,' parallel structure, repetition, and 'attractive cadence'" (ibid., 7). For a recent study of Roosevelt's speech patterns in a media theory context, see the aforementioned text by Edward Miller.

45. Loviglio, *Intimate Public*, 8.

46. Ibid., xvi.

47. Brinkley, *Voices of Protest.*

48. In *Unspeakable Sentences* (1982), Ann Banfield argues that her system "sees narrative fiction as linguistically constituted by two mutually exclusive kinds of sentences, optionally narratorless sentences of pure narration and sentences of represented speech and thought. Both these sentences are 'unspeakable'" (185). Alternatively, Brian McHale argues that sentences in *U.S.A.* break supposedly unbreakable rules in Banfield's system ("Talking," 254–55, 257).

49. Three years after Dos Passos published "The Radio Voice," his friend Archibald MacLeish wrote *The Fall of the City*, a radio drama directed by Irving Reiss and broadcast on CBS on April 11, 1937. Neil Verma has written eloquently about how MacLeish's "word" and Reis's "microphone" organize the play's acoustic, narratological and political space in a struggle between MacLeish's "pessimistic liberalism" and Reis's populism (*Theater*, 55). The distinction pertains to what Verma names the play's "audioposition," or where listeners locate themselves in the drama: "MacLeish's correspondent is a hero because he guides the listener away from the folly of the people in the square, while Reis's is a hero because he brings us together with them. Given that the play is expressly concerned with the politics of communication to begin with, perhaps the best way to read the broadcast is as a meditation on contradictions lurking in how New Deal culture processed the idea of addressing multitudes" (55). Noting how Dos Passos worked with the "word" in order to engage listeners to think through how their listening situated them politically—as either an intimate friend of the president or a skeptic aware of the gulf of wealth and power that separated speaker and addressee—we can ponder how *The Fall of the City* draws on Dos Passos's writing about listening in order to create a new acoustic environment working through the same political and aesthetic issues in formulating different conceptions of "the voice of the people."

50. This notion of "acoustic properties" imagines an alternative to the mere commodification of the acoustic that the term describes in Stephen Best (see *Fugitive's Properties*, 59). As I move throughout my argument, I treat the alienability and inalienability of the acoustic dialectically.

51. For more on FDR and his speeches, see Brinkley, *Roosevelt*; Hilmes, *Radio Voices*; Craig, *Fireside Politics*; and Levine and Levine, *People*.

52. "A Propos de Dos Passos et de '1919,'" in *Critiques littéraires (Situations, I)* (Paris: Gallimard, 1947), 94. The essay was originally published in August 1938.

53. Dos Passos, *The Big Money*, 59.

54. Dos Passos, *42nd Parallel*, 255.

55. Denning recognizes a similar feature of Dos Passos's narrative but associates it with "an anti-populist Taylorized colloquialism," in which "Dos Passos's compound words are not a device of defamiliarization but a streamlined and efficient elimination of unnecessary parts and motions" (*Cultural Front*, 177). He goes on to identify Dos Passos's use of indirect speech as "the 'about how' construction: 'Agnes had told her all about how getting drunk was something that men did and that they hadn't ought to'; 'he tried to break it up by telling about how he'd been taken for Charles Edward Holden in that saloon that time, but nobody listened.' The text creates a continuous present, a constant ticking of the clock, so regular that you hardly notice it, with little use of flashback, internal monologues, or disrupted time sequences: its storytelling is as regular and relentless as an assembly line" (177).

56. In many ways Sartre's notion of the "populist" style in Dos Passos comes closest to what other narrative theorists have identified as characteristic of the epic. See, for instance, Eric Auerbach's chapter "Odysseus' Scar," where he writes, "a problematic psychological condition . . . is impossible for any of the Homeric heroes, whose destiny is clearly defined and who wake every morning as if it were the first day of their lives: their emotions, though strong, are simple and find expression instantly" (*Mimesis*, 12); or, "the Homeric poems conceal nothing . . . Homer can be analyzed . . . but he cannot be interpreted" (13). Similarly, Mikhail Bakhtin's comments that "the epic world of the absolute past is inaccessible to personal experience and does not permit an individual, personal point of view or evaluation . . . [I]t is impossible to experience it, analyze it . . . [T]he important thing is this formal constitutive characteristic of the epic as a genre . . . its reliance on impersonal and sacrosanct tradition, on a commonly held evaluation and point of view—which excludes any possibility of another approach" (*Dialogic*, 17).

57. Pizer, *Modernist Style*; Beal, *Networks*; Denning, *Cultural Front*.

58. Denning, still usually the novel's most perceptive reader, describes the Camera Eye as "the dreamwork of the trilogy, a series of symbolist prose poems . . . [that] seek to unite private memory and public history" (*Cultural Front*, 171). The Newsreels, on the other hand, are not "the work of public history" but "an emblem of industrial society's 'idiot lack of memory" (171). Finally, "If the Newsreels offer the official history of the New Century, and the Camera Eye offers a memorial to its victims, the grand narrative of the Lincoln republic is outlined in the twenty-seven biographical portraits of actual historical figures" (171).

59. Dos Passos, *1919*, 238.

60. Ibid., 239.

61. Ibid., 246.

62. Beal, *Networks*, 96. For more on the way in which the "network" provides an alternative to existing models of relationality between the fragment and the totality, see ibid., 101. Patrick Jagoda understands "network aesthetics" as the relation between a particularly historicized physical infrastructure, social science, and set of narrative imaginaries.

63. Hirschkop, *Mikhail Bakhtin.*

64. Lukács, *Realism,* 24. In one of Bertolt Brecht's responses to Lukács, he notes: "Dos Passos himself has given an excellent portrayal of 'close interactions between human beings in struggle,' even if the struggles he depicts are not the kind Tolstoy created, or his complexities those of Balzac's plots. Secondly, the novel certainly does not stand or fall by its 'characters,' let alone with characters of the type that existed in the 19th century . . . We know something about the bases on which the cult of the individual, as practiced in class society, rested. They are historical bases" ("Brecht against Lukács," 77).

65. Ngai, *Ugly Feelings*, 41. Ngai's reading of Melville's *The Confidence Man* (1857) follows this inclination, joining tone's sonic and affective connotations to reveal how sound can redirect attention to discursive form and instantiate new political awareness. When two of the novel's characters interrupt their statements with what Ngai calls "non-semantic" "sonorous patterning"—alliteration, assonance, and onomatopoeia—their utterances index a change in the novel's tone: away from "confidence," which she identifies as "the tone of capitalism" (62), and toward a negative affect she describes as the "*nonfeltness*" (ibid., 76; emphasis in original) of confidence: the radical insecurity that capitalism's confident mood needs to cover. Textual sound's nonsemantic sensuousness leads, in Ngai's argument, to the reader's perception that the story "*runs on a feeling that no one actually feels*" (ibid., 69; emphasis in original). Against the backdrop of the novel's larger themes, this statement amounts to a critique of the credit economy. See also Adams, "Boom Years," 2; and Lukács, "From *Realism in Our Time*," 759–64. See also Frederic Jameson's argument that realist fiction becomes increasingly abstracted as it moves from the time of narration toward affect from the nineteenth to the twentieth century (*Antinomies*).

66. Denning writes: "This anti-populist Taylorized colloquialism led Edmund Wilson to complain to Dos Passos that 'you tend to make your characters talk clichés'" (*Cultural Front*, 177). Jacques Rancière comments: "Literary purity can't undo the tie that binds it to the democracy of writing without eliminating itself. Its own process of differentiation leads it to the point where its difference becomes indeterminable. The other side of this paradox will show itself the following century when the will to transform itself into a means of intervention will drive an accusatory literature to welcome into its pages the standardized messages of the world. The most representative work of the political literature of the twentieth century, Dos Passos's *USA* cycle, testifies to this" (*Politics*, 27).

67. Sartre, "A Propos," 88. Reviewers of the trilogy in the 1930s generally praised the book but consistently bemoaned the lack of individually wrought characters. Even a 1997 review in the *New York Times* of the Library of America's edition of *U.S.A.* mistakes Sartre's praise for criticism (see Gilman,

"Many Lives"). A collection of reviews from the 1930s can be found in Maine, *Dos Passos*.

68. Sartre, "A Propos," 95; emphasis in original.

69. de Beauvoir, *Prime*, 113.

70. Lucey, "Simone de Beauvoir," 103.

71. Sartre, "Prière."

72. Sartre, *Sursis*, 53. "Le docteur Schmidt comprit que le temps des interprètes était revenu.

Hitler s'approcha de la table et la belle voix grave monta dans l'air pur; au cinquième étage de l'hôtel Massilia, une femme qui prenait le frais à son balcon l'entendit, elle dit : 'Gomez ! viens écouter le nègre, c'est charmant!' . . . Et le nègre chantait . . . Chamberlain disait : 'En ce qui concerne le premier paragraphe, je présenterai deux objections' et le nègre chantait: *Bei mir, bist du schön*; cela signifie: 'Vous êtes pour moi la plus jolie' (ibid., 794).

73. "Heaving along from side to side, like the sailor on the unsteady deck of a ship, Lincoln Agrippa Daily, familiarly known as Banjo, patrolled the magnificent length of the great backwater of Marseilles, a banjo in his hand" (McKay, *Banjo*, 1).

74. The nonfiction corollary to this scene from *Le sursis* is Sartre's theorization of radio collectivity as a "practico-inert ensemble" in the chapter "Collectives" in *Critique of Dialectical Reason, Vol. 1*.

75. Verma, *Theater*, 68.

76. Ibid., 67.

77. Ibid., 67, 67, 68.

78. Sartre, "Dos Passos," 93; emphasis added.

79. Dos Passos, *Forty-Second Parallel, in U.S.A.*, 298.

80. Ibid.; emphasis added.

81. Father Coughlin, Aimee Semple McPherson, and other religious orators brought religion to radio in fiery, ecstatic and, in Coughlin's case, often hate-filled speeches. My argument is not that religion and radio failed to mix. My claim is rather that radio and other aural technologies also created an alternative model for the transferable popular speech associated with the religious movements of the eighteenth century. Dos Passos confronts Huey Long's radio oratory as the main antagonist of his novel *Number One*.

82. For more on *Amos 'n' Andy* and the politics of race on 1930s radio in the United States, see Hilmes, *Radio Voices*.

83. Denning, *Cultural Front*, 169.

84. Qtd. ibid.

CHAPTER 2

1. Sartre, "A Propos," 24.

2. Verma, *Theater*, 55.

3. Given Welles's ongoing interest in Latin America, beginning with at least his plan to film an adaptation of Joseph Conrad's *Heart of Darkness* in the Panama Canal Zone as early as 1938, it is worth noting that this first starring radio role was for a play that MacLeish claims was inspired by a visit to

Tenochtitlan in 1929, and his thoughts about Cortés's 1521 takeover of that city (see Drabeck, Ellis, and Wilbur, *MacLeish*, 106–7).

4. The definitive critical work about Welles's film is Catherine Benamou's *It's All True: Orson Welles's Pan-American Odyssey*. A number of excellent studies trace the hemispheric cultural politics associated with the Office of Inter-American Affairs (OIAA) (see esp. Fox, *Panamerican*; Sadlier, *Americans All*; and Cohn, *Literary Boom*).

5. Paley's grandson, the son of CBS longtime chief executive William S. Paley, has recently restarted the La Palina brand that Samuel Paley popularized (see "The Original La Palina" on the La Palina website: www.lapalinacigars.com/). For more on the iconography and politics of cigars in Cuba, see Lane, "Smoking."

6. Bergreen, *Look Now*.

7. Callow, *Orson Welles*, 159.

8. Of the many possible criticisms of Welles's optimistic representation of the urban melting pot in the United States, perhaps the most appropriate would be Tony Schwarz's 1955 audio documentary *Nueva York: A Tape Documentary of Puerto Rican New Yorkers* (see Stoever-Ackerman, "Splicing").

9. Writing about the *War of the Worlds* broadcast, Michael Denning comments, "It captured the contradictions of anti-fascist culture: at once trying to create a 'people's culture' with the new means of communication, while fearing that the 'people' constituted by the apparatuses of mass culture were less a participating audience than a dispersed series of panicked individuals" (*Cultural Front*, 384).

10. Welles contributed the voice-over to the 1937 propaganda film *The Spanish Earth*, written by Dos Passos and Ernest Hemingway. Some versions of the film have Welles's voice, while the film that premiered featured Hemingway instead. Welles was also supposed to perform an adaptation of Dos Passos's novel *Number One* (1943) on the April 3, 1945, production of the CBS radio program *This Is My Best*, but it was replaced at the last minute by F. Scott Fitzgerald's "The Diamond as Big as the Ritz."

11. Woloch, *The One*, 19; emphasis in original.

12. Auerbach, *Mimesis*; Spitzer, "Perpectivism"; Lukács, "Balance"; Clark, *Image*.

13. See Radford, *Modern Housing*, 177–98. For more on the racial segregation policies under the Federal Housing Administration (FHA), see Sitkoff, *New Deal*, 50. For more on the Communications Act of 1934, see Craig, "A New Deal for Radio? The Communications Act of 1934," in *Fireside Politics*, 78–91.

14. USHA was again amended by the Housing Act of 1949, and, in 1965, USHA, along with the Public Housing Administration and the House and Home Financing Agency, became part of the United States Department of Housing and Urban Development (HUD).

15. For more on Mumford's place in media history, see Mindell, *Human and Machine*, 1–16, 112, 318–21.

16. Hammond, "Interior," 181; emphasis added.

17. Cather's essay "The Novel Démeublé" was initially circulated in the *New Republic* 30 (April 12, 1922): 5–6. However, it reached readers in book form with Cather's essay collection *Not under Forty*.

18 Jameson, "Synoptic Chandler," 36.

19. Apter, *Feminizing*; Brown, *Material Unconscious* and *Sense of Things*; Freedgood, *Ideas*.

20. If, as Adorno complained, one can always turn off the radio, even muted it carries the resonance of its extraterritorial purpose (Adorno, "Symphony").

21. Freedgood, *Ideas*, 8.

22. Chandler, *Later Novels,* 51.

23. Jameson, "Synoptic Chandler," 40.

24. Ibid.

25. Chandler, *Later Novels,* 80; emphasis added. Chandler's observation seems to anticipate what the new media theorist Howard Rheingold has named "sentient things." Rheingold, writing in 2002, speaks of "technical architecture" (*Smart Mobs*, 86) and quotes the researcher Mark Weiser's notion of "ubiquitous computing," which "is invisible, everywhere computing that does not live on a personal device of any sort but is in the woodwork everywhere" (87). Rheingold goes on to mention software designer Intel's expectation "that in the near future, Intel will include radio transponder circuitry in *every chip* Intel manufactures" (86; emphasis in original), thus securing a future in which "shirt labels gain the power to disclose what airplanes, trucks, and ships carried it, what substances compose it, and the URL of a webcam in the factory where the shirt was manufactured. *Things tell you where they are*" (85; emphasis added). One could argue that Jameson and Bill Brown's notions of living spaces becoming work spaces and Things expressing human relationships have as much to do with the laptop computer, RSS tags, and digital wireless technology as with radio and the other technologies that paved its way.

26. Jameson, "Synoptic Chandler," 43.

27. Chandler, who, in the introduction to his essay "The Simple Art of Murder," singles out Henry James's *The Spoils of Poynton* as a "graceful and elegant evocation" (*Later Novels*, 1019), was a keen reader of James. Thus, we might consider his work with real estate's privilege over the individual as a critical response to the ventriloquism Bill Brown identifies in James's *The American Scene*. Brown borrows the notion of objects longing for rights in his *The Sense of Things* from Miguel Tamen's *Friends of Interpretable Objects*.

28. Perhaps it's not by chance that the apartment's apparent owner, Mrs. Fallbrook, later revealed to be Muriel Chess, responsible for murdering the tenant, lectures Marlowe on the difference between personal and real property: "'Well of course it's much easier for you,' she said. 'About the car, I mean. You can just take it away, if you have to. But taking a house with nice furniture in it isn't so easy. It takes time and money to evict a tenant'" (*Later Novels*, 82). In Chandler's case one might also think of how his literary production functioned in less than aesthetic ways in his own life. As Judith Freeman writes in her book about Chandler and his wife, Cissy, *The Long Embrace*, Chandler learned to write in order to pay the bills, and, it should be noted, to allow him and his wife to move to more than thirty-six different addresses in and around Los Angeles, Santa Barbara, and San Diego. In other words, Chandler's writing directly served his own survey of Southern Californian real estate. One could be forgiven, if, like Birdie Keppel in *The Lady in the Lake*, one were

to also ask of Chandler, "Are you in the real estate business, Mr. Marlowe?" (*Later Novels*, 51). For more on "portable property" in particular, see Plotz.

29. For more on the gothic, realism, and real estate in nineteenth-century American fiction, see Michaels, "Romance." For a trenchant critique of the housing crisis in the 2000s and its connection to realism and the gothic, see McClanahan, "Dead Pledges."

30. Brown, *Sense of Things*, 184.

31. Chandler, *Early Novels*, 767.

32. Ibid., 768.

33. Ibid., 770. Despite the novel's keen attention to the sound of space, Marlowe has a tin ear when it comes to racial difference. Not only can he not hear the voices at Florian's as anything but a "chant," his later encounters with an Asian woman and an American Indian man result in caricatured dialect.

34. Ibid., 784; emphasis added.

35. Ibid., 847.

36. Ibid., 849.

37. Ibid., 922.

38. Ibid.

39. Ibid.; emphasis added.

40. Ibid., 834.

41. Ibid., 976.

42. Ibid., 865.

43. Ibid., 976.

44. Ibid., 947.

45. While hunting for clues about Malloy's whereabouts in chapter 35, Marlowe looks out at the gambling boat, the *Montecito*, where he will eventually find a man, Brunette, who will lead him to Malloy. As he looks at the ocean, he observes "two stumpy masts just high enough for a radio antenna. There was light on the *Montecito* also and music floated across the wet dark sea" (*Early Novels*, 947). As Lawrence Soley discusses in his analysis of "free radio": "The term 'pirate radio' is often used to describe unlicensed radio stations because the first unlicensed stations that broadcast popular music to the United States and Europe were located aboard ships in international waters. The earliest unlicensed, offshore station appears to have been RKXR, which was based on the S.S. *City of Panama*, a floating gambling casino anchored off the California coast in 1933" (*Free Radio*, 53).

46. Chandler's letters, essays, and novels emphasize his sophisticated understanding of tone. In his essay "Notes (Very Brief, Please) on English and American Style," he writes: "The tone quality of English speech is usually overlooked. This tone quality is infinitely variable and contributes infinite meaning. The American voice is flat, toneless, and tiresome. The English tone quality makes a thinner vocabulary and a more formalized use of language capable of infinite meanings. *Its tones are of course read into written speech by association.* This makes good English a class language, and that is its fatal defect" (*Later Novels*, 1014–15; emphasis added). Marlowe, the character meant to combine Arthurian chivalry with cowboy curtness, makes this observation, "It just goes to show that we talk different languages to different

people" (ibid., 153). Chandler's dialogue-driven style and ear for tone no doubt facilitated the adaptation of almost all of his novels to the screen and radio, revealing yet another dimension of his reciprocal stylistic relationship with radio and voice-over.

47. Chandler, *Early Novels*, 976.

48. For instance, we encounter this reversible relationship in *The Big Sleep* (1939) when Geiger's house serves as his workplace and Harry Jones's office becomes his resting place. Jameson argues that this exchangeability of "dwelling" and "office" derives from the novel's "radio aesthetic." Of course, Chandler is not alone in associating the radio set as instituting the oscillation between public and private life. For a cinematic example contemporaneous with Chandler's novel, see the opening scenes of Jean Renoir's *La règle du jeu* (1939), in which the aviator's public revelation of his private feelings on the radio is received on the radio sets in the private boudoirs of his possible mistress and her husband. The effect, according to Christopher Faulkner's commentary in the Criterion Collection version of the film, is that Renoir does not so much just invert the public and private, and thus "place differing sets of values in opposition to one another, as ask us to sustain the possibility of their interchangeability" (*La règle du jeu*).

49. For more on overhearing and the novel, see Gaylin, *Eavesdropping*.

50. Chandler, *Early Novels*, 976–77.

51. Ibid., 983.

52. Ibid.

53. Vale, *Puritans*, 17.

54. Ibid.

55. Ibid., 16.

56. McCullers, *Heart*, 37–38.

57. Cather, *Not under Forty*, 6.

58. Although *Dick Tracy* began as a comic strip in 1931, the radio version had a popular run from 1934 to 1948, including a stint in primetime from 1938 to 1939, the years just preceding the publication of *The Heart Is a Lonely Hunter* (1940) (see http://otrcat.com/dicktracy-p-1175.html, and http://en.wikipedia.org/wiki/Dick_Tracy#Radio).

59. For an interpretation of this scene more concerned with Mick's potential exclusion from a hierarchy of male power, see Spivak, "Three."

60. Friedrich Kittler comments, "In 1900 . . . 'word art' became synonymous with literature" (*Discourse*, 185). Without merely mapping Kittler's "discourse network" onto McCullers's scene, we should note that Mick, in a turn representative of New Deal fiction, converts the avant-garde impulse of Dadaist collage into the demotic list of radio names. One might begin to think of a "popular vanguard" aesthetic learned from the simple montage of turning the radio dial to different types of programs on one machine.

61. Brown, *Sense*, 141.

62. McCullers, *Heart*, 37.

63. Ibid., 102, 52.

64. Ibid., 163.

65. Loviglio introduces his theory via a reading of Saul Bellow's memoir *Add It Up*, where the latter recalls walking down Michigan Avenue in Chicago

as Roosevelt's voice came out of car windows and drifted into the street. Unlike the national community Bellow's anecdote is meant to imply, Mick's listening is an intimate yet anonymous shared listening that occurs through and across borders without the desire to assimilate to any national narrative.

66. McCullers, *Heart*, 163.

67. Ibid., 52–53.

68. Ibid., 210.

69. Ibid., 353.

70. See Sterne, *Audible Past*, 23–52, 83; and Mills, "Deaf Jam." Singer reacts against the deaf pedagogy practiced by Bell and that sought to make deaf people speakers: "At the school he was thought very intelligent. He learned the lessons before the rest of the pupils. But he could never become used to speaking with his lips. It was not natural to him, and his tongue felt like a whale in his mouth. From the blank expression on people's faces to whom he talked in this way he felt that there was something disgusting in his speech. It was painful for him to try to talk with his mouth, but his hands were always ready to shape the words he wished to say. When he was twenty-two he had come south to this town from Chicago and he met [the deaf] Antonapoulos immediately. Since that time he had never spoken with his mouth again, because with his friend there was no need for this" (McCullers, *Heart*, 11).

71. McCullers, *Heart*, 94.

72 Wright, "*The Heart*," 195.

73. In this they are not entirely wrong: "At first he had not understood the four people at all. They talked and they talked—and as the mouths went on they talked more and more. He became so used to their lips that he understood each word they said. And then after a while he knew what each one of them would say before he began, because the meaning was always the same" (McCullers, *Heart*, 206). As Singer's room becomes similar to a psychoanalyst's office, it not only recalls the oscillation between dwelling and office Jameson associated with the 1930s "radio aesthetic" but also helps tell psychoanalysis's own media history. For, while the telephone was developed through work with deaf subjects like Singer, Freud himself modeled the psychoanalyst's method on the telephone (see Kittler, *Discourse* 283–84; and Picker 100–107).

74. McCullers, *Heart*, 200, 223, 232.

75. Ibid., 209.

76. Ibid., 210, 211, 212.

CHAPTER 3

1. The definitive text about 7 Middagh Street, which was also known as "February House" because so many of its initial inhabitants were born in that month, is Sherill Tippins's *February House*.

2. See Kiuchi and Hakutani, *Wright*, 137.

3. Wright, "Inner Landscape."

4. Muldoon, *Meeting*, lines 7–21.

5. For the historical super's complaints, see Kiuchi and Hakutani, *Wright*, 137.

6. Ulin, "Astonishing," 209.

7. Ulin eloquently asserts that "for Wright issues of housing were always issues of democracy" ("Talking," 164).

8. Wright, *Native*, 499.

9. Although the guidebook's collective sense of authorship means that no single author's byline is attached to each entry, multiple sources attest to Wright's final authorship of the "Negroes" section in the *Panorama*, and he is listed as one of twenty-four writers to contribute to the nine-hundred-page *New York City Guide*. J. J. Butts notes, "An archived production chart notes that this tour [the Harlem River Houses] was assigned to Richard Wright, but essays in the guidebooks were published anonymously and they were often heavily rewritten by both local and federal editors" ("Projects," 131n1; see also Kiuchi and Hakutani, *Wright*, 58, 70, 71).

10. Butts has done the most of any critic to document the interrelationship between the Federal Writers' Project and the literary production by writers affiliated with the guides (see Butts, "New World," 649–50).

11. In addition to his work for the FWP and the *Daily Worker*, and his well-known books, like *12 Million Black Voices*, Wright also contributed to smaller projects. His pamphlet essay "The Negro and *PARKWAY* Community House" (April 1941) demonstrates his keen attention to and care for the lived experience of housing for African Americans in Chicago, struggling to find their way in the northern industrial cities in the wake of the Great Migration. A letter from Ralph Ellison to Wright on May 11, 1940, also underscores how Wright's fiction entered the housing debates of the day: "*Look* magazine this week is carrying photos of housing in Harlem along with a snapshot of you and references to N.S. And Horace R. Clayton has a pamphlet on Negro housing in Chicago which is highlighted by quotations from the book. I'll try to send you a clipping from the Defender which tells of a man named D. Ellison being saved from an eviction when the judge was reminded of N.S" (Ralph Ellison to Richard Wright, May 11, 1940, Box 97, Folder 1314, Richard Wright Papers, Beinecke Rare Book and Manuscript Library, Yale University).

12. "The FWP shaped the work of African American authors in several ways: first, it reinforced the importance of spatial narration as a means of imagining community; second, it focused on the slum dynamics of Harlem, providing a directed reading of the environment and a call for redress; and third, it reconceptualized the relationship between folk culture and modern urban communities, a shift that was central to ideas about the intersection of African American and American culture, thus authorizing Harlem's claim to social justice by advancing an inclusive notion of national community" (Butts, "New World," 650).

13. Federal Writers' Project, *New York City Guide*, 258. Butts documents how both Wright's *12 Million Black Voices* (1941) and Roi Ottley's *New World A-Coming* examine Harlem as both the capital of black culture in the United States and also an unexceptional model of the "urban slum" present throughout the United States during the New Deal years (Butts, "New World").

14. Federal Writers' Project, *New York City Guide*, 392.

15. Although the Harlem River Houses were championed by New Deal advocates as a great advance in housing, historians of public housing have

detailed the restrictions placed on the complex's early inhabitants, the higher income and detailed background check required to secure an apartment there, and the ways in which other major public housing projects were quickly repurposed in the 1940s for wartime families whose incomes were often three to four times higher than the neediest residents, for whom the apartment houses were ostensibly designed. As early as the summer of 1937 Wright wrote an article about spending cuts slowing admission to housing and tightening the already absurdly competitive list of applicants. He noted: "There is at present a waiting list of 11,534 applicants for the 574 apartments, indicating that this project does not even begin to fill the urgent need of Harlem housing . . . Though the Harlem River Houses represent a good step forward in some measures it brings into relief the obstacles which prevent workers generally from raising their standard of living: Low wages, job insecurity, and the fact that for every decent apartment available there are thousands of workers who need them but who can't pay the price" (Wright, "WPA Slashes"). From Butts's perspective, Wright turned against the "futurological discourse" of the New Deal guides in his later writing to insist on the failed promise of public housing (see Butts, "New World"; Goetz, *Ruins*; Vale, *Purging*; and Wright, "WPA Slashes").

16. Federal Writers' Project, *New York City Guide*, 394.

17. "Catherine Bauer, the leading spokesperson of the [modern housing] movement, placed enormous faith in a new design style to change or influence personal values" (Simon, "Housing," 687).

18. Walker, "Contested Rhetorics," 292. In the section on the Harlem River Houses, the Federal Writers' Project *New York City Guide* underscores how a building's success depends on its social life: "Transcending the physical aspects of the development are the social, and one item in the first year's record spoke eloquently: not a single case of delinquency or crime or social disorder was reported for Harlem River Houses. Apartments and courts were maintained with scrupulous care by young and old. A compact, progressive community had emerged, and its very success made the plight of the less fortunate residents of Harlem seem by contrast more bitter than ever" (394).

19. Drawing on Charles Abrams's 1947 ACLU- and NAACP-published pamphlet *Race Bias in Housing*, Ulin comments: "This text outlined the intersection between housing discrimination and the democratic project. As Charles Abrams argues, 'It was in housing that segregation received its greatest impetus and momentum. Once rooted there the segregation pattern spread unarrested' (Abrams 7). Abrams sets the inequalities of the housing system against the broad democratic claims of the United States. 'Present policy,' he states, 'will either lastingly fix the segregated pattern, with all its hazard for the future, or recreate, at long last, an environment worthy of a democratic nation. In that sense federal housing policy involves more than housing; it has become a proving ground on which the practical validity of a great ideal can be demonstrated' (9). Recalling Richard Wright's discourse of the danger posed by the neighbor, Abrams points out that 'the Negro servant once afforded a convenience to those he served by living in a backstreet nearby. *But welcome as a servant he became intolerable as a neighbor*' (Abrams 10, emphasis

mine). Since a servant does not exist as an owner, s/he is not a full citizen, and is therefore insignificant. Ownership in America has always signified power, with homeownership in particular providing not only a possession of property but a sense of economic security, familial stability, and a basis for long-term investment" (Ulin, "Astonishing," 198).

20. Wright, "Blueprint," 106. Wright recognized the necessity of making writing equivalent in value to other public professions and of retaining what he called "the autonomy of the craft," and "the sensory vehicle of imaginative writing": "Writing has its professional autonomy; it should complement other professions, but it should not supplant them or be swamped by them" (ibid., 105). When Abrams met with Robert Moses in 1952, "Abrams had been engaged in a quixotic attempt to hamper Moses's and others' brandishing of the word *slum*. Abrams understood the rhetorical power of *slum*—he himself tried to legally define it early in his career (A. S. Anderson, *Housing*, 56). Furthermore, "[Abrams] lamented urban living conditions for minorities and the poor, but as their advocate, he realized that glowing institutional terms like *slum clearance* and *urban renewal* would not likely benefit them . . . The *slum* label ignited fervent action to check the declining condition of housing in the early twentieth century, priming Moses's rise to power through an active riddance of designated slums" (Walker, "Contested Rhetorics," 291). For more on the politics of New Deal housing debates, and the difference between progressives, who primarily supported "slum clearance," and the modern housing advocates, like Bauer, who sought to institute Bauhaus-style governmental housing for a broad swath of the population, see Goetz, *Ruins*, esp. chap. 1. For more on the New Deal debates about slum clearance, see Vale, *Purging*. For more on the debates between Abrams and Moses, see A. S. Anderson, *Housing*; Lavine, "Slum Clearance"; and Walker, "Contested Rhetorics."

21. In a remark that has proved influential to later theorists of the phonograph, African American subjects, and ethnography, Thomas Alva Edison described the phonograph's mission in 1878 as a project to capture "fugitive sound waves" ("The Phonograph and Its Future").

22. William Beverley argues that "the slave narrative also functions as the crucial precursor to modern lam stories, prefiguring the tactics of flight, their strategies of surveillance, and to some degree their narrative structure. Furthermore, the light these two forms shed on each other illuminates the implications of criminality within the racial coding of American mobility" (*On the Lam*, xvi).

23. Wright, *Lawd!*, 216.

24. Ibid., 218.

25. Ibid., 219.

26. Ibid.

27. Ibid.

28. Ibid., 4.

29. Ibid., 5; emphasis in original.

30. Ibid., 105.

31. Ibid., 200.

32. Ibid., 106.

33. Ibid., 128.

34. A number of psychoanalytic theories find the ideal form of acousmatic authority in the Jewish *shofar*: "The law itself, in its pure form, before commanding anything specific, is epitomized by the voice, the voice that commands total compliance, although it is senseless in itself" (Dolar, *A Voice*, 53).

35. Wright, *Lawd!*, 158.

36. Ibid.

37. Willihnganz, "Voice," 131, 130–31.

38. Ibid., 126. The exclusion or alienation from the national network is recognized by one of Jake's friends in the post office. In a conversation that anticipates Studs Terkels's *Working*, the friend remarks, "I wouldn't get so damn tired if I knowed where some of this mail was going" (150). For more on the post office in connection to histories and theories of media, see Henkin, *Postal Age*.

39. Willihnganz, "Voice," 134.

40. Willihnganz further connects such exclusion to the sonic communications network when he observes Jake "hugging tightly at a steel telephone pole, opposite his home" (*Lawd!*, 216) as an emblem for Jake's lack of access to the broadcast voice and its (white) power: "Jake stands outside communication in two ways here. He embraces part of a communication network but is not a participant. And he looks on a space that he has, partly out of the experience of disembodied voices, transformed into dumb violence" ("Voice," 139).

41. Ulin's careful archival research and deft readings of Wright's unpublished "Black Hope," and his posthumously published radio play *Man of All Work* argue for a critical reassessment of Wright's representation of gender. Moreover, her summary of "Black Hope," which concludes with an African American domestic servant inheriting a fortune from her employer, who leaves her mansion to the Domestic Workers Union, reveals Wright finally providing the utopic image of interracial solidarity in housing (see Ulin, "Astonishing").

42. Wright, *Lawd!*, 216.

43. Michel Fabre observes that Wright "read a few Faulkner novels, including *Absalom! Absalom!* and *Sanctuary*, before 1940" (*Books*, 50).

44. Faulkner, *Absalom*, 296.

45. Richard Godden writes, "'Bon,' not as a person (fictional or real) but as an associative path through a collection of words, leading back to Sutpen's act of naming, that is, to the owner's translation of a nominal son into real property (Sutpen believes, quite literally, that he has paid in acres and slaves for Eulalia's labor and 'goods')" ("Unreadable Revolutions," 272).

46. Willihnganz, "Voice," 125.

47. "Wright said that Waldo Frank's writing was important to him because of its lyricism and social analysis and the portrayal of the tragedy of racial hatred in *Holiday, City Block, Dawn in Russia*. He saw in the work a desire similar to his own for writing a communal life" (Webb, *Wright*, 403n8).

48. Zadie Smith, for one, would disagree. See her essay "*Their Eyes Were Watching God*: What Does Soulful Mean?"; and Wright, "Between."

49. Hurston, "Saturday Review," 32; emphasis added.

50. Farrell, "*Partisan Review*." William J. Maxwell, drawing on the work of Carla Cappetti and Werner Sollors, has identified this split as emerging

from the sociology department at the University of Chicago (where Farrell was an undergraduate) and the anthropology of Columbia University's Franz Boas (with whom Hurston studied) (see Maxwell, *New Negro*, 159–60; see also Cappetti, *Writing Chicago*, 188; and Sollors, "Modernization").

51. Again, for more on this disabling binary, see Maxwell, *New Negro*, 159–60. Stein's own work with listening and writing has been detailed in Bennett.

52. North, *Dialect*, 9.

53. Eugene Miller, "Wright and Stein," 111.

54. Qtd. in Eugene Miller, *Voice*, 67.

55. North, *Dialect*, 76.

56. North draws here on Eugene Miller's aforementioned work, which has done the most to archive Wright's statements of Stein's influence not only in his posthumously published autobiographic sequel to *Black Boy*, *American Hunger*, but also in his articles "Gertrude Stein's Story Is Drenched in Hitler's Horrors" published in *P.M.* magazine on March 11, 1945, "Why I Chose 'Melanctha' by Gertrude Stein" in *I Wish I'd Written That*, edited by Whit Burnett in New York by McGraw Hill in 1946, and the unpublished paper "Memories of My Grandmother." Gavin Jones has criticized North's work for taking African American speech as a monolithic whole. *Strange Talk: The Politics of Dialect Literature in Gilded Age America*, Jones's study of dialect and vernacular writing in the late nineteenth-century writing of the United States, emphasizes the multiplicity of African American dialects represented in fiction, as well as the class distinctions between different forms of vernacular speech.

57. See Best, *Fugitive's Properties*; Biers, "Syncope"; Gitelman, *Scripts*; Goble, *Beautiful*; Wagner, *Disturbing*; and Weheliye, *Phonographies* and "Feenin." For more on the phonograph and the history of acoustic technology more generally, see Sterne, *Audible Past*. Published on the back of the first edition of *Three Lives* was Georgiana Goddard King's review, which read, "The patient iteration, the odd style, with all its stops and starts, *like a stubborn phonograph*, are a part of the incantation" (qtd. in Charters, "Introduction," xviii). While most critics emphasize Stein's method as developing out of her association with Picasso's cubism and the two artists' shared passion for Cézanne, Ann Charters observes that "in one of her later essays [Stein] said the key to understanding her books was realizing that she wrote 'by ear' rather than 'by eye'" (ibid., xvi). Eugene Miller also points to the fact that in her essay "Portraits and Repetition" Stein connected her use of repetition to "oral tradition: 'When I first really realized the inevitable repetition in human expression that was not repetition but insistence . . . was when . . . I . . . lived . . . principally with a whole group of very lively little aunts who had to know anything . . . [A]s there were ten or eleven of them they did have to say and hear said whatever was said and any one not hearing what it was they said had to come in to hear what had been said . . . I began consciously listening . . . You listen as you know'" (*Voice*, 69). Friedrich Kittler also notes the kinship between Stein's prose, the phonograph, and her research on "Normal Motor Automatism," which she published in 1896 and 1898 while a student of William James at Harvard (*Discourse*, 225–29). Reading through a different oral

technology, Sarah Wilson notes that in Stein's writing "the radio broadcast conveys a sense of an immediate and concentrated present; it begins again and again, as Stein's characteristically insistent phrasings indicate to us" ("Stein," 263). My comments here hope to answer what has proved to be a contentious and dissonant debate about Stein's influence on Wright. As Eugene Miller observes, the available scholarship "encourages among students of Wright two views: that a literary snobbishness, however defensively and unconsciously developed during his formative years, led him to claim connection with Stein; or the seemingly kinder attitude that he did not know what he was talking about. For example, how could he condemn Zora Neale Hurston's *Their Eyes Were Watching God* because it presented 'quaint' views of Afro-Americans, the way white Americans liked to see them, and yet call 'Melanctha,' with its stereotypical references to the 'simple, promiscuous unmorality of the black people' and 'the wide abandoned laughter that makes the warm broad glow of negro sunshine,' 'the first realistic treatment of Negro life I'd seen when I was trying to learn to write'?" ("Wright and Stein," 108).

58. It is worth remembering that Zora Neale Hurston was a student of Franz Boas at Columbia and that she tuned her writing ear producing ethnographic phonograph recordings. Wright's turn away from the ethnographic model through the phonograph presents us with two very different notions of the "black phonographic voice." For more on Hurston and Boas, see Butts, "New World" 652; and Luis-Brown, "Rising Tides of Color': Ethnography and Theories of Race and Migration in Boas, Park, Gamio, and Hurston," *Waves of Decolonization.*

59. Best, *Fugitive's Properties*, 19.

60. Ibid., 52.

61. Ibid., 19.

62. In a summary of his sophisticated connection between the phonograph's effect on copyright laws and the technology's legal and historical implications for rethinking slavery in the United States, Best writes:

> Alongside any analysis of the formal legal protections of voice's mechanical reproduction we must assess analogous visions of personhood's theft—specters of theft that must serve a heuristic function, as analogue and precursor to the pilfering of another's persona through the imitation of his or her voice.
>
> We move, then, to a single example of the law's administration of specious exchange, a case in which the violation of an already suspect property right is reimagined as a species of credit fraud—a transformation of prediction into prodigality and futures into fugitiveness, which reconceives threats to the security of property as the evasions of a guilty, responsible, and obligated person . . . I refer, of course, to Dred Scott's challenge to John F.A. Sanford's reasonable expectation of stable market value. (*Fugitive's Properties*, 64–65).

63. Wagner, *Disturbing*, 194. Elsewhere, Wagner is more emphatic about this process as particularly embodying: "Rather than alienating the voice from the body, the phonograph appears to have grounded the black voice in

the body" (290n9). In contrast to Wagner, Weheliye writes, "Paradoxically, black voices are materially disembodied by the phonograph and other sound technologies, while black subjects are inscribed as the epitome of embodiment through a multitude of U.S. cultural discourses" ("Feenin," 26). Wagner also implicitly connects what Biers wants to imagine as the liberating force of the black voice in James Weldon Johnson to the frustrations of nineteenth-century music collectors "preoccupied with transcription's impossibility" (Wagner, *Disturbing*, 211).

64. "Collectors had long complained that the trajectory of black speech could not be stabilized to allow for transcription in the time when they were face-to-face with an informant. In an ethnographic exchange, black voices did not sound like they could be indexed to their speakers. This was a voice whose 'odd turns' broke the diatonic scale and whose speech had to be transcribed in tortured syntax and effusive misspelling if its timbre were even to be approximated" (Wagner, *Disturbing*, 210).

65. Ibid., 219.

66. Ibid., 194. For another version of this argument in relation to the Lomaxes' recordings in particular, see ibid., 219–20. Goble also understands race as an effect of technology. "Race, as I explore its fascination here," he writes, "emerges less as the sign of what remains natural and authentic within the circuitous mechanics of recorded sound, and more as the material result of modernism's own commitment to the pleasures of both technology and technique" (*Beautiful*, 158).

67. Wright, "Ledbetter." For more on the relationship between Wright and Ledbetter, see Stoever, *The Sonic Color Line*.

68. Qtd. in Eugene Miller, *Voice*, 67.

69. Qtd. in Eugene Miller, "Wright and Stein," 111.

70. In comparing Wright's critic Eugene Miller's reading of Stein with Stein's early reviewer we can hear the same exchange of folk for phonograph. Attempting to reconcile Wright's connection between his grandmother's speech and Stein's writing, Miller argues that "Stein's prose is certainly folklike in its most famous characteristic, its *repetition*" (*Voice*, 69), thus substituting the "folk" for what Stein's early reviewer identified as "the patient iteration, the odd style, with all its stops and starts, *like a stubborn phonograph* . . . a part of the incantation" (qtd. in Charters, "Introduction," xviii; emphasis added). Differing from Miller's criticism of Stein, Michael Denning claims *Three Lives* represents "a vernacular without the condescension of dialect, a patient detailing of ordinary acts with studied objectivity" (*Cultural Front*, 242).

71. Wright, "Long Black Song," 103.

72. Ibid., 104. Psychoanalytic critics from Klein, Winnicot, and Lacan onward have labored the etymological implications of the "infant," the *infans*, or the one who can't, or can't yet, speak (see Dolar, *Voice*; Kelly, *Post-Partum*; and Silverman, *Acoustic Mirror*). To borrow from Dolar's gloss on Lacan's *Seminar XI*, infants "do not address a definite interlocutor at hand, but their solipsism is nevertheless caught into the structure of address" (*Voice*, 27). In addition to the poststructuralist and psychoanalytic approach favored by these critics, my own reading of this particular infant voice in Wright's story

acknowledges the overdetermined infant cry as participating in a structure of address and, moreover, signaling in each of its iterations, unique historical factors that pertain as much to a history of technological mediation, copyright law, and slavery as they do to any ontological status of the "*cri pur*" or "*cri pour.*" Indeed, the story's play with signification and "waiting" expresses a historical logic of deferral that might supply a critique of indeterminacy as participating in rather than resisting the means to keep dreams deferred.

73. Wright, "Long Black Song," 107.

74. Ibid., 108.

75. Ibid.

76. Ibid.

77. Wright, *Lawd!*, 4. Goble reads the scene similarly, focusing on how "Wright outlines the allegory of this encounter between white (technology) and black (nature)" (*Beautiful*, 220).

78. Sterne, *Audible Past*, 267.

79. Goble, *Beautiful*, 223. For more on the theoretical implications of the famous RCA Victor advertisement, "His Master's Voice," see Dolar, *Voice*, 74–78.

80. Recalling Wright's extensive narrative and professional work with radio, the traveling graphophone salesman, who travels the region bringing music to people's doors, seems like a physical instantiation of a radio network. For more on the history of the graphophone, see Sterne, *Audible Past*, 179–80, 256–60.

81. Wright, "Long Black Song," 109. As Weheliye observes, the spiritual was historically appropriated by white subjects: "As both Ronald Radano and Jon Cruz have shown, spirituals, once transcribed and compiled, served both white and black abolitionist purposes as embodiments of black humanity. Black sacred and later secular music took on two simultaneous functions: proving black peoples' soul and standing in for the soul of all U.S. culture, keeping the racially particular and national universal in constant tension. Thus spirituals ushered in a long history of white appropriations of black music" ("Feenin," 28).

82. Wright, "Long Black Song," 110.

83. Ibid., 113.

84. Ibid., 103.

85. Benjamin, "Theses on the Philosophy of History," in *Illuminations*. Werner Sollors links Wright with Hurston around the question of the clock and graphophone as emblems of modernity and "the intrusion of the capitalist ethos as a sexual seduction" (qtd. in Maxwell, *New Negro*, 168) in "Long Black Song" and Hurston's "The Gilded Six-Bits." Advancing Sollors's observation, Maxwell adds that "the seducers from the North" should be analogized "with the seductions of the Great Migration" (ibid.). Sollors and Maxwell fail to address how this "seduction," which seems more like rape, has led many feminist critics to attack Wright's misogyny. Barbara Johnson, in an essay that primarily focuses on *Native Son*, and in a critique she connects to Wright's inability to appreciate Hurston's work with the black vernacular, criticizes the story's misogyny at the level of plot: "When the black woman

[Sarah] does attempt to take control of her own plot in Wright's short story, 'Long Black Song,' the black man dies in an apocalyptic fire. The unavailability of new plots is deadly" ("The Re(a)d," 154). While considering how Wright's paternalistic narrative partially circumscribes Sarah in the figure of the "simple peasant woman," Cheryl Higashida supplies a Marxist reading of the story that includes the historical specificity of the Great Migration, and finds in Sarah a more sophisticated rendering of a "folk" perspective than previous critics have attributed to Wright. Acknowledging Sarah's isolation derives from her husband's concern with "accumulating land and commodities [rather] than with human relations" ("Aunt Sue's," 404), and her "communal-familial ethos" (408) hints at "the possibility of interracial cooperation" (406). Yet, her alienation from her husband, and her longing for her ex-lover Tom's return from the war resists a simple nostalgia for a folk way of life, involving her and "the rural South [as] inextricably part of the world system. International warfare and the economic structures of Southern agriculture shape Sarah's social consciousness, as African American men are drafted to fight in Europe, raise crops for exchange in the market, and leave their homes in search of better pay in industry" (407).

Brent Hayes Edwards, writing about Duke Ellington's use of sound to resignify literary texts by composing a "musical 'parallel,'" describes something similar to the way Wright works with sound here: "This operation privileges the sound of words over the particular ways they are written on the page. Again, it underlines the specific parameters of a musical 'parallel,' an interpretive mode that reads by 'hearing' phonemically at a certain distance from the literary source text . . . It brings sound to the fore, as it were, places sound before sense, in a spirit of semantic disturbance or 'fugitivity' that Nathaniel Mackey, among others, has argued is endemic to black traditions of literate and musical expressions alike" ("Ellington," 11).

86. To argue that this scene somehow represents the "authentic" black voice of Silas's dialect fighting against the "inauthentic" recording would misread the framework of my analysis and its links to Wagner's work. Furthermore, as Wright's debate with Hurston makes clear, the story's acoustics resist precisely the desire to imagine an "authentic black voice" derived from an originary folk culture.

87. Wright, "Long Black Song," 117–18; emphasis added. Looking back to the ethnography of the time, Silas's status as a property holder also seems to exclude him from the type of life and song that would make him ethnographically identifiable as "authentic" in his blackness. As Wagner observes: "In 'Some Current Folk-Songs of the Negro' (1912), Will Thomas agrees that the ideal informant is someone who tries to 'live in this world without working.' 'So far as my observation goes,' Thomas writes, 'the property-holding negro never sings.' Because 'property lends a respectability' and 'respectability is too great a burden' for any tradition to bear, the only 'negro' who 'sings' is the one who is losing, or has never found, his 'economic foothold.' A good worker makes a bad informant because he is too respectable to waste time singing, and a bad worker makes for the best kind of informant because the dearth in industry and respectability is richly compensated by musical genius.

Collecting authentic black expression became a matter of knowing where to look" (*Disturbing*, 30).

88. Wright, "Long Black Song," 125.

89. Goble, *Beautiful*, 224.

90. Wright, "Long Black Song," 125.

91. Goble, *Beautiful*, 224.

92. Wright, *Lawd!*, 218.

93. Wright's careful work with time in his prose might be thought alongside recent attempts to grant attention to narrative time that does not work at the solely macro levels of history, or the patterned metrical rhythms of poetry (see Yahav, "Sonorous").

94. Wright, "Long Black Song," 128.

91. Ibid.

96. Wright, who had read Freud before the publication of *Uncle Tom's Children*, might have had the "fort/da" game, the repetition compulsion, and the "death drive" in mind when he linked the baby's "bang bang bang" to Sarah's erotic fantasies about Tom, the salesman's aggression, and Silas's death at the story's end (see Freud, *Beyond*).

97. My reading departs from Allen Weiss's readings of Poe and Mallarmé. Likewise, Wright's work with acoustics and housing offers a historically contingent pessimism that counters the utopian desires of Gaston Bachelard, ("Radio et Rêverie").

98. Wright, "Long Black Song," 110.

99. Butts, "Projects," 123.

100. Wagner, *Disturbing*, 237.

101. Goble, *Beautiful*, 172.

102. Wright, *Native*, 277–78.

103. Ibid., 466. Max's words also carry the weight of other, earlier thoughts of Bigger: "He committed rape every time he looked into a white face. He was a long, taut piece of rubber which a thousand white hands had stretched to the snapping point, and when he snapped it was rape. But it was rape when he cried out in hate deep in his heart as he felt the strain of living day by day. That, too, was rape" (263).

104. Tanner, "Uncovering," 143.

105. Ibid., 142.

106. In "Everybody's Healing Novel," Alessandro Portelli summarizes and then critiques the critical literature that reads this scene as an example of Bigger's linguistic mastery. For an example of Bigger as the hyperarticulate hero, see Reilly. Tanner provides the most sophisticated reading of Max and Bigger's exchange, although she, too, ultimately hears Bigger's speech as proof of his prowess in "the storytelling competition that governs the political world and the novel itself" (145).

107. For a recent attempt to rethink Descartes's connection to an acoustic philosophy, see Erlmann, *Reason*.

108. Wright, *Native*, 270.

109. Ibid., 501.

110. Reilly, "Giving" 60. "Marking Bigger's freedom by the power of self-expression, Wright makes Bigger's voice the emblem of his novel signifying

that through the brilliant complex of linguistic acts we know as *Native Son* freedom also comes to black writers" (ibid.).

111. Biers, "Syncope," 100.

112. Wright, *Native*, 500.

113. Ibid., 500–501.

114. Although Bigger also stutters in this closing encounter, it is Max's stutter that breaks the generic expectations established through American literary form, and it is for that reason that I focus on his words here. Nonetheless, we might hear Max's stutter as a "contagious" influence in which the dividing lines between lawyer and client begin to break down.

115. Deleuze, "Stuttered," 25.

116. Ibid., 24, 24; emphasis added.

117. Barthes, *Rustle*, 76.

118. Ruttenburg, *Democratic*, 345–46.

119. Mackey, *Discrepant*, 252.

120. Ibid., 249.

121. Ibid., 253.

122. See Deleuze and Guattari, *Kafka*.

123. Melville, *Billy Budd*, 357.

124. Ruttenburg, *Democratic*, 363.

125. In Melville's book the stutter inspires violence in the face of the law; Billy kills because he can't speak. While Bigger's "impulsion to try to tell was as deep as had been the urge to kill" (*Native*, 358), his legal representative, speaking for him, stutters in the face of the outlaw. In this sense, the narrative displaces the stutter onto Max, which then allows Bigger to finally speak what he has thought throughout the novel. As should be clear from my argument, the dividing lines in Bigger and Max's dialogue elaborate an entire system of differences at play. However, they do not merely repeat the separation between the referential and performative aspects of speech Barbara Johnson elaborates in her reading of *Billy Budd*. Rather, the stutter in *Native Son* organizes a discursive universe that includes Billy Budd's innocence, fugitive slave narratives, Emersonian ideals of poetic inarticulacy, and radiophonic articulation in order to identify each as inappropriate models for imagining democratic speech at the end of the New Deal. For more on the stutter in *Billy Budd*, see Johnson, "Melville's Fist"; and Ruttenburg, *Democratic*, 344–78.

126. In this staging of a problem of address between Max and Bigger we should remember Wright's fundamental disagreement with Hurston: "In the main, her novel is *not addressed to the Negro*, but to a white audience" (Wright, "Between," 23; emphasis added). What takes place here is a property struggle over voice but also an implicit hope in the possibility of sharing voice rather than subsuming it to another or laying claim to one voice as one's sole property. This is not a disembodiment, but an example of the limits of visuality, and the importance of embodied listening at a distance.

127. Webb, *Wright*, 251. Wright recounted the conversation with Stein to his biographer Constance Webb: "I've been working toward an idea—and hope, somehow to reveal it in my next book, taking off, only as far as style, from the middle section, 'Flight,' of *Native Son*; that Negro self-discovery, the facts of

Negro life, constitute a great body of facts of importance about mankind in general. But white Americans probably won't accept it. I think Europeans will accept and understand what I'm saying and then it will filter back to America from Paris and London and Rome. And then the Negro may be better known. Maybe the Negro must present his case to the world, to Europe" (ibid.).

128. Wright, *Native,* 473; emphasis added. The narrative repeats this claim later: "He did want to see Max and talk with him again. He recalled the speech Max had made in court and remembered with gratitude the kind, impassioned tone. But the meaning of the words escaped him" (488–89).

129. Ibid., 498.

130. Ibid., 499.

131. To move from Bigger's existential crisis of consciousness to a Marxist explanation and justification for his crimes through Max's court room speech: Bigger murdered because of his social and economic conditions, because, as Max rhetorically asks Mr. Dalton in court: "Do you think that the terrible conditions under which the Thomas family lived in one of your houses may in some way be related to the death of your daughter?" (ibid., 379).

132. For more on the legal and literary discourse regarding the rights of objects, see Tamen, *Friends*; and Brown, *Sense of Things*,184.

133. Attention to the fugitive voice builds on previous theories of the fugitive in *Native Son.* Earlier critics have understood *Native Son* as a novel whose generic originality resides in the narrative's fugitive border-crossing and the story's manipulation of space. Recalling my comments regarding Wright's Steinian influence and his writing on the Harlem housing projects, Michael Denning joins tone and neighborhood space in plot as the twin problems of the 1930s novels he names "ghetto pastorals": "For the plebeian writers, modernism meant two things: on the one hand, a way to use a vernacular that was not an 'ethnic' dialect, always already a minstrel exercise in misspelling, broken grammar, and comic solecisms; on the other hand, a freedom from plot, a way to avoid the well-crafted intrigues and counterplotting of the novel proper. The most striking aspect of the ghetto pastorals is their lack of unifying narrative, their sketchiness" (*Cultural Front*, 243). Left without a "unifying narrative," the writers Denning describes adopted "the crime stories of the popular thriller," or chose "to lay a melodramatic plot over the sketches of everyday life. But for the most part the writers of the ghetto pastorals were left conjuring fiction out of worlds without narrative" (243). While *Native Son* seems to follow this generic path, it proves exceptional in surpassing one of the generic problems in representing the neighborhood: "Despite the intricacy of the neighborhood's internal boundaries, few of the ghetto pastorals take their characters out of the neighborhood . . . The one novel constructed around the crossing of the border is Richard Wright's *Native Son*" (246). As I will explain, this border crossing, which occurs because of Bigger Thomas's attentive listening, and eventually leads to his own fugitivity and his exploration of fugitive voices, elaborates the acoustic logic of "fugitive sound" already found in "Long Black Song." What might appear as a "world without narrative," emerges instead as a meticulous acoustic plot reckoning with the place of the black fugitive voice within American literature. The narrative's work

with neighborhood boundaries, boundaries constructed in the novel by the real estate mogul Charles Dalton, prepares the novel's explicit interrogation of the limits of the acoustic imagination in the face of real property. Reinforcing Denning's sense of narrative claustrophobia in *Native Son*, Barbara Johnson writes: "What Wright's writing demonstrates again and again is the deadly effect both of overdetermination and of underdetermination in storytelling. It is because the 'rape' plot is so overdetermined that Bigger becomes a murderer. It is because there are so few available models for the plots of Indian maidens that Wright's heroine 'has to die.' And it is because the 'rape' plot about white women or the 'idealization' plot about Indian women are so overdetermined that the plot about black women remains muffled beyond recognition. When the black woman does attempt to take control of her own plot in Wright's short story, 'Long Black Song,' the black man dies in an apocalyptic fire. The unavailability of new plots is deadly" ("The Re(a)d," 154).

134. Butts, "New World," 658.

135. Qtd. in Phu, "Bigger," 46–47.

136. Ibid., 47.

137. For more on the changing landscape of the Argentine film industry during the Good Neighbor Policy, see Falicov, "Rogue."

138. Fabre, *Quest*; Brunette, "Two"; Burks, "Cinematic."

139. Falicov, "Rogue," 248–49.

140. See also Lopéz and Alvarado, *Mediating*. Falicov quotes a State Department memorandum on the film policy in Latin America promoted by the Office of the Coordinator of Inter-American Affairs (OCIAA) from 1942 that states, "It was the greatest outpouring of propagandistic material by a state ever" ("Rogue," 249).

141. Argentina under Juan Perón was not free of racial bias. For more on the representation of black characters in Argentine film, see C. Thompson, "Margins."

142. E. Scott, "Blacker," 95.

143. "On November 4 the first premier of *Native Son* was held on board a Pan American strato clipper between Buenos Aires and Nice" (Webb, *Wright*, 302).

144. In "Blacker Than Noir," Ellen Scott adds that the flashing Sunkist sign used to illuminate Bigger and Bessie in their tenement hideout might be taken as an allusion to Pablo Neruda's poem "La United Fruit Co." from his *Canto General* (1950).

145. "The motivation behind Wright's promotion of Bessie's character to protagonist status may be found in his increasing sense of the complexity and international relevance of jazz" (E. Scott, "Blacker," 99).

146. A review of the uncut film discovered and restored over the last decade concludes: "Its literary pedigree aside, *Native Son* is genuine in its moral ambiguity and existential despair. It's fascinating to see all the genre conventions and stylistic flourishes we associate with noir being brought to bear in a racially charged drama about African-Americans—made contemporaneously with the classic noir era in Hollywood. The recovery of this film gives us an essential and previously missing link in mid-20th-century cinema" (see Krebs, "Noir Thing"; see also E. Scott, "Blacker," 93–94).

147. Richard Wright Papers, Beinecke Rare Book and Manuscript Library, Yale University.

148. Pierre Chenal to Richard Wright, July 31, 1951, Box 96, Folder 1257, Richard Wright Papers, Beinecke Rare Book and Manuscript Library, Yale University.

149. See Phu, "Bigger," 41–45.

150. E. Scott, "Blacker," 101.

151. Ibid., 103. Scott's article provides an exhaustive list of changes demanded by the NYMPD.

152. "I made the mix with the tapes [les bandes] *following* [*conformes*] the Argentine tape. This way, even if they intend to make use of [exploites] the American version, they . . . will not be able to!" (2) (Pierre Chenal to Richard Wright, May 17, 1951, Box 96, Folder 1257, Richard Wright Papers, Beinecke Rare Book and Manuscript Library, Yale University).

153. Webb, *Wright*, 294.

154. E. Scott, "Blacker," 115.

155. Borges, "On Dubbing." See also Yampolsky, "Voice Devoured." Puig, after failing to finish two English-language screenplays at the neorealist Centro Sperimentale di Cinematografia in Rome, found work in London writing subtitles for Spanish-language films. We might think of this early training in writing *on* the screen, rather than *for* the screen, as a type of writing that is bound to the writer and reader's listening.

156. Not all cuts came from the censors. Ellen Scott adds that Gould "exacerbated its censorship, further cutting the trial sequence to align it with an action/suspense pacing consonant with the exploitation aesthetic preferred by the 'buyers' that remained after the film's emasculation. Gould also may have altered the film's music in order to accelerate its pacing, a move that further collaged its already awkward soundscape" ("Blacker," 104). Briefly connecting Bigger's fate to Jake Jackson's exclusion from the nation's communication network, a partially restored version of the film introduces the montage with a quote from the police chief Britten: "Just say the word Mr. Dalton. I have 500 hundred friends who will take that coon out and string him up to the first telephone pole."

157. E. Scott, "Blacker," 104.

158. Copjec, "Phenomenal."

159. Silverman, *Acoustic Mirror*, 57.

160. The film also begins with a voice-over that excerpts Carl Sandburg's "Chicago" over images of the city to contrast the urban center's heroic image with "its secrets which are seldom written or sung": the intense poverty and what the voice calls "a hidden world . . . Chicago's Black Belt."

161. Kozak, "Tesoros."

162. Krebs, "Noir Thing."

CHAPTER 4

1. Qtd. in Luis López, *La radio*, 27. ITT was formed out of the Puerto Rican Telephone Company owned by the U.S.-born brothers Sosthenes and Hernán

Behn. In his *Culture and Customs of Cuba*, William Luis claims that "[Zayas's] speech, in English, was meant to be heard in the United States, since Cuba had fewer than 100 radios at that time" (*Culture,* 68). Oscar Luis López writes that the Cuban president "pronounces his discourse *first in English*, and then in Spanish. This last fact has not been proven" (*La radio*, 34; emphasis in original). I have found no more definitive archival proof that Alfredo de Zayas spoke in either English or Spanish; however, all historians appear to agree that Raúl P. Falcón's introduction to the president's speech was delivered in English and Spanish (ibid., 33). Zayas might have also had in mind the listeners in the room adjacent to his office in the presidential palace, an audience that included the U.S. general consul in Cuba, Frank Steinhart, a member of Theodore Roosevelt's "Rough Riders," and Enoch Crowder, the personal envoy of U.S. president Warren G. Harding.

2. Qtd. in Luis López, *La radio*, 29.

3. Damien Keane, in a book about radio and literature in Ireland, brilliantly expounds on the importance of recognizing radio as a material medium that was both able to cross national borders and impeded by physical and technological limitations.

4. Reports about the broadcast came from listeners as far away as New York, but no other nation in the Americas, aside from Argentina, had access to radio, and there are no records indicating the broadcast reached listeners there. Of course, there were also listeners in Cuba—the unofficial station LRC had already been in operation for three months before the president's address. In an appendix to his history of Cuban radio Luis López lists a number of these amateur operators on the air before the official beginning of broadcast in Cuba. In a sense uniting both points in a chapter on "Radio Talk," Erving Goffman observes: "One of the basic resources of the announcer (perhaps even more than of the ordinary speaker) is that of conveying something that listeners will be privy to but which cannot stand as something they openly have been given access to. *The audience is, as it were, forced into the role of overhearers*, but of messages the announcer is sending only to himself or not to anyone at all" (*Forms*, 303; emphasis added). He goes on to explain that through certain shifts in tone of voice, "the announcer makes his audience privy to his own feelings (not the station's or sponsor's or any generalized 'we'), shifting the audience's status to that of overhearers" (305). Thus, while all radio communication is not, by necessity, *overheard*, the medium does allow the announcer to force his audience into the role of overhearing.

5. Based on the Platt Amendment, in 1902 the U.S. leased Guantánamo Bay and Bahía Honda, the latter of which was released in 1912 with the expansion of the lease on Guantánamo. The United States signed a lease to pay two thousand dollars per year with no stipulated end date (Thomas, *Cuba*, 502). The Platt Amendment's doctrine of intervention was supplemented by President Theodore Roosevelt's 1904 "Corollary" to the Monroe Doctrine, in which "the United States gave itself the right to preemptive military intervention [in Latin American countries]" (Schoultz, *Beneath*, 290; see also Thomas, *Cuba*, 476). As a result of these adjustments to U.S. foreign policy, the United States sent the marines to Cuba to address labor disputes and apparently fraudulent elections in 1906, 1912, 1917, and 1920.

6. For more on this period see Pérez, *Platt.*

7. Schoultz, *Beneath*, 295. Schoultz claims Hoover had already talked about being a "good neighbor" in his 1928 goodwill tour of Latin America (290). FDR's eventual policy borrowed from Hoover's early policy, and the legacy of 1928 Secretary of State Frank Kellogg's "Kellogg-Brian Pact," which opposed a (Teddy) Rooseveltian reading of the Monroe Doctrine, known as the Roosevelt "Corollary." Kellogg wrote an official document to these ends, but Hoover suppressed it in 1929 and 1930 in order to pass the "London Naval Treaty"; he then published it quietly in 1930 as a 236-page government document (ibid., 291–92). As the Coolidge administration's secretary of commerce, Hoover was responsible for authorizing radio licensing and became a powerful advocate of the Radio Corporation of America's (RCA) lobbying interests to monopolize the radio waves (Soley, *Free Radio*, 35).

8. As Lawrence Soley points out: "In 1933, encouraged by President Franklin D. Roosevelt's criticisms of Cuban leader Gerardo Machado, young Cubans organized demonstrations and protests against the Cuban dictator. In the midst of the protests a free radio station appeared, broadcasting anti-Machado commentaries and calling upon the Cuban army to join the rebellion. Although the government located the transmitter and arrested its operators, other freedom stations soon appeared. These stations also called upon the army to rebel, which it did, and Machado was ousted and forced into exile" (*Free Radio*, 8).

Surprisingly, in early 1933 the same Welles had "sent the President-elect three pages of draft language about Latin America for his inaugural address, and suggested a prohibition on 'the dispatch of the armed forces of the United States to any foreign soil whatsoever.' This commitment to military nonintervention became the defining characteristic of FDR's Good Neighbor policy" (Schoultz, *Beneath*, 299). However, later in the same year, after Welles's friend Carlos Manuel de Céspedes, the U.S.-born former Cuban ambassador to the United States, was deposed by a group led by General Fulgencio Batista and replaced with Ramón Grau San Martín, Welles sent repeated requests to Washington "for 'a strictly limited intervention,' which he said would require 'the landing of a considerable force at Habana and lesser forces in certain of the most important ports of the Republic.' Realizing that the literal-minded might interpret this as a proposal for military intervention, Welles argued that U.S. troops would be only 'an armed force lent by the United States as a policing power' and therefore 'would most decidedly be construed as well within the limits of the policy of the "good neighbor"'" (ibid., 300).

9. A decade after FDR refused to send the U.S. military (although he did decline to recognize Grau's government, ringed the island with gunships, and sent Welles back to undermine the Cuban government), "when FDR welcomed a rehabilitated Grau San Martín to Washington, he remarked that 'the President-Elect is largely responsible for the good neighbor policy' because he forced the 1933 decision not to send U.S. troops into Cuba. Because the term 'nonintervention' was defined narrowly to encompass only a military invasion, the outcome in Cuba was considered a major victory for the new administration's nonintervention policy" (Schoulz, *Beneath*, 303). However, in 1933 Grau had felt differently and "responded that America's non-recognition policy was

'a new type of intervention—intervention by inertia'" (Dallek, *Roosevelt*, 64). In Robert Dallek's view, "while the President rejected intervention in Cuba, he believed that pressing domestic need and preservation of the Good Neighbor idea required him to interfere in its affairs . . . [T]hough political interference . . . deprived Cuba of the freedom fully to decide its own political fate, it impressed Roosevelt as the best means to both immediate domestic and long-term foreign goals" (ibid., 65). Dallek also points out that Roosevelt's idea of the Good Neighbor carried through diplomatically. He met with the Argentinian, Brazilian, Chilean, and Mexican envoys to discuss his decision about intervention in Cuba, making him "the first President in the nation's history to discuss current United States policy in the Hemisphere with Latin diplomats . . . [and] he remained convinced that direct military involvement in Cuba would distract from pressing domestic concerns and all but destroy the Good Neighbor idea" (63).

10. Gellman, *Roosevelt*, 113.

11. Schoultz, *Beneath*, 305.

12. Gellman, *Roosevelt,* 115. Gellman makes clear the connections between the abrogation of the Platt Amendment's policy of intervention and Cuba's newly strengthened dependency on the U.S. economy: "This political alignment was closely associated with economic actions. The Eximbank loans, the Jones-Costigan Act, the lowering of the Cuban sugar tariff, and the reciprocal trade agreement have given authors reason to praise Roosevelt for his assistance in alleviating Cuba's abysmal economic situation, but in fact, Cuba paid a tremendous price. *Signing the agreement damaged the island's diversification program severely. The lowering of duties on 426 products from the mainland made Cuba even more dependent on American goods.* The Cuban government's freedom to sign commercial treaties with other nations was also greatly impaired. Finally, the trade agreement helped to continue the status quo against any reformist alternatives" (ibid., 120; my emphasis).

13. Dallek, *Roosevelt*, 93. In a more recent history of the New Deal in a global context, Kiran Klaus Patel observes: "In 1934, the United States lowered its protectionist tariffs on sugar imports, but simultaneously introduced a quota system. Liberalization did not go far. Cuban participation in the US market nonetheless increased, expanding from 25 percent in 1933 to 31 percent by 1937, while Cuban sugar output rose from 2.2. to 2.9 million tons during the same period. Besides its beneficial effects for the Cuban sugarcane business, this expansion of production and trade hampered economic diversification and industrialization efforts on the island. The policy exacerbated the ecological degradation of the island, and helped keep Cuba largely dependent on a single crop and its huge neighbor to the north. Both economically and politically, the country became more vulnerable; in fact, the Good Neighbor policy led to a new phase of North American primacy in the Cuban economy, after years of increasing independence since the early 1920s. The new growth in trade did little to alleviate the situation of most Cubans. Sugarcane was in the hands of just a few owners, many of whom were *norteamericanos.* The State Department was by no means blind to these ambivalent effects. In 1935, vice consul Hernan C. Vogenitz stressed that most of the benefits of the trade

agreements landed in the pockets of American sugar interests: while 'a few American corporations might gain larger profits . . . the policy will most certainly lead to internal disorder in Cuba'" (*New Deal*, 153).

14. Shoultz, *Beneath*, 305. Writing of the international importance of this economic policy, David Haglund points out that the Reciprocal Trade Agreements, established by Secretary of State Cordell Hull, were "considered an adequate solution to the problem of extrahemispheric economic incursions" in 1936, but Hull's efforts "on behalf of free trade would later come under scathing attack from critics who feared that the Reciprocal Trade Agreements were an impotent response to the German challenge" (Haglund, *Latin America*, 52; see also 47–48).

15. Soley, *Free Radio*, 37.

16. Commenting on the growth of radio listeners worldwide, Soley writes: "In 1934, for example, there were 42 million radio receivers in the world. Only 100,000 of these were in Africa, 2 million in Asia, and 900,000 in Latin America. By 1949 there were 160 million receivers worldwide with 2 million in Africa, 8 million in Asia, and 6.5 million in Latin America. Radio penetration had grown quickly during World War II because it was an efficient, safe, and inexpensive method for obtaining information, which could be easily assessed for accuracy by comparing the information provided by one station . . . with that of another" (ibid., 17).

17. Fox, Sadlier, and others have made strong arguments for the influence of state-supported Good Neighbor radio programs broadcast through NBC and CBS under the control of Nelson Rockefeller's Office of Inter-American Affairs. While I discussed one such program, Orson Welles's *Hello Americans*, in chapter 2, my intent in this chapter is to draw the focus away from U.S.-based English-language programs and consider instead the complicated radio politics developed in Cuba.

18. Salwen observes that Goar's initial interest in radio ownership came after RHC canceled an advertising contract with him when the station realized they could make more money using the airtime for other purposes. Realizing that CMQ and RHC had monopolized the airwaves, Mestre decided that he and his brothers should take over financial control of one of the stations. They were able to do so only after joining with the advertising firm of Augusto Godoy, whose accounts included Pepsi Cola ("Origins," 318). Salwen notes that "during the war years, RHC–Cadena Azul and CMQ together accounted for as much as 80% of the audience in Havana's 30-plus station market" (ibid., 317). Regarding the stations' North American affiliations, he writes: "The United States government encouraged American radio networks to establish Latin American affiliates to counter what it viewed as hostile Axis propaganda in the region. The American networks needed little encouragement. They went into pan-American broadcasting as much for profit as for patriotism . . . NBC and CBS, with 124 and 76 Latin American affiliates, respectively, led the way in pan-American broadcasting in 1941. Pan-American broadcasting turned out to be far less profitable than the American networks hoped" (ibid., 316–17). Howard Frederick points out that the United States began to counter Nazi radio propaganda to Latin America with the 1940 appointment of Nelson

Rockefeller as coordinator of inter-American affairs: "President Roosevelt organized a collation of broadcasters to set up a government shortwave service to Latin America. By July 1941, *Business Week* reported that WRUL in Massachusetts was broadcasting 50 kilowatts to Latin America. It was put on the air with $200,000 in contributions from the National Broadcasting Company (NBC), the Columbia Broadcasting System (CBS), Westinghouse Corporation, and others 'to counteract the deluge of shortwave propaganda emanating from Europe" (*Radio Wars*, 16).

19. Salwen, "Origins," 138. Writing more generally about the commercial relationship between the United States and Cuba, Michael Chanan observes: "The island became an offshore testing laboratory for U.S. penetration of Latin American media markets from the end of World War I, a bridgehead for companies like ITT and RCA—the latter took to promoting Cuban music and radio productions throughout Central America, supported by advertising agencies that trained up Cuban artists and copywriters" (*Cuban Cinema*, 17).

20. Salwen, "Origins," 315. For the definitive Cuban account of U.S. commercial influence and melodramatic radio programs, see González, *Llorar.*

21. González, *Llorar*, 66.

22. In *The Female Complaint*, Berlant writes, "Publics are affective insofar as they don't just respond to material interests but magnetize optimism about living and being connected to strangers in a kind of nebulous *communitas*" (xi). Berlant also argues that femininity itself is a genre (3–4).

23. Melodrama, detective fiction, thrillers, and adventure stories—those genres known as "middlebrow"—made up the primary fictional content on the radio in Cuba from *Chan Li-Po* in 1934 up through the Revolution in 1959. This twenty-five-year period also included adaptations of print novels. *Playing the Race Card: Melodramas of Black and White from Uncle Tom to O.J. Simpson*, Linda Williams's extraordinary study of what she calls the melodramatic "mode" that traverses media and centuries of cultural production in the United States, considers how melodrama structures the possibilities and limitations of melodrama to address racism in the United States.

24. Berlant, *Complaint*, xi.

25. Ibid., 5, viii.

26. Luis-Brown, *Waves*, 5–6.

27. Ibid., 36.

28. "*El derecho de nacer* de Félix B. Caignet sigue siendo el más fiel espejo de cierta realidad sensible, inmediata, de la América Latina."

29. "Lo único que me devolvió el sosiego fueron los amores contrariados de *El derecho de nacer*, la novela radial de Don Félix B. Caignet, cuyo impacto popular revivió mis viejas ilusiones con la literatura de lágrimas . . . Inspirado por el éxito de *El derecho de nacer*, que seguí con atención creciente durante todo el mes, había pensado que estábamos en presencia de un fenómeno popular que los escritores no podíamos ignorar" (*Vivir*, 494, 495).

30. Luis-Brown, *Waves*, 38.

31. L. William, *Race Card*, 6.

32. For the implications of such a voice in the transnational space of the Caribbean, and Puerto Rico in particular, see Rivero, *Tuning Out*.

33. "*El derecho de nacer* es la pieza donde confluyen las concepciones estilísticas y éticas del melodrama radiofónico en su período de auge, con auténtica y aleccionadora comunicación" (González, *El más humano*, 157).

34. "una caricatura de la pronunciación bozal, estereotipada para la caracterización radiofónica" (ibid., 199).

35. Lane, *Blackface Cuba*, 47–49.

36. Ibid., 16, 2.

37. Ibid., 13.

38. Ibid., 10–14. For more on a nineteenth-century postracial identity in Cuba, see Ferrer, *Insurgent Cuba*.

39. Antonio López, *Unbecoming*, 67.

40. "Minstrelsy brought to public form racialized elements of thought and feeling, tone and impulse, residing at the very edge of semantic availability, which Americans only dimly realized they felt, let alone understood" (Lott, *Love and Theft*, 6).

41. Kutzinski, *Worlds*.

42. Masiello, "Rethinking"; Moore, *Nationalizing*.

43. My translation here does not attempt to mimic the *bozal* form. For a contemporary rendering of "la poesía negra" into a U.S. poet's idea of black vernacular, see Langston Hughes's translation of Nicolás Guillén's poetry in *Cuba libre*.

44. T. Anderson, *Carnival*, 169.

45. Dupree, *Science*, 389–90.

46. Kunheim, *Beyond*, 48, 47.

47. González, *El más humano*, 36.

48. Antonio López, *Unbecoming*, 98.

49. Ibid., 64.

50. Qtd. in T. Anderson, *Carnival*, 163; emphasis added. "[E]l noventa por ciento del mito de mis versos está en tu interpretación . . . ¡Mis versos lo son cuando los dices tú!"

51. Maguire, "Eusebia Cosme Show," 5. Antonio López points out that in some cultural and legal ways Cosme became black in her journey to the United States from Mexico.

52. González, *El más humano*, 200–203.

53. Batista's own racial status is a complex subject in Cuban political history (see Argote-Freyre, *Batista*, 3–5).

54. In an interview with Luis López, the radio actress Sol Pinelli recalls working for one of Alejo Carpentier's *Dramas de la guerra* programs at CMQ where the "studio, divided by an enormous glass window, which was next to us, allowed you to see what they were rehearsing or broadcasting on the other side. On more than one occasion, Carpentier had to call us, the actresses above all, to attention, because unconsciously we were all entertaining ourselves looking at the neighboring participants [*participantes vecinos*], whose broadcast started an hour before our rehearsals. We knew about the enormous popularity and charisma of the space's director, the leader of the Ortodoxo Party Dr. Eddy Chibás, who was often accompanied by a young revolutionary and student leader . . . Fidel Castro Ruz" (*Carpentier*, 79).

55. Birkenmaier, *Carpentier*, 209. The link between sentimentalism and both Latin American politics and the struggle for racial equality in the nineteenth-century Americas has been treated in Luis-Brown. Citing Nancy Armstrong, Elizabeth Barnes, Lauren Berlant, Ann Douglas, Juli Ellison, Arthur Riis, and Gillian Silverman, as well as primary documents from Harriet Beecher Stowe and Helen Hunt Jackson, Luis-Brown argues that as "sentimental melodrama" and "romantic racialism" travel from the United States and England to "politically contested zones of empire" they become the conflicted but politically persuasive means to imagine a more expansive community to contest colonial power. Among those influenced by sentimentalism, according to Luis-Brown, were W. E. B. Du Bois and José Martí, the latter of whom saw the U.S. sentimental tradition as fundamental to moving toward what he called "Our America" (Nuestra América).

56. Ibid.

57. Thomas, *Cuba*, 766.

58. Salwen, *Radio*, 73.

59. Ibid., 75.

60. Ibid. According to the collective memoir of Radio Rebelde, these were the last words of Chibás on the CMQ: "¡Compañeros de la Ortodoxia, adelante! ¡Por la independencia económica, la libertad política y la justicia social! ¡A borrar a los ladrones del Gobierno! ¡Pueblo de Cuba, levántate y anda! ¡Pueblo cubano, despierta! ¡Este es mi último ALDABONAZO!" (Martínez Víctores, *7RR*, 23).

61. As Reynaldo González points out: "La agitación desde la radio tiene su climax en marzo de 1933, cuando se lanza una falsa noticia desde una emisora portátil: '¡Pueblo cubano, eres libre! ¡Ha caído la dictadura de Gerardo Machado!' La población habanera se lanzó a la calle, a celebrar el acontecimiento, y se produjo un lamentable disturbio'" (*Llorar*, 93). In 1957, "armed groups belonging to the Directorio Revolucionario led by José Antonio Echeverría, together with some of Carlos Prio's *auténticos*, had launched an audacious daylight assault on the Presidential Palace, and temporarily seized the twenty-four-hour Radio Reloj station in Havana. But the assaults failed, and in the shoot-outs that ensued, over forty people had died" (J. Anderson, *Guevara*, 246). As Hugh Thomas describes the scene: "Echevarría had captured the radio station and broadcast an excited message: 'People of Havana! The Revolution is in progress. The presidential palace has been taken by our forces and the Dictator has been executed in his den!' But after he had blown up the central control panel in the radio station, Echevarría was shot dead in the street by police" (*Cuba*, 929).

62. Castro, *La historia*, 31.

63. Ibid., 30. The belief that the recorded sound can propel the masses to draw blood recalls Friedrich Kittler's media chronology in which the radio derives from the phonograph, the medium from which Edison "found a way to transfer the functions of his ear to his sense of touch" (*Gramophone*, 28), drawing blood with his voice as he touched the needle.

64. See McEnaney, "Digital."

65. Allen S. Weis and Douglas Kahn's writings on radio, and even Anke Birkenmaier's readings of Carpentier and Desnos's radio experiments in Paris

are typical examples of this trend to hear radio as ghostly. While I tend to oppose this analytical method for failing to attend to the medium's materiality, I invoke it here because of Castro's own anxiety about bodies, history, and control. Derrida's *Specters of Marx* seems instructive in this case and might help indicate why the *animus* represents an especially troubling concept for Castro's status as a commander intent on galvanizing a unified community: "the specters of Marx. Why this plural? Would there be more than one of them? *Plus d'un* [More than one/No more than one]: this can mean a crowd, if not masses, the horde, or society, or else some population of ghosts with or without a people, some community with or without a leader—but also the *less than one* of pure and simple dispersion. Without any possible gathering together. Then, if the specter is always animated by spirit, one wonders who would dare to speak of a spirit of Marx, or more serious still, of a spirit of Marxism. Not only in order to predict a future for them today, but to appeal even to their multiplicity, or more serious still, to their heterogeneity" (*Specters*, 3). See Derrida, *Specters*; Weiss, *Breathless*; Weiss, *Experimental*; and Kahn, *Noise*. For more on haunting and media more generally, see Sconce, *Haunted Media*.

66. Thus, Denis Hollier's insistence on radio's "live" ontology as similar to the Derridean *anarchival* becomes complicated by this scene of mixed media.

67. Commenting on the production of liveness in radio, Jonathan Sterne writes: "To borrow Roland Barthes's language, the point of the artifice is to connote denotation, to construct a realism that holds the place of reality without being it. Like the descriptive speciality, the live radio broadcast did not so much capture the event as it became an event itself . . . The goal in reproducing live events was not reproducing reality but producing a particular kind of listening experience. Early sound reproduction—whether live or wholly contrived—sliced up reality in order to fashion a new aesthetic realism" (*Audible Past*, 245–46). Chibás's suicide and Castro's speech make clear that the construction of this "aesthetic realism" could migrate into the political sphere with powerfully nonaesthetic consequences on real bodies.

68. Franqui, *Mito*, 258. "Yo no dejé que [Fidel Castro] metiera por Radio Rebelde la mala poesía ni incluso la radio novela, como era su intención."

69. Franqui, *Family*, 65.

70. Martínez Víctores, *7RR*, 9; emphasis added. "La *gran novela* era la epopeya de retar a una tiranía oligárquica, defendida por un ejército profesional armado, asesorado y entrenado por Estados Unidos."

71. Ibid., 10. "El país en pleno (cuando digo *país* hablo de *pueblo*), en las horas de transmisiones nocturnas de 'Radio Rebelde', le volvió la espalda . . . a las lloronas novelas radiofónicas, para pegar su oído a aquel llamado de encendida prédica martiana y de absoluta honestidad informativa (Martínez Victores, *7RR*, 10).

72. The replacement of the *radionovela* with Martí unwittingly hearkens back to Martí's own borrowing of melodramatic narrative to construct his own rhetorical power (see Luis-Brown).

73. Guevara, *La Guerra*, 137–38.

74. Soley and Nichols, *Clandestine*, 169.

75. Debray, *Revolution*, 108.

76. Ibid., 109.

77. Soley and Nichols, *Clandestine*, 14. Recalling the station's military role, Fidel Castro says: "Through Radio Rebelde we communicated with the different fronts and columns. Thus, as a communications center it was supremely important for the military, in addition to having been an instrument of mass broadcasting that played a greatly transcendental political role during the entire war" (*Rebelde*, 3). Radio Rebelde also developed an international audience: "In May [1958], Jules Dubois, the *Chicago Tribune* correspondent, used Radio Rebelde's newly boosted transmitter links with the outside world to conduct an interview with Fidel from Caracas" (J. Anderson, *Guevara*, 322).

78. This reorganization of radio's meaning comes close to Fanon's description of Algerian rebels' use of radio to oppose French colonialism.

79. Thomas, *Cuba*, 1029.

80. Castro, *Rebelde*, 78.

81. Ibid., 14.

82. J. Anderson, *Guevara*, 370.

83. Hamberg, *Construction*, 3.

CHAPTER 5

1. Qtd. in Tyson, *Dixie*, 225. Robert Williams also recalls this comment in his remarks before the Bay Area Fair Play for Cuba Committee on March 11, 1961. Williams laid out a broad "inter-nationalist" position, writing in the first issue of the *Crusader* (3, no. 8 [April 1962]) he published from Cuba: "We must be willing to join hands with forces throughout the world that are now fighting to rid the universe of the evil enemy that has plagued the Afro-American for so long in the racist U.S.A. We must realize that the people of Latin America, Asia and Africa are engaging the same enemy in the same liberation struggle" (2). Elsewhere in the same issue, criticizing the resistance to his platform of armed self-defense, he writes, "All of the passive morality in the world will not save [the Afro-American] if he fails to make friends on an international plane" (5).

2. Although these charges were eventually dropped, four others were wrongfully charged and sentenced. Among them was Williams's friend Mae Mallory, a single mother who was sentenced to sixteen to twenty years in prison (Cohen, *Crusader*, 268). The *Crusader* featured repeated appeals for legal help with Mallory's case. In one of these, under the headline "Hands across the Mason-Dixon," Williams compares Ohio governor DiSalle's decision to extradite Mallory to Monroe to the time of the fugitive slave law. He adds: "It is an indictment of American justice to have a Northern state collaborate with the South in a legal lynching. The Mallory case proves that even a Northern State like Ohio is not half as humane as integrated Cuba. Realizing how Fidel Castro feels about racial injustice, the racist U.S. Government has not even bothered to ask Cuba to return an Afro-American to North Carolina lynch justice" (R. Williams, *Crusader* 3, no. 9 [May 1962]: 3).

3. Williams cabled Mao Tse-Tung, Algeria's Ben Bella, Ghana's Nkrumah, and Indonesia's Sukarno in 1963, after the Birmingham church bombings, asking for a statement of support for African Americans against the racist

violence in the United States (Cohen, *Crusader*, 250). Of the four, only Mao issued a statement, which Williams published in The *Crusader*'s 1964 special edition of October. For more on Williams's time in China, see ibid., 253–66, 279–81, 319–30; Tyson, *Dixie*, 96; Robin D. G. Kelley and Betsy Esch, "Black Like Mao: Red China and Black Revolution," Fred Ho, "The Inspiration of Mao and the Chinese Revolution on the Black Liberation Movement and the Asian Movement on the East Coast," and Diane C. Fujino, "The Black Liberation Movement and Japanese American Activism: The Radical Activism of Richard Aoki and Yuri Kochiyama," in Ho and Mullen, *Afro-Asia*.

4. Of course, by the "enunciative function," Foucault means something more specific and more general than geographical place, but so too does Williams. Albeit in very different language than Foucault, Williams's statements acknowledge that he has found himself positioned within a national discourse that seems to forbid certain types of statements when those statements emerge from particular racialized bodies in certain (politically and geographically) "foreign" contexts. His radio program seizes a particular technological form—broadcasting—in an attempt to reorganize the discursive space and his place within it. In his chapter "La fonction énonciative" from *L'archéologie du savoir*, Foucault writes: "Il est une place déterminée et vide qui peut être effectivement remplie par des individus différents . . . Décrire une formulation en tant qu'énoncé ne consiste pas à analyser les rapports entre l'auteur et ce qu'il a dit (ou voulu dire, ou dit sans le vouloir), mais à déterminer quelle est la position que peut et doit occuper tout individu pour en être le sujet" (125–26).

5. See Berlant, *Complaint*, 270–71, for more on the ways in which realism and melodrama coexist.

6. "On [*sic*] March 1960 Goar Mestre abandoned the island and headed for New York . . . Abel Mestre, Goar's brother, stayed in Cuba to oversee CMQ's business, but on 1 April he also left the country. CMQ and Radio Progreso were later taken over by the government. Even Castro's popular commentator, José Pardo Llada, who believed that there was room for the Communist party in the revolution, defected in March 1961, recognizing that an independent position was no longer possible. By the time of Pardo Llada's defection, the government had gained control of radio and television" (Luis, *Culture*, 74–75).

7. Thomas, *Cuba*, 1273–74. Thomas writes that after the closure of CMQ, "[t]he only effective vocal opposition to the government could now come from the old newspapers, *Diario de la Marina* and *Prensa Libre*, and from the pulpit, since Channel 12, the other big television network, was under a government intervener and since all the small radio stations had by force or intimidation been grouped together in a new government-inspired corporation ironically known as FIEL (*Frente Independiente de Emisoras Libres*), under the virulent-voiced commentator Pardo Llada" (1274). Wryly pointing to the irony of this and subsequent transitions, Thomas describes the soundscape in Cuba in 1961: "In the streets the revolutionary songs still, almost continuously, blared from gramophones, and, above all the *Hymn of the 26 July*, still jingled to their seaside rhythm. One of the best-known voices of the revolution, however, Pardo Llada, the chief of the radio and television corporation FIEL, nicknamed the 'Minister of Hate,' defected to Mexico late in March,

screaming 'betrayal' in the same voice as he had until then screamed 'imperialism'" (1343).

8. Frederick, *Radio Wars*, 6.

9. Ibid., 6. In the meantime, the Argentine journalist, activist, and crime writer Rodolfo Walsh had already decoded a cable from the head of the CIA in Guatemala informing Washington of the preparations carried out for the invasion of Cuba, including the Guatemalan training camps for the attack's mercenaries. Walsh, who was working for *Prensa Latina*, was given the mysterious coded cable by the news agency's director, Jorge Ricardo Masetti, deciphering the accidentally intercepted message in his apartment at the Focsa building in Havana (see Arrosagaray, *Walsh*).

10. Frederick, *Radio Wars*, 17. Recognizing that shortwave transmissions could only reach the limited number of shortwave receivers in Cuba, Kennedy's press secretary, Pierre Salinger "devised a plan whereby a powerful network of American commercial AM stations from Florida and the Gulf could flood the Cuban AM band with American reports" (ibid.). Ten stations volunteered their services, from Miami, Atlanta, Chicago, New Orleans, Cincinnati, and two shortwave stations in Scituate, Massachusetts, and Redwood City, California (ibid., 18). In addition, the United States established a 50-kilowatt transmitter at the Dry Tortugas in the Florida Keys, which was on the same frequency as WHO (AM) in Des Moines, Iowa, and another at the city of Marathon, also in the Keys, which has been "interfering with WHAM [Rochester, New York] illegally for 19 years . . . in violation of Federal Communications Commission (FCC) allocations and international treaties. Radio Marathon can be heard throughout Cuba and the Caribbean, and WHAM complains that Cuban jamming of that frequency interferes with its signal. The violation of international treaties referred to is the fact that Radio Marathon was not reported to the International Frequency Registration Board, part of the International Telecommunication Union in Geneva until 1981" (ibid.).

11. Washburn, *Broadcasting*, 87.

12. Ibid., 39. The issue of officially intentional or unintentional domestic propaganda has remained relevant. For a more recent discussion regarding the United States involvement in Iraq, see Barstow, "Two Inquiries."

13. Frederick, *Radio Wars*, 7–9. Frederick notes: "In the beginning, RHC's broadcasts were in Spanish and were repeated many times. In midevening, there were also broadcasts in English. Later, broadcasts were transmitted in French, Portuguese, Arabic, Guarani, Quechua, and Creole. By 1963, 188 hours were transmitted per week. By the end of the year, RHC's weekly programming had reach 267 hours" (9). He adds that "by 1963, Cuba held fourth place among Communist international broadcasters in hours per week beamed abroad. By 1965, Cuba was broadcasting an estimated 160 hours per week to Latin America. Programming emphasized the coordination of revolutionary action programs. Relatively simple broadcasts were beamed in Creole to Haiti; guerrilla warfare manuals were read over the air" (10).

14. Ibid., 10–11.

15. Heel, "Radio y TV."

16. Writing in the pages of his broadsheet the *Crusader* (4, no. 2 [August 1962]) from Havana, Williams argued, "The only Alliance for Progress for the black man is a close alliance with the struggling peoples of South America, Asia, and Africa" (2). In a later issue (*Crusader* 4, no. 3 [September 1962]), he wrote, "The U.S. Government and the Ku Klux Klan's Alliance for Terror is becoming a menace to the entire world" (7). For more on Kennedy's "Alliance for Progress," see Taffet; and Rosen. Williams supposedly told Manuel Piñeiro, the Cuban rebel known as "Barbaroja," or "Red Beard," "What I propose is that the Revolution provide me with radio facilities for a regular program aimed at Black America" (Cohen, *Crusader*, 209). Tyson writes: "Despite its range, Radio Free Dixie 'was aimed at the South, primarily,' [Williams] recounted, 'because the black people in the South didn't' have any voice" (*Dixie*, 287). Cohen adds that Williams, tuning in to Radio Havana Cuba from Monroe, told the visiting Bob Taber of the Fair Play for Cuba Committee in 1960, "if the Cubans were to use some English in their long-wave programs, they'd be able to get through to millions of listeners in the United States who have ordinary receivers" (*Crusader*, 209). By 1963, FBI agents had tuned in to the program: "Reminding Hoover that state authorities in Alabama had complained about the program, the agents noted that Radio Free Dixie was 'directed to the Negroes in the southern part of the United States, calling upon the oppressed Negroes to rise and free themselves'" (Tyson, *Dixie*, 287).

17. For more on the complex ways in which African American performance in the 1930s could both dissent from and comply with U.S. imperialism, see Batiste, *Darkening Mirrors*.

18. Writing about the literary exchanges between Nicolás Guillén and Langston Hughes, and the cultural climate in which they occurred, Vera Kutzinski notes: "In the postwar US, a cultural and political environment still steeped in racial binarisms and anxieties about intermarriage—this is still well before the last antimiscegenation law was repealed in Virginia—the very idea of conceding, let alone celebrating, the impact of racial mixing on the national culture would have been anathema to prevailing social sensibilities on both sides of the color line" ("Fearful," 124).

19. Emily Apter defines a "translation zone" as "a linguistic hot spot," similar to a "military zone," and, ultimately, "a translation no-fly zone, an area of border trouble where the lines dividng discrete languages are muddy and disputatious; where linguistic separatism is enforced by high surveillance missions; or where misfired, off-kilter semantic missiles are beached or disabled" ("Balkan," 69). Linking translation to black internationalism and what he names "the practice of diaspora," Brent Hayes Edwards writes, "Diaspora is a term that marks the ways that internationalism is pursued by translation" (*Practice*, 11). As I have already written, Williams does not make translation, whether from dialect to dialect, or major language to major language, part of his political platform. Although he played his friend Langston Hughes's spoken-word pieces on his radio program, the broadcast eschewed dialect. Uninterested in Hughes's translations of Nicolás Guillén, Williams did not confront the racial politics Kutzinski describes inhered in the translation of the Spanish word "negro," for instance ("Fearful," 115). Nor did he seek to

broadcast translations of other authors from Cuba, Martinique, or any of the African nations. However, the pages of the *Crusader* did include translations of Fidel Castro's speeches, and telegrams from Mao Tse-Tung, and Chinese and Vietnamese schoolchildren.

20. Amy Kaplan and David Luis-Brown describe, in different ways, how cultural production intertwined the two countries in the late nineteenth and early twentieth centuries. Kaplan, in particular, demonstrates how the "domestic" and "imperial" spheres of U.S. government and private life influenced each other.

21. Amy Kaplan, *Anarchy*, 138; emphasis added.

22. Listeners who wrote to the Williamses often mentioned the tagline for the show (see Richard C. Laslo to Robert F. Williams, February 23, 1964, "Correspondence," Box 1, Robert F. Williams Papers, Bentley Historical Library, University of Michigan).

23. The first notice for *Radio Free Dixie* appeared in the pages of the *Crusader*'s August 1962 edition and stated, "This will be the first completely free radio voice that black people have had to air their case against brutal racial oppression" (5). Speaking of *Radio Free Dixie*, Williams claimed, "This was really the first true radio where the black people could say what they want to say and they didn't have to worry about sponsors, they didn't have to worry about censors" (qtd. in Tyson, *Dixie*, 287–88). For more on the Mexican-U.S. history of border radio, see Fowler and Crawford, *Border Radio.* Brian Ward points out that "the two most militant black voices regularly heard on radio in the South during the late 1950s and early 1960s emanated from outside the country. The Black Muslims weekly *Muhammed Speaks* shows covered much of the United States, thanks to a 250,000-watt transmitter at XERF in northern Mexico. The proto-black power *Radio Free Dixie* programs produced by North Carolina's controversial NAACP leader and armed-self-reliance advocate Robert Williams were beamed in from Cuba" (*Radio and the Struggle*, 124).

24. Williams, *Crusader* 3, no. 8 (April 1962): 5.

25. Tyson, *Dixie*, 288.

26. Williams, *Crusader* 3, no. 8 (April 1962): 5.

27. The advertisement for records was framed by images of undulating radio waves and repeated "S.O.S." signals, and asked readers to send their records to an address in Canada, which sent them on to the Williamses around the U.S. blockade to Cuba (see *Crusader* 4, no. 3 [September 1962]: 1–8). In a letter dated December 18, 1962, Leroi Jones (soon to change his name to Amiri Baraka) told Williams: "I sent some records sometime ago, direct, through events of the last few months have shown me how dumb that was. Anyway, I've now sent them to the forwarding address Marc gave me, so you should get them soon . . . I sent: Bags & Trane (milt jackson& [*sic*] john Coltrane); Clyde McPhatter, Billie&Dede Pierce, which is new orleans trad.; yng trumpeter, Richard Williams; clark Terry; Dizzy Reece; Rocky Boyd, yng saxophonist w/ Kenny Dorham. 7 records in all to start, will get more there as soon as. Yr book is out. I'm supposed to review it for the Village Voice if they don't chicken out. Called: *Negroes With Guns*. I've already read it. Marc did a good

job. You probably didn't get a copy for the same reason you didn't get records. These cats are always tampering with my mail, slicing envelopes and such. Letters from you and Mae are always of special interest in this free open society. I manage to pick up RFD, it seems, when the wind is right" (Leroi Jones to Robert F. Williams, December 18, 1962, "Correspondence," Box 1, Robert F. Williams Papers, Bentley Historical Library, University of Michigan).

28. This quote is taken from a recording of the broadcast available on the website for the Public Broadcasting System's documentary, *Negroes with Guns: Rob Williams and Black Power* (Independent Lens, May 3, 2011).

29. Barlow, *Voice Over*, 204; emphasis added.

30. "What surprised Williams most of all was [*sic*] the letters from White Southerners who said they were happy someone had the courage to tell the truth about what was going on in the United States" (Cohen, *Crusader*, 212). Ward notes the importance radio programs like Williams's played in shifting the political landscape through the soundscape: "By helping to publicize the goals and methods of the early movement beyond Dixie, radio contributed to a national process of legitimization that by 1965 had persuaded even some southern stations to allow pro-civil rights programming" (*Radio and the Struggle*, 124).

31. Robert F. Williams to J. A. Lumpp, January 30, 1964, "Correspondence," Box 1, Robert F. Williams Papers, Bentley Historical Library, University of Michigan.

32. Early in the program's history, Williams received a letter dated August 30, 1962, from Sammy J. Kents, who claimed solidarity with Williams. Kents wrote: "I used the Kenya Broadcasting Service as my personal tool. This year in March, I went in the Kenya Radio then announced to all people in Kenya that European Farmers must leave Farms to Africans, and if they did not like they had better leave Kenya, otherwise Revolution was to take place and that could make the to run away then leave all the farms to Africans. I want you to settle my case with the Cuban Officers and let me know the outcome of it. I want very much to come to Cuba then join you in your Free Radio of America (RADIO HABANA) as a speaker of English and if possible Swahili" (Sammy J. Kents to Robert F. Williams, August 30, 1962, "Correspondence," Box 1, Robert F. Williams Papers, Bentley Historical Library, University of Michigan).

33. *Radio Free Dixie* Broadcast Transcript, July 27, 1962, Box 10, Robert F. Williams Papers, Bentley Historical Library, University of Michigan. See also U.S. Justice Department, Federal Bureau of Investigation, "Robert Franklin Williams, Field Office File # 100–7670," Box 8, Robert F. Williams Papers, Bentley Historical Library, University of Michigan.

34. Jim Cumbie to Robert F. Williams, November 30, 1963, Robert F. Williams to J. A. Lumpp, January 30, 1964, "Correspondence," Box 1, Robert F. Williams Papers, Bentley Historical Library, University of Michigan; Richard C. Laslo to Robert F. Williams, February 23, 1964, "Correspondence," Box 1, Robert F. Williams Papers, Bentley Historical Library, University of Michigan.

35. Robert F. Williams to Jack Martina, January 30, 1964, "Correspondence," Box 1, Robert F. Williams Papers, Bentley Historical Library,

University of Michigan; Jo Salas to Maria Cordova, December 19, 1963, "Correspondence," Box 1, Robert F. Williams Papers, Bentley Historical Library, University of Michigan.

36. Tom Todd to Robert F. Williams, February 17, 1964, "Correspondence," Box 1, Robert F. Williams Papers, Bentley Historical Library, University of Michigan; William Johnson Jr. to *Radio Free Dixie*, August 22, 1963, "Correspondence," Box 1, Robert F. Williams Papers, Bentley Historical Library, University of Michigan; James H. Williams to Robert F. Williams, October 12, 1962, "Correspondence," Box 1, Robert F. Williams Papers, Bentley Historical Library, University of Michigan.

37. David Rattray to Robert F. Williams, September 24, 1963, "Correspondence," Box 1, Robert F. Williams Papers, Bentley Historical Library, University of Michigan.

38. Clark M. Nobil, August 26, 1963, "Correspondence," Box 1, Robert F. Williams Papers, Bentley Historical Library, University of Michigan.

39. An August 19, 1963, letter from Kay Greaves at radio station KPFA in Berkeley, California, reports: "WE are pleased to advise that KPFA will rebroadcast 'RADIO FREE DIXIE' at 2pm to 3pm on Thursday, September 5. This program is a recording, made by our sister-station—WBAI FM in New York City—of your broadcast on AM radio made in May." However, perhaps more impressive for its grassroots improvisation and belief in Williams's message, a letter dated December 20, 1962, from Robert Perkins in Los Angeles, California, informed Williams: "Enclosed are two copies of a leaflet that we had printed up after receiving a tape of the interview that you gave in Havana. After the playing of the tape, we took up a collection and sent it to the Mae Mallory defense committee. I am letting my other nationalistic friends make copies (of the TAPE) and telling each of them to let someone make a copy of theirs. I am also attempting to get in touch with all of your followers . . . P.S. Let me know if you received the letter sent thru Mexico." A listener himself, Williams also contributed to this tape and radio network. In a November 5, 1963, letter from the Peking Hotel in Peking, China, Williams writes Mr. Richard Elman at Radio Station WBAI in New York City to tell him: "I have on hand a number of tapes. One is a 2½ hour recording of a Peking rally attended by 10,000 persons in support of the Negro struggle against racial discrimination in the U.S.A. It includes speeches of noted persons from many parts of the world in many languages, simultaneously translated into english [*sic*]. It has also been broadcast to many parts of the world. I also have a tape of the songs, 'Old Man River' and 'We Shall Overcome' sung in Chinese, plus Revolutionary guerrilla songs of the Angolan Freedom Fighters in their actual voices. In as much as I shall be returning to Havana soon, if you are interested in these tapes, please rush exchange tapes to Havana" ("Correspondence," Box 1, Robert F. Williams Papers, Bentley Historical Library, University of Michigan).

40. Cohen, *Crusader*, 211.

41. Qtd. ibid., 215.

42. In the October–November 1962 edition of the *Crusader*, Williams writes that his pamphlet "has no political affiliation and does not reflect

any official policy. It is printed in Cuba by workers who want to help the Afroamerican struggle in the racist USA . . . It is not a project sponsored by the Cuban Government, therefore, the workers themselves make its continued publication possible . . . THE CRUSADER enjoys a freedom of the press that the racists of the USA could never bring themselves to tolerate" (1). Williams repeated this message in the many letters he wrote to listeners inquiring about the program's political stance. However, even late into his stay in Cuba, he made clear that the country's politics did not produce the terrifying racial hatred he constantly faced when living in the United States: "RADIO FREE DIXIE does not follow any political party line. We stick to the race issue. I've never belonged to any political group and I don't belong to one now, however, I am very grateful to this 'Communist country' for saving this black American veteran from a lynch mob of racist savages. My predicament is living proof that the Communists believe more in justice for black folks than any of those self-righteous hypocrites who deceptively speak of defending the world from 'oppressive communism.' I have seen and know real oppression and terror in the so-called 'Free World.' There is no racial discrimination or lynch mobs in 'Communist Cuba'" (Williams to Lumpp, 2–3).

43. See Gleijeses, *Conflicting*, 185–213. The PAIGC almost accepted African American fighters as well. Gleijeses reports that the Black Panther leader Stokely Carmichael went to Cuba in June 1967 and "told his Cuban hosts that he wanted African Americans to fight alongside the PAIGC in Guinea-Bissau to stress their solidarity with Africa and to atone for their participation in the war of aggression against Vietnam" (193). The PAIGC leader Amílcar Cabral met with Carmichael in Conakry at the Cubans' behest and told the latter that he would support them as long as the fighters were trained in another country. Nyerere in Tanzania agreed to allow the men to train there, but Carmichael then "met the South African singer Miriam Makeba, and his plans changed. 'He dropped us!' Dreke remarked. He married Makeba, became a citizen of Guinea, and settled down" (ibid., 193–94).

44. See Sarduy, *Los matadores de hormigas*, 1077–93. All translations from the play are mine, although I have consulted Philip Barnard's excellent translation, *The Ant Killers*, in *For Voice*, 111–36 (Pittsburgh: Latin American Literary Review Press, 1985).

45. The printed Spanish text reads "alemanes" (Germans), but Philip Barnard's English translation changes their nationality to French. The production of the play on Radio France on July 10, 1981, says "touristes françaises" (French tourists).

46. In the introduction to Sarduy's radio play *La Playa* (The Beach; 1969), he observes that the blank spaces of concentric circles between songs on a vinyl record are known as "playas" and that each one of these concentric circles marks the moment of starting again: "A cada 'playa' la sequencia es reestructurada, uno de sus detalles varía o se transforma en su contrario; a cada playa el retrato parte de cero, borra, desdice, niega lo anteriormente escrito e impone una nueva versión. No hay resultado final, puesto que las versiones son equivalentes" (15). *Los matadores de hormigas* uses this narrative premise as its basis, but rather than transforming the event each time, it slowly builds and

accrues significations, creating a layered event that allows for movement back and forth across those circles as they are linked into networks. Sarduy, *Para la voz.*

47. See Fanon, "Ici la voix de l'Algérie . . ."

48. Sarduy, *Los matadores*, 1077.

49. See Brian Edwards, "Fanon's *al-Jazáir*; and Mowitt, "Breaking Up."

50. Commenting on Delpech's song as it plays over the speakers from a bar at a Portuguese port near Faro in sequence one, "woman one" says, "It was a French song, with astute references to the revolution" (1080). The quoted lyrics include these lines: "Oh, Marianne was so pretty / when she sang from the rooftops / It will be fine, it will be fine" (1080). The reference to "Marianne," the symbolic name for the French Republic, and the expression "ça ira" (it will be fine), which references the song "Ah, ça ira" popular during the French revolution, encode revolutionary sentiment into the play's soundscape. It is a song whose message seems more directed to the French radio audience than the German, Portuguese, and Angolan characters in the play. The song's reference recalls how songs functioned as hidden signals during the civil rights movement, and in anticolonial wars. An impressively direct musical "encoding" occurred during the 1963 Birmingham civil rights demonstrations, where the local gospel show host, the Reverend Erskine Fausch, converted one song into an actual marching order. In order to confuse Bull Connor's police force, which had placed barricades around the city to prevent marchers from organizing, young activists were positioned with portable radios in basements throughout Birmingham. When Fausch played the song "All Men Are Made by God," the marchers emerged en masse and from all different directions (Barlow, *Voice Over*, 207–8). This kind of large-scale mobilization through radio and social networking embodied the promise Williams claimed for radio.

51. Sarduy, *Los matadores*, 1082.

52. Ibid., 1089.

53. Ibid., 1079.

54. In fact, Sarduy's inclusion of music from Goa and Brazil to frame the sequence goes far beyond the limited category of the "black Atlantic." From the middle half of the sixteenth century until 1752, Portuguese colonizers referenced the "Estado da Índia Portuguesa," or the "State of Portuguese India," to include imperial possessions from Southern Africa to Southeast Asia, including Goa and Mozambique. Portuguese colonists also employed African slaves in Goa, thus forcing the current of the "black Atlantic" to run eastward as well. Geographically, Goa and Brazil "frame" Africa, indicating the two directions toward which African slaves were displaced by the Portuguese empire. There is a robust amount of literature in "Atlantic studies," much of it responding to John Gilroy's *The Black Atlantic*. For a critique of Gilroy's frame of reference in defining the black Atlantic, see Goyal, "Theorizing Africa."

55. Whatever bodily history we might want to hear in Holiday's voice should not forget the important role the microphone played in making the intense intimacy of her delivery audible (see Spellman, "Billie Holiday."). For more on the importance of recording in developing singing style, see Katz, *Capturing.*

56. Michael Denning writes that "like 'world music,' the 'world novel' is a category to be distrusted; if it genuinely points to the transformed geography of the novel, it is also a marketing device that flattens distinct regional and linguistic traditions into a single cosmopolitan 'world beat'" ("Novelists," 703).

57. The radio dial's ability to flatten out generic difference reveals the darker side of Peter Hallward's more positive imagining of horizontal relations in Sarduy's work with the voice: "Sarduy explores a univocal regime of 'constant variation', a singular plane of immanence in which 'everything is real' and everything coheres on the same level, the same exclusive (or all-inclusive) scale of intensity" (*Postcolonial*, 254).

58. Sarduy, *Los matadores*, 1077.

59. Ibid.

60. Sarduy, "Juana de Arco," 30.

61. Ibid.

62. Ibid.

63. Ibid. Notably, the voices he identifies here and elsewhere in the essay are voices that come from somewhere else: Tangier, a female African American singer ("her voice was 'whitening' [*blanqueando*] . . . very close to a *degree zero of the voice*"), another female singer from Bahía in Brazil ("the noise of the Earth . . . the sound of my land"), and a journalist from the Canary Islands ("the voice of the islands . . . with its hot and ashy stratus"). The voice's "grain," the material difference or "bumpiness" that derives from tone, accent, and timbre attaches to bodies by means of their ability to carry with them some residue of "earthiness." Perhaps embarrassed by the simple exoticized desire he expresses in listening to these voices, which he names only reluctantly, Sarduy warns his reader that it would be better if "each listener . . . could imagine them [*fantasmarlas*], and embody them [*encarnarlas*] in the body [*el cuerpo*] that they wanted" (31). However, this fantasy only furthers the exotic eros Sarduy attaches to the voice (see Barthes, "Grain").

64. Harkness, *Songs*, 77. The first "is the site in which the phonic production and organization of sound intersect with the sonic uptake and categorization of sound in the world," and the second "describes an alignment to, or taking up of, a kind of perspective or moral stance in respect of semiotic text. It is the expression of an interest in relation to, or explicitly against other interests in the social world" (ibid.).

65. Sarduy, *Los matadores*, 1077.

66. Ibid.

67. Ibid.

68. Qtd. in Lucey, "Never," 15.

69. Pushing Ducrot's point further, Erving Goffman perceives, in Michael Lucey's gloss, that different uses of the first-person pronoun "involve sufficiently different aspects of a same speaker that it could be argued . . . that the referent of the pronoun—even though it is the 'same' pronoun—is not the same" (Goffman qtd. in Lucey, *Never*, 18). However, even this referential flexibility remains part of "interactional discourse" (8), which Sarduy considers "colonial."

70. Birkenmaier, *Carpentier*, 170.

71. Qtd. in Brennan, "Introduction," 27; emphasis added. "[El *speaker*] debe tratarse de deshumanizar su voz en todo lo posible. Dotarlo de una voz neutra, uniforme, que se destaque sobre el conjunto como la de una especie de máquina parlante" (Carpentier, "Nuevas posibilidades," 393).

72. Deharme, "Proposition," 408; emphasis in original.

73. Ibid., 409.

74. Goffman, "Forms."

75. Sarduy, *Los matadores*, 1077.

76. Ibid., 1078.

77. Hallward, *Postcolonial*, 307.

78. Goffman, "Forms," 124–61.

79. Hallward, *Postcolonial*, 257.

80. Sarduy, *Los matadores*, 1080.

81. Ibid., 1081.

82. My argument here departs from "sound symbolism," the denotative function of phonic utterances. A good example of this other use of sound comes from Silverstein, who observes that sonic changes in the Wasco-Wishram Chinookan language can shift the phrase, "I skate on (some location)" to the otherwise structurally identical form "I (go to) squat someplace [sc. For defecation]" ("Wasco," 52). Sound, in this case, changes the referential meaning of the phrase, rather than the nonreferential, but indexical function I hear in *The Ant Killers*.

83. Ibid., 53.

84. Lucey, *Never*, 61.

85. Ibid., 64–66.

86. Sarduy, *Los matadores*, 1081.

87. Ibid., 1084.

88. Ibid., 1088.

89. Ibid., 1092.

90. Ibid., 1084.

91. See McLuhan, *Understanding*, 298–307. In a chapter that largely borrows from Frankfurt school theories of radio, McLuhan describes the radio as "a profound archaic force" (301), "a mighty awakener of archaic memories, forces, and animosities" (306) with "a tribal magic" (302) that allowed the German people, "brooding upon the resonating Africa within" (301), to "dance entranced to the tribal drum of radio that extended their central nervous system to create depth of involvement for everybody" (298).

92. Ibid., 301.

93. Sarduy, *Los matadores*, 1084. In addition to calling to mind the alternative society of male same-sex misogyny depicted in that novel, the reference to Burroughs in this radio play obsessed with the radio recalls his earlier novel *Naked Lunch*. Set in Tangier during the years in which the United States "was protecting the radio transfer stations built in the Tangier Zone (Mackay, RCA, and Voice of America)," *Naked Lunch* develops "the connection between Tangier as international radio relay station, as a site where the voices of the world mingle din static, and as a historical crossroads on which imperial and emergent nations encounter one another in proximity (both geographical and

historical). In his journals, Burroughs writes: 'It is frequently said that the Great Powers will never give up the Interzone because of its value as a listening post. It is in fact the listening post of the world" (Brian Edwards, *Morocco*, 180). Sarduy quotes from *Naked Lunch* in his radio play, *Relato*, published alongside *Los matadores de hormigas* in his collection, *Para la voz*. The link between Burroughs's novel and anticolonial radio becomes clearer when Edwards points out: "Burroughs makes frequent reference in *Naked Lunch* to radio signals, whether to the 1920 crystal set I mentioned before, or to the noise of Cairo stations blaring through Tangier cafés . . . Like Frantz Fanon, whose writing about the ongoing Algerian revolution in the late 1950s brought him to advocate a broken signal and a clear message emanating from jammed radio signals as the true voice that would break France's hold on the Algerian nation, Burroughs saw a ruptured 'voice' in Tangier as a means to disrupt American hypocrisy, both global and domestic" (179).

94. Sarduy, *Cartas*, 121. "Ya me estoy convirtiendo en el autor del Derecho de Nacer francés." This quote appears in a letter from November 1963 where he tells his family that he is in the midst of composing "un sainete radiofónico" (ibid., 104). This radio farce, which he titled "Dolores Rondón," became both a chapter of his first novel *De donde son los cantantes*, and a critical hit on German radio two years later. As the accolades continued, and his more than half-dozen radio works were produced in Germany, France, England, and the United States—although never in Spanish—he wrote back to Havana again. For more on Sarduy's radio play and "Dolores Rondón," see Birkenmaier, "Juana de Arco."

95. Sarduy, *Los matadores*, 1093.

96. Ibid., 1082.

97. Ibid., 1089.

98. Ibid., 1090.

99. Ibid., 1092.

100. Ibid., 1093.

CHAPTER 6

1. Venegas, *Digital Dilemmas* 39.

2. Luis, *Culture* 71.

3. Sirvén, *El rey*; Amato, "Goar Mestre."

4. For more on the event's significance in changing radio's popularity in Argentina, see Ricardo Gallo, *La radio*; Ulanovsky, *Días*; Matallana, "*Locos*," 83; Sarlo, *La imaginación técnica*; and Claxton, *Perón*.

5. The nationalist fervor occurred on both sides of the airwaves. Daniel Fridman and David Sheinin cite Boyden Sparkes of the *New York Tribune*, who "suggested that only a duel between George Washington and Simon Bolívar might have generated equivalent attention . . . As a result of the fight, he went on, Latin Americans would begin to buy 'American automobiles, American drugs, American hardware, [and] American clothing" ("Wild Bulls," 57). The American realist painter George Bellows's depiction of the scene in his 1924 painting *Dempsey and Firpo* used to hang at the entrance to the Whitney

Museum of American Art. For a discussion of the political implications of this decision, see Wood, "Madness."

6. Thinking through the Kittlerian framework of the typewriter as the technology of the "symbolic" and the phonograph and radio as representatives of the "real," we can recognize how Dempsey's collision with the writing machine (*máquina de escribir*) signals the type of immediate contact with the Real desired in the writings of militant political authors in the Americas of the 1960s and 1970s.

7. For one of the early debates about this turn, see Collazos, Cortázar, and Vargas Llosa, *Literatura*.

8. See the aforementioned histories of radio in Argentina for the contentious debates regarding when and where "broadcast" officially began in the world. Spanish vocabulary more easily distinguishes between two types of "radio": "*radiodifusión*," or "broadcast," and *radiotelefonía*, the kind of point-to-point wireless systems developed by Lee De Forest and others earlier in the century.

9. Qtd. in Matallana, "*Locos*," 13–14; emphasis added. It is worth noting the acoustic difference of broadcasting. In the reporter's words, the radio "sounds [*resuena*] clearer than the phonograph, and without the noise [*rumor*] of the needle on the disc." Radio brought with it not only a new way of hearing—"listening without seeing"—but also, in the ears of our reporter, a more refined sound. Yet, given the attention to sound, historians of acoustics will find it strange that radio began broadcasting from the Teatro Coliseo, when Buenos Aires's more famous Teatro Colón, like Boston's Symphony Hall, remains among the top five acoustic spaces in the world (see Long, "Shoebox Halls"). Notably, Long singles out the Teatro Colón as the only one of the top five music halls whose shape, the European shoehorn, distinguishes itself from the shoebox model. Long cites Beranek, *Concert Halls*.

10. Sarlo, "Oralidad," 30.

11. Ibid., 32.

12. Ibid., 33.

13. Ibid., 36.

14. Karush, *Culture of Class*, 10, 17.

15. Matallana, "*Locos*," 35–36.

16. See Matallana for the codevelopment of U.S. and Argentine broadcasting regulations (ibid., 24).

17. Qtd. in Fraga, *La prohibición*, 39.

18. Matallana, "*Locos*," 46.

19. Qtd. ibid., 49.

20. Ibid., 61.

21. Qtd. ibid., 49.

22. Ehrick, *Gendered*, 138. For Puig's interest in Marshall, see S. Levine, *Spider*, 68. Marshall was a particular target of Borges's criticism. As Juan José Sebreli explains: "Borges decía en 'Las alarmas del Dr. América Castro,' (*Otras Inquisiciones*, 1952) refiriéndose a los filólogos 'poseen fonógrafos, mañana trascribirán la voz de Catita [la famosa personaje de Marshall].' En la edición de las *Obras Completas* Borges suprimió este párrafo que fue, no

obstante, predictivo y efectivamente años más tarde la universidad conservará esas grabaciones. Ana María Barrenchea ilustraba sus clases de lingüística con ejemplos de los manierismos de Catita y de Mónica como expresión de dos clases sociales" (n12).

23. S. Levine, *Spider*, 68.

24. Qtd. in S. Levine, *Subversive*, 123–24; emphasis added.

25. Sarlo, "Oralidad," 37.

26. Ibid.

27. In this sense, Puig aligns himself with Borges's contemporary Roberto Arlt, whose central European name, monolingualism, and propensity for reading and incorporating "bad translations" into his writing, along with his keen ear for *lunfardo*, has made him the pillar of "lo popular" in late twentieth- and twenty-first-century Argentine literary culture. Perhaps more than any other writer and critic, Ricardo Piglia has helped immortalize Arlt as a radically nonproprietary writer, most infamously through Piglia's own story collection *Nombre falso*, wherein he passed off one of his own imitations of Arlt as a missing Arlt story. The "falsely attributed" story, "Luba," was first discovered by an academic critic years after the anthology was published, at which time Piglia revealed the story was not an imitation of Arlt at all but rather of the Russian writer Leonid Andreyev's "The Dark," whom Arlt had read in translation. Thus, Piglia challenged the entire concept of authorship as property by revealing Arlt's own writing as an imitation of another (see Speranza, "Ricardo Piglia"). In an analysis of Piglia's fictional character Renzi's analysis of Arlt, Brett Levinson observes: "Arlt, Renzi argues, strives in his writings to produce or reproduce something like an 'Argentine language' . . . [,] to capture the immigrant dialects of Argentina, as well as the lower classes. In seeking to accomplish such a goal Arlt does not recur to the technique employed by the realist or social realist writers. He does not listen to the speech of 'actual immigrants' and then try to duplicate that *jerga* in his fiction. In fact, he tracks down these speech patterns . . . in the poor Spanish translations of European or Russian novels. Arlt's 'wretched writing' is actually the bad translation of those bad translations. It is a doubly imperfect translation that Arlt posits as both the language of immigration and as Argentina's 'national tongue'" (*Ends*, 58).

Piglia, "Luba," *Nombre falso*; McCracken, "Metaplagiarism"; Levinson, *Ends*. Although Puig's writing favors such challenges to property, he himself complained about Arlt's overuse of *lunfardo* and dialect in general.

28. Pauls, *Manuel Puig*, 22.

29. Commenting on Warhol's *a, A Novel* (1968), Gustavus Stadler writes: "Warhol's interest in sound was always fundamentally an interest in broader forms of mediation and, in particular, commercial mass mediation. He was fascinated by the ways sounds grew big by traveling through everyday publicity channels like vinyl records, radio, and the gossip and chatter of celebrity culture. More literally, he embraced the agglutination of individual particles of human sound—especially words and voices—into an aural mass (Tavel again: '[I]t was human sound itself he wanted rather than any particular words.')" ("My Wife," 429).

30. S. Levine, *Spider* 143; Kolesnicov, "Primeras joyas."

31. Castelli, "Entrevista," 22.

32. Suzanne Levine notes McCullers's influence on Puig when she writes that "motivated by the film of a play he had seen, starring a young new actress named Julie Harris, he read Carson McCullers's *A Member of the Wedding*. McCullers's autobiographical treatment of a young girl's coming-of-age in a small Southern town and her vivid use of regional speech brought to life the story of an eccentric 'misfit' with humor and compassion. The literary American South had produced this play, Manuel wrote Malé, 'about a twelve-year-old girl, but it can be enjoyed just the same by those who are allergic to stories about children.' He would soon begin the 'failed' screenplay that became his own autobiography of childhood, *Betrayed by Rita Hayworth*" (*Spider*, 121). See also the December 3, 1959, letter from Puig in Puig, *Mi querida* 195. In a 1976 interview, F. B. Claire notes that Puig has Frank McShane's *The Life of Raymond Chandler* on his bookshelf, and Puig responds, "Suspense makes the reader become a detective, a super-active reader" (*City 5* [Winter 1976–77]). Suzanne Levine also notes Chandler's influence on Puig (*Spider*, 70).

33. Pauls says gossip is "anti-juridical" and "against *property*" because it is "impossible to attribute it to an enunciator, an author. Gossip doesn't tolerate *copyrights* . . . We have to stop talking about gossip as if it had a subject: there is only a *carrier* [portador] (as we say that someone carries a virus)" (Pauls, *Manuel Puig*, 50).

34. Chad Bennett has argued that in her play *Ladies' Voices*, "Stein positions gossip as a style of listening" ("Ladies' Voices," 319).

35. Pauls, *Manuel Puig*, 86. Francine Masiello makes a different point about Puig's work with and against property. For Masiello, Puig's writing focuses on questions of sexual identity, spectacle, and monetary exchange related to the Chicago school's introduction of neoliberal economic policies in Latin America: "Much like the play of words suggested, in Spanish, by the cognates *invertido* and *inversión*, Puig links the interplay between 'deviant' sexuality and money. Invertir: to turn outside in or to clothe or envelop. Inversión: an economic practice whereby one hopes that an original sum will expand to yield a greater return . . . [T]he play of *inverted* identities becomes Puig's *investment* in a form of social exchange" (*Transition*, 84–85).

36. "For the first part of my life, I had been submerged in the reality of the first Peronist regime, and later on in that of the second and third" (Puig and Corbatta interview, 169).

37. Puig, who participated in the translations of his own works, paid detailed attention to the epigraphs in the English-language version of *Boquitas pintadas*, translated as *Heartbreak Tango*. As his main English translator and biographer Suzanne Jill Levine recounts: "Puig's vast knowledge of North American mass culture was invaluable to our creative collaboration. The solution we finally came up with was to translate some tango lyrics that were *essential* to the plot (I'll soon explain), but to replace at least half of the epigraph quotations with either tag lines from Hollywood films or Argentine radio commercials, originally borrowed from Madison Avenue inventions. That is, artifacts relevant to the original context but that rang a funny, familiar, exaggerated bell for American readers . . . These substitutions added a

new dimension of interpretation, or more precisely, they emphasized certain elements in the text that had been more implied than explicit. The highlighted advertisements as in the epigraph to episode II underscore Puig's implied comment on the commercialization of culture as well as his intent, in writing the books he writes, to show how popular media have dissipated traditional distinctions between consumerism and art" (S. Levine, *Subversive*, 128).

38. "In Spanish American popular music there are two genres that have crossed national boundaries and become continuing sources of common myths and emotional expression about relationships: the tango and the bolero. Of the two, the bolero has been identified with romantic feelings, lyricism, and persistent hope in love" (Campos, "Poetics," 637). More particularly, the bolero traveled to Argentina because of its popularization on Mexican radio in the 1930s, and its boom in Mexican cinema in the 1940s. Mexican cinema's own rise resulted, in part, from Argentina's weakened film industry during the 1940s, which, as I explained in chapter 3, came about through tensions with the United States over Argentina's refusal to accept the political and economic program of the New Deal.

39. McCracken, "Heartbreak," 30.

40. Puig, *Boquitas*, 78. All quotes are from Manuel Puig, *Boquitas pintadas* (Barcelona: Seix Barral, 1972).

41. Ibid., 163.

42. Ibid., 170–71.

43. Ibid., 171.

44. Ibid., 174.

45. Ibid., 132–33.

46. Ortíz , *Eva Perón*, 43.

47. Monegal, "Los sueños." For a more recent, and archivally inclined analysis of the connections between Eva Duarte's radio dramas and her political radio speeches, see Ehrick, "Savage."

48. See Monegal and Sebreli, *Comediantes*; and Ehrick, *Gendered*.

49. Ehrick, *Gendered* 124; Karush, *Culture of Class*.

50. Ehrick, *Gendered,* 118–20. "Evita's dreams were the materialization—beneath the form of a voice that performed pleasure and pain on the radio—of the dreams of all the *cabecitas negras*, of all those women from the countryside, who lived in a desert, only relieved by the radio voice . . . that they (like Evita) also dreamed to become one day" (Monegal, "Los sueños," 34).

51. Qtd. in Monegal, "Los sueños," 34.

52. Ehrick, *Gendered*, 122–23.

53. Ibid., 122; Cramer, "Word War."

54. For more on Marinetti's writing and the radio, see Campbell, *Wireless*.

55. Thus, Puig does not repeat the anti-Peronist politics of Cortázar's "Casa Tomada" (House Taken Over), in which a wealthy brother and sister are driven from their home by mysterious sounds, often read by critics as the populist overthrow of the bourgeois.

56. Piglia, *Pedazos*, 116.

57. "Puig's work is the obsessive and systematic unfolding of the same single question: How do we live together? (Which is literally equivalent to: How can we be contemporary?)" (Link, *Ñ*, 9).

58. Giordano, *Manuel Puig*, 65

59. Goldchluk, "El archivo Puig," 10–11.

60. Danto, *Transfiguration*.

61. Romero, "Las traiciones," 16.

62. Amícola, *Camp*, 89. "Puig's *La traición* is a question of place: it situates itself in the middle road between the trivial and the consecrated space of *high* literature" (Romero, "Las traiciones," 16).

63. Speranza, "Ricardo Piglia," 218. Speranza identifies Puig's existence as between social worlds: "The always double experience of that limbo in which he lives (two languages, two cultures), the movement between two realities equally lived (inside and outside of the dark room) and the shift between two sexual identities (inside and outside of the family) pressurize the opposition and encourage a new aesthetic form capable of suturing them" (ibid., 212). Logie and Romero write: "Graciela Speranza has described that place, neither peripheral, nor central occupied by Puig, in the following terms: Puig goes beyond narrative modernization in Latin America and thus breaks with a long tradition with respect to the literatures of the center" ("Extranjeridad," 33).

64. Link, *Cómo se lee*, 268. In the weekly cultural supplement Ñ dedicated to the twentieth anniversary of Puig's death, Link writes of "an admirable equilibrium" in the influence of Puig's characters on the author or viceversa (see "La obra").

65. Pauls, *Manuel Puig*, 56–57.

66. Ludmer, *El cuerpo*, 133.

67. Sarduy, "Notas," 556, 557.

68. Ibid., 556. I have consulted Suzanne Jill Levine's translation of Sarduy's essay here (see Sarduy, "Note to the Notes to the Notes . . . A Propos of Manuel Puig," trans. Levine).

69. Sarduy, "Notas," 565.

70. Ibid., 567. Sarduy writes that the use of the *radionovela* produces a sense of "being-in-the-past. The action in *Boquitas pintadas* takes place ten years before it is recounted by its own characters; the events in the radio soap opera 'occur' twenty years before the soap opera" (566).

71. Qtd. in Puig and Corbatta interview, 168.

72. Qtd. in S. Levine, *Subversive*, 124; emphasis added.

73. Puig, *Boquitas*, 194. Later in the scene Mabel asks Nené's opinion of the program, and the latter criticizes the actress: "Nené was afraid to praise the actress's work. She remembered that her friend didn't like Argentine actresses. 'Why no, she's very good, I like her.' Mabel replied, remembering that Nené never did know how to judge movies, theater, and radio" (199). In this way, the aesthetics of listening enter into the social pragmatics of organizing the power relations in the room.

74. Ibid., 195 (both quotes).

75. Ibid., 196–97.

76. Brian Edwards draws on Deleuze and Guattari's notion of "deterritorialization" to demonstrate how the radio's mixed signals lead to a mixed linguistic "signal" in Fanon's writing that undoes French as a single, national language (*Fanon's al-Jazáir*).

77. Sartre, "A Propos," 93.

78. Auerbach, *Mimesis*, 12, 13.

79. Bahktin, *Dialogic*, 17.

80. These forms of "epic" or "mythic" mimesis are not equivalent with Adorno and Horkheimer's notion of "myth and enlightenment" in *The Dialectic of Enlightenment.* However, their use of the Odyssean resistance to the Sirens through a self-imposed temporary deafness, and its relation to their opposition to radio in their chapter "The Culture Industry: Enlightenment as Mass Deception," contrasts with the aesthetic I find in Puig. Indeed, Puig's writing approaches the extreme limits of Adorno's "critical mimesis," which focuses on mimesis as a relation that acknowledges difference rather than an identification between subject and object that would transform representation into a relationship similar to exchange value. Puig's later novels attempt to renegotiate representation by turning the apparent object, his interviewee, into the subject, thereby undoing the mimetic categories.

81. Speranza, *Fin*, 74,

82. "Just as in the work of pop art, through the apparent anonymity and minimal intervention, Puig achieves the effect of a deceiving impersonality. The sentimental and emotional outbursts of his characters openly contrast with the objectivity—the distance—of the technique of reproduction. The original discourse transforms itself into art without losing the objective property that links it to daily experience and without producing, therefore, aesthetic or moral judgments" (*Fin*, 114). Sarlo, "El brillo."

83. Speranza, *Fin*, 96. Speranza quotes Lucy Lippard, who writes: "In the '60s, male artists entered feminine spaces and pillaged them without punishment. The result was Pop Art, the most popular art movement in U.S. history . . . If the pop artists had been women, the movement never would have left the kitchen" (*Fin*, 93).

84. Ibid., 108.

85. Speranza writes that Puig's novels "offer fragments of a lover's discourse that work like interpretative keys with which characters can communicate affective relationships" (ibid., 106).

86. "Nos encontrábamos dos horas cada vez. El no quería grabadora, así que todo lo que iba diciendo yo lo escribía a máquina. El diálogo en el que yo me ponía en el lugar del Viejo, es la novela" (Castelli, "Entrevista," 25).

87. "Puig ficcionaliza lo testimonial y borra sus huellas" (Piglia, *Pedazos*, 116).

88. Sarlo, *Tiempo*, 62.

89. Indeed, Walter Benjamin's distinction between *Erlebnis* and *Erfahrung*, most particularly in his essay "On Some Motifs in Baudelaire," remains one of the key categorical issues in approaching questions of narrative representation or aesthetic work of any kind. For more on the politics of experience in a different context, see J. Scott, "Evidence."

90. Moreno, "Doble casetera," 3.

91. Beverley, "Margin."

92. Speranza, *Fin*, 112.

93. Moreno adds, "the corrections to the transcripts that I could see are typos and the transcriptions were done by professionals" (3).

94. Moreno, "Doble casetera," 1.
95. S. Levine, *Spider*, 294.
96. Speranza, *Fin*, 31.
97. Puig, *Curse* 3.
98. Logie and Romero, "Extranjeridad," 41.
99. Masiello, *Transition*, 87; emphasis added.
100. For more on the comparison between literary and sonic fidelity, see McEnaney, "Realismo."
101. Puig, *Curse*, 15.
102. As Daniel Link comments: "[I]f certain genres like the interview fictionalize a conversation among socially significant interlocutors . . . in Puig we attend the spectacle of always insignificant speakers." As opposed to Chandler's stories, in which "important characters in the story make significant utterances," Puig recognizes that "conversation . . . is a thematic and discursive drift: speech without a reason and a specific destination; life itself . . . The most scandalous thing about Puig's writing is the rigorous stubbornness with which he constructs a literature from conversational waste" (*Cómo se lee*, 267–68).
103. Puig, *Curse*, 29.
104. Ibid., 32.
105. Ibid., 83.
106. Ibid., 60.
107. Piglia, "Writing and Translation," 64.
108. Puig, *Curse*, 190.
109. S. Levine, *Subversive*, 299.

CHAPTER 7

1. Puig, interview by Wheaton.
2. Moreno, "Doble casetera."
3. Puig, interview by Wheaton.
4. Stadler, "My Wife."
5. Costa and Perreault, "Introduction."
6. Moreno, "Doble casetera."
7. González-Echevarría, *Biografía*, 110.
8. Piglia, *La ciudad ausente*, 145. While all translations here are my own, I have consulted Sergio Waisman's excellent translation, *The Absent City*.
9. See Guevara, *Guerrillas*, 137–38. See also Fanon, "La voix." See Soley and Nichols, *Clandestine*; and Soley, *Free Radio*.
10. Matthew Fuller comments: "Any radio receiver can, with a modicum of fiddling, be converted to a transmitter . . . The inverse correlation between loudspeaker and microphone continues this existence of the transmitter as mirror-world of the receiver" (*Media Ecologies*, 25).
11. Allusions to and citations from Brecht abound throughout Piglia's work, most obviously in the epigraph to the latter's novel *Plata quemada* [Money to Burn], where he quotes Brecht: "What is robbing a bank compared to founding one?" (¿Qué es robar un banco comparado con fundarlo?).
12. Brecht, "Radio," 42 (all quotes in this paragraph).

13. To take one popular example, see "Here Comes Everybuddy," Andy Merrifield's critical response to Clay Shirky's new media utopianism (*Here Comes Everybody*); and Malcolm Gladwell's critique of "weak-tie" online activism. For more academic studies of the relationship between old and new media, see Gitelman and Pingree, *New Media*; Chun, Fisher, and Keenan, *New Media, Old Media*; and Marvin, *Old Technologies*.

14. For the most influential recent critiques of the testimonial genre, see Sarlo, *Tiempo*; and John Beverley's response to Sarlo's work in Beverley, *Latinamericanism after 9/11*.

15. Barthes, "Listening," 259.

16. For more on hard-boiled fiction in Argentina, see Ludmer, *El cuerpo*.

17. For more on the role of overhearing in the novel, see Gaylin, *Eavesdropping*.

18. Williams, *Other Side*, 159.

19. Piglia, *Absent City*, 11.

20. For more on the Argentine civil catastrophe known as the "Dirty War" (1976–83), see Daniel Balderston et al., *Ficción y política*; Feitlowitz, *Lexicon*; and Corradi, Fargen, and Garretón, *Fear at the Edge*.

21. See Adorno and Horkheimer's, "The Culture Industry: Enlightenment as Mass Deception," in *Dialectic*. Adorno, *Current*; Brecht, *On Film*; McLuhan, "Radio: The Tribal Drum," in *Understanding Media*; Kaplan, *Reproductions*; Apostolidis, *Stations*; and Campbell, *Wireless*. While the dominant reading of radio's political, historical, and cultural impact continues to be derived largely in reference to fascism following the early Frankfurt school theory, a number of volumes have begun to widen the scope of the theoretical understanding of radio. See, in particular, Hilmes, *Radio Voices*; Hilmes and Loviglio, *The Radio Reader*; Rubén Gallo, *Mexican Modernity* 117–68; Birkenmaier, *Carpentier*; the articles and translations in the special issue of Modernism/*modernity* 16, no. 2 (2009); and Verma, *Theater*.

22. Hollier, "Death," 19.

23. Kittler, *Gramophone*.

24. Hanrahan, "Index."

25. See Rozitchner, *Vías Argentinas*; and Misemer, *Moving Forward*.

26. Piglia, *Absent City*, 11–12.

27. Piglia, *Absent City*, 12.

28. Hirschkind, *Ethical*, 19

29. Avelar, *Untimely*.

30. Adorno and Horkheimer comment that under the culture industry in the United States "the radio becomes the universal mouthpiece of the Führer; his voice rises from street loud-speakers to resemble the howling of sirens announcing panic—from which modern propaganda can scarcely be distinguished anyway. . . . The gigantic fact that the speech penetrates everywhere replaces its content. . . . No listener can grasp its true meaning any longer, while the Führer's speech is lies anyway. The inherent tendency of radio is to make the speaker's word, the false commandment, absolute" (*Dialectic*, 159).

31. Mowitt, "Breaking Up"; Edwards, "Fanon's al-Jazáir."

32. See Donghi, *Peronist*; Lejtman, *Perón vuelve*; and Turner and Miguens, *Reshaping*.

33. Piglia, *Absent City*, 101.

34. I borrow the term "remediation" from Jay David Bolter and Richard Grusin's book *Remediation: Understanding New Media* (1999). As their subtitle implies, Bolter and Grusin themselves borrow the logic of remediation from McLuhan's *Understanding Media*, where the latter already presents the idea that every medium takes another medium as its content. Kittler's term for a similar process is "medial ecology." Citing Kittler, N. Katherine Hayles remarks, "Like the phonograph, audiotape was a technology of inscription, but with the crucial difference that it permitted erasure and rewriting" (*Posthuman*, 209). Piglia explicitly incorporates the logic of remediation into his character Emil Russo's explanation to Junior about the origins of the storytelling machine: "To invent a machine is easy, as long as you can modify the pieces of an earlier mechanism. The possibilities of converting something that already exists into something else are infinite" (*Absent City*, 140). The connection between the machine and radio is also implied elsewhere in the narration: "[Macedonio] thought about perfecting the 'apparatus' (that's what he called it), with the intention of entertaining the peasants in the villages. It seems to me a more entertaining invention than the radio, he said, but it's still too early to declare victory" (42). Gitelman, *Scripts*; Sterne, *Audible Past*.

35. Piglia, *Absent City*, 73.

36. "Mac," referenced throughout the novel, carries multiple, overlapping signifieds. While the name is short for Macedonio Fernández, it also obviously references the Macintosh computer, most famous in the year of the novel's publication for Ridley Scott's *Blade Runner*–inspired television commercial in 1984, as well as, in a more literary vein, "the man in the Macintosh," the mysterious character from Joyce's *Ulysses*. (This last reference would seem more obscure if not for the fact that Piglia devotes an entire chapter in *The Absent City* to describing a utopian community whose bible is *Finnegans Wake*.) Furthermore, the man Elena identifies as Mac when she emerges from an underground tunnel into an electronics repair shop seems to reference not only Macedonio or Russo, the machine's two inventors, but also Rodolfo Walsh, Piglia's esteemed forebear in Argentine political detective fiction, testimonial literature, and underground political activism. Walsh, as the Argentine author and journalist María Moreno told me, was an amateur electronics repairman who converted his television into a radio receptor to tune into the police's radio communications during the Dirty War. A powerful voice of the leftist rebellion against the dictatorship, Walsh was murdered in 1977, a day after publishing his "Carta abierta a la junta militar" (Open Letter to the Military Dictatorship) (María Moreno, in conversation with the author, August, 5, 2010).

37. Piglia, *Absent City*, 71; emphasis added.

38. Ibid., 65. Elena's forgetting here resonates with another, less overtly political scene of forgetting associated with Elena's husband, Macedonio, the cocreator of the cyborg that houses her consciousness: "He had discovered the existence of the verbal nuclei that keep remembrances alive, words they had

used that brought all the pain back into his memory. He was removing them from his vocabulary, trying to suppress them, and establish a private language without any memories attached to it. A personal language, without memories, he wrote and spoke English and German, so he would mix the languages, in order to avoid even grazing the skin of the words he had used with Elena" (148).

39. Sartre, *Dialectical,* vol. 1.

40. Considering Piglia's turn to the free market as a model for reimagining radio collectivities, it is worth noting Frederic Jameson's comment in the introduction to the first volume of Sartre's *Critique of Dialectical Reason*: "Certainly seriality also knows its paradoxes: it is indeed no accident that Sartre's principal extended illustrations are those of radio broadcasts and of the free market. In fact, I believe, the notion of seriality developed here is the only philosophically satisfactory theory of public opinion, the only genuine philosophy of media, that anyone has proposed to date: something that can be judged by its evasion of the conceptual traps of collective consciousness on the one hand, and of behaviorism or manipulation on the other. If this were all Sartre had managed to do in the *Critique*, it would already mark a considerable achievement (*neglect of which may reflect the turn of current media theory towards a problematic of the image* as much as anything else)" (xxviii; emphasis added).

41. Piglia, *Absent City*, 75.

42. Ibid.

43. I borrow the term "footing" from Erving Goffman's sociolinguistic analysis, *Forms of Talk*.

44. For more on the relation between Piglia's fiction and economic forces, see Levinson, *Ends*.

45. Piglia, *Absent City*, 166, 80.

46. In Argentina, and in Piglia's work especially, "plata" points to multiple meanings. Literally "silver," via metonymy the word indicates money in general. However, in Buenos Aires, located on the Río de la Plata, an entire linguistic culture of "castellano rioplatense" exists. In his subsequent novel, *Plata quemada* (Burnt Money; 1997), Piglia makes use of "plata" to signify both the literal burning of stolen money, and the destruction and corruption of the *rioplatense* region.

47. Piglia, *Absent City*, 166.

48. What Colleen Lye has identified as "Asiatic racial form" in the context of fiction in the United States seems to function in Piglia's novel as Korean and Japanese characters are reified into a stereotype of "Asianness," scapegoated as the sole causal force behind global capital as well as a threatening inhumanness or mechanization of thought that Argentine political insurgents identify as both the origin of an unthinking adherence to state discipline and, paradoxically, the means by which to resist the nation's police state. In the novel's conclusion, Russo tells Junior: "The Japanese model of feudal suicide, with its paranoid courtesy and its Zen conformity, was Macedonio's main enemy. They make electronic devices [*aparatos*] and electronic personalities and electronic fictions, and in all the States of the world there is a Japanese

brain that gives orders" (142). And yet the narrator describes "[a] Korean man, Tank Fuyita, [who] has been the keeper and the guard for years. He came with the second generation of immigrants, smugglers of cheap watches on the free market. They wore the watches on their arms, ten or twelve Japanese watches, and spoke in their Oriental whispers" (166). Fuyita, commodity smuggler and museum guard, two activities that help continue the rebels' diffusion of narrative history against the State's desire for total erasure, also rescues Elena and the Tano in "The White Nodes" by bringing them into the museum from the streets, where Elena's narration reads: "The more criminal activity found among Asian refugees, the more Asian refugees that the police must recruit as informants. Insanity of resemblance is the law, the Tano thought. To look alike in order to survive" (77). Thus, the same rule of simulacra that allows the insurgents to escape the State through their exchange of identities is modeled on the racist logic of identifying all Others as the Same. The novel, therefore, desires to exclude Asianness as a category against which to define the insurgent subject, and yet this subject finds itself parasitic of this same model. In this, however, the novel merely follows the "techno-orientalism" that makes up the generic rules of the cyberpunk literature of William Gibson and Philip K. Dick that it borrows from throughout "The White Nodes" chapter. As Stephen Hong Sohn comments: "David Morley and Kevin Robbins assert that cyberpunk representations embody a kind of techno-Orientalism. They note: '[w]ithin the political and cultural unconscious of the West, Japan has come to exist as the figure of empty and dehumanized technological power. It represents the alienated and dystopian image of capitalist progress. This provokes both resentment and envy. The Japanese are unfeeling aliens; they are cyborgs and replicants' (170). Morley and Robins suggest that this offshoot of Saidian Orientalism manifests through ambivalence due to both a desire to denigrate the unfeeling, automaton-like Alien/Asian and an envy that derives from the West's desire to regain primacy within the global economy" ("Introduction," 7). Piglia's "techno-Orientalism," as well as his more Saidian Orientalist invocation of Scheherazade, is at least as important as Joyce's *Finnegans Wake* and Fernández's *Museo de la novela eterna* (Museum of the Eternal Novel) in helping him develop a possible outer limit to the governmental censorship and surveillance against which he writes (see Lye, *America's Asia*; "Racial Form"; and "Afro-Asian"; see also Sohn, "Introduction").

49. Piglia's technique undoubtedly borrows from William Burroughs's *Naked Lunch* (1959), itself an experiment in cutting up tape recordings and typewritten pages and then pasting them back together at random to create a narrative order beyond the author's control. The connection becomes obvious when, explaining the identity exchanges of the underground, the narrator cites Burroughs's novel without acknowledging it: "In New York half of the addicts are detectives" (76–77). I mention the allusion here because of the way in which Burroughs's work with identity exchange and the cut-up technique relates to radio in particular. See, for instance, Brian Edwards's chapter on Burroughs in *Morocco Bound*: "Steven Shaviro has suggested that one of the 'lessons' Burroughs teaches his readers is that language is a virus . . . Language's function in this frame is not to communicate but to replicate itself.

From this Shaviro derives the observation that individuals/authors do not try to express a 'self' but rather speak in tongues, channel messages as the radio does" (181).

50. The political ambition in this network is to follow the model of the rebel contact Grete Müller, another inhabitant of the "Korean sector" who, realizing her consciousness "had been infiltrated . . . buried her past and adopted a fictitious one" (75). This expedient maneuver for political purposes then leads to an aesthetic ideal in the production of photographs that reveal her ability to "look at the world with eyes that weren't her own" (75).

51. Piglia, *Absent City*, 54. The rush of female literary figures here includes Amalia, the eponymous character of the Argentine novelist José Marmol's 1851 novel about Juan Manuel de Rosas's police state, Molly Bloom from *Ulysses* (1922), Hipólita from Roberto Arlt's 1929 anarchist and revolutionary novel *The Seven Madmen* (Los siete locos), and Temple Drake from William Faulkner's 1931 novel *Sanctuary*.

52. Piglia, *Pedazos*, 115–16.

53. In her article "Rousseau's *Emile ou de l'education*: A Resistance to Reading," Janie Vanpée writes that "Rousseau's masterpiece . . . gives priority to one of the most fundamental problems of all pedagogical discourses—the problem of transmission, and in particular, of their own transmission" (156). Piglia's Emil Russo places the failure of transmission as the basis for his resistance to the censorious state against which he and the other rebels strive.

54. Hirschkind, *Ethical*, 55.

55. Ibid., 50–51.

56. Ibid., 27.

57. Piglia, *Absent City*, 42.This quotation echoes Benjamin's investment in "experience" (*Erfahrung*) as the governing problem that links technology and storytelling.

58. Ibid., 142.

59. Ibid., 57.

60. Rancière, *Schoolmaster*, 2; emphasis added. The similarities between Rancière's thought, Hirschkind's reading of Benjamin, and my own analysis of the storytelling cyborg in Piglia's novel derive from an insistence on the democratic quality of storytelling. As Kristen Ross relates in her introduction to her translation of Rancière's *Le maître ignorant*: "Storytelling then, in and of itself, or *recounting*—one of the two basic operations of the intelligence according to Jacotot—emerges as one of the concrete acts or practices that verifies equality. (Equality, writes Jacotot, 'is neither given nor claimed, it is practiced, it is *verified*.') The very act of storytelling, an act that presumes in its interlocutor an equality of intelligence rather than an inequality of knowledge, posits equality, just as the act of explication posits inequality" (ibid., xxii). If we read this lesson back into *The Absent City*, the rebels' listening network might be seen to represent an especially radical commitment to equality.

Works Cited

Abrams, Charles. *Race Bias in Housing*. American Civil Liberties Union, National Association for the Advancement of Colored People, American Council on Race Relations, 1947.

Adams, J. Donald. "John Dos Passos Pictures the Boom Years in *The Big Money*." *New York Times Book Review*. August 16, 1936.

Adorno, Theodor. *Current of Music*. Frankfurt am Main: Suhrkamp, 2006.

———. "The Radio Symphony: An Experiment in Theory." In *Radio Research 1941*, edited by Paul F. Lazarsfeld and Frank N. Stanton, 110–39. New York: Duell, Sloan and Pearce, 1941.

Adorno, Theodor, and Max Horkheimer. *The Dialectic of Enlightenment*. New York: Continuum, 1999.

Agha, Asif. "Voice, Footing, Enregisterment." *Journal of Linguistic Anthropology* 15, no. 1, (2005): 38–59

Allen, Mike, and Jonathan Martin. "Obama Unveils 21st Century New Deal." *Politico*, December 6, 2008.

Amato, Alberto. "Goar Mestre, padre de la TV latinoamericana, dos veces despojado." *Clarín*, January 19, 2014.

Amícola, José. *Camp y posvanguardia: Manifestaciones culturales de un siglo fenecido*. Buenos Aires: Paidós, 2000.

Anderson, A. Scott. *Housing and the Democratic Ideal: The Life and Thought of Charles Abrams*. New York: Columbia University Press, 2000.

Anderson, Benedict. *Imagined Communities: Reflections on the Origins and Spread of Nationalism*. London: Verso, 1983.

Anderson, Jon. *Che Guevara: A Revolutionary Life*. New York: Grove, 1997.

Anderson, Thomas F. *Carnival and National Identity in the Poetry of Afrocubanismo*. Gainesville: University Press of Florida, 2011.

Apostolidis, Paul. *Stations of the Cross: Adorno and Christian Right Radio.* Durham: Duke University Press, 2000.

Apter, Emily S. "Balkan Babel: Translation Zones, Military Zones." *Public Culture* 13, no. 1 (Winter 2001): 65–80.

———. *Feminizing the Fetish: Psychoanalysis and Narrative Obsession in Turn-of-the-Century France.* Ithaca: Cornell University Press, 1991.

Argote-Freyre, Frank. *Fulgencio Batista: The Making of a Dictator.* New Brunswick, N.J.: Rutgers University Press, 2006.

Arnheim, Rudolf. *Radio.* London: Faber and Faber, 1936.

Arrosagaray, Enrique. *Rodolfo Walsh en Cuba: Agencia Prensa Latina, milicia, ron, y criptografía.* Buenos Aires: Catálogos, 2004.

Auerbach, Eric. *Mimesis.* Translated by Willard R. Trask. Princeton: Princeton University Press, 1953.

Avelar, Idelber. *The Untimely Present.* Durham: Duke University Press, 1999.

Bachelard, Gaston. "Radio et Rêverie." In *Le doit de rêver.* Paris: Presses universitaires de France, 1970.

Bakhtin, Mikhail. *The Dialogic Imagination.* Translated by Caryl Emerson and Michael Holquist; edited by Holquist. Austin: University of Texas Press, 1981.

Balderston, Daniel, et al., *Ficción y política: La narrative argentina durante el proceso militar.* Buenos Aires: Alianza, 1987.

"Bandera de Pirátas." *Revolución* (Havana, Cuba), August 27, 1963.

Banfield, Ann. *Unspeakable Sentences.* Boston: Routledge and Kegan Paul, 1982.

Barlow, William. *Voice Over: The Making of Black Radio.* Philadelphia: Temple University Press, 1999.

Barnard, Rita. *The Great Depression and the Culture of Abundance: Kenneth Fearing, Nathanael West, and Mass Culture in the 1930s.* New York: Cambridge University Press, 1995.

Barnouw, Erik. *A Tower in Babel: Broadcasting in the United States to 1933, Vol. 1.* Oxford: Oxford University Press, 1966.

Barstow, David. "Two Inquiries Set on Pentagon Publicity Effort." *New York Times,* May 24, 2008. Web.

Barthes, Roland. "The Grain of the Voice." In *Image-Music-Text.* Translated by Stephen Heath. New York: Hill and Wang, 1977.

———. "Listening." In *The Responsibility of Forms.* Berkeley: University of California Press, 1991.

———. *The Rustle of Language.* Translated by Richard Howard. Berkeley: University of California Press, 1989.

Barthes, Roland, and Frédéric Berthet. "Présentation." *Communications* 30 (1979): 2–5.

Batiste, Stephanie. *Darkening Mirrors: Imperial Representation in Depression-Era African American Performance.* Durham: Duke University Press, 2012.

Bazin, André. "The Ontology of the Photographic Image" (1945). Translated by Hugh Gray. *Film Quarterly* 13, no. 4. (Summer 1960): 4–9.

Beal, Wesley. *Networks of Modernism: Reorganizing American Narrative*. Iowa City: Iowa University Press, 2015.

Beauvoir, Simone de. *The Prime of Life*. 1976. Boston: Marlowe, 1992.

Benamou, Catherine. *It's All True: Orson Welles's Pan-American Odyssey*. Berkeley: University of California Press, 2007.

Benjamin, Walter. *Illuminations*. Translated by Harry Zohn. New York: Schocken, 1969.

Bennett, Chad. "Ladies' Voices Give Pleasure: Gossip, Drama and Gertrude Stein." *Modern Drama* 53, no. 3 (Fall 2010): 311–31.

Beranek, Leo. *Concert Halls: How They Sound*. Woodbury, N.Y.: Published for the Acoustical Society of America through the American Institute of Physics, 1996.

Bergeron, Katherine. "A Bugle, a Bell, a Stroke of the Tongue: Rethinking Music in Modern French Verse." *Representations* 86 (Spring 2014): 53–72.

Bergreen, Laurence. *Look Now, Pay Later: The Rise of Network Broadcasting*. New York: Doubleday, 1980.

Berlant, Lauren. *The Female Complaint*. Durham: Duke University Press, 2008.

———. *The Queen of America Goes to Washington City: Essays on Sex and Citizenship*. Durham: Duke University Press, 1997.

Bernstein, Susan. *Housing Problems*. Stanford: Stanford University Press, 2008.

Best, Stephen. *The Fugitive's Properties: Law and the Poetics of Possession*. Chicago: University of Chicago Press, 2004.

Beverley, John. *Latinamericanism after 9/11*. Durham: Duke University Press, 2011.

———. "The Margin at the Center: On Testimonio (Testimonial Narrative)." In *The Real Thing: Testimonial Discourses and Latin America*, edited by Georg M. Gugelberger, 23–41. Durham: Duke University Press, 2001.

Beverly, William. *On the Lam: Narratives of Flight in J. Edgar Hoover's America*. Jackson: University Press of Mississippi, 2003.

Biers, Katherine. "Syncope Fever: James Weldon Johnson and the Black Phonographic Voice." *Representations* 96 (Fall 2006): 99–125.

Birkenmaier, Anke. *Alejo Carpentier y la cultura del surrealismo en América Latina*. Madrid: Iberoamericana, 2006.

———. "'Soy una Sor Juana de Arco electrónica': Severo Sarduy's Radio Play 'Dolores Rondón.'" *La Habana Elegante* 57 (November 2015).

Bolter, Jay David, and Richard Grusin. *Remediation: Understanding New Media*. Cambridge: MIT Press, 1999.

Borges, Jorge Luis. "The Argentine Writer and Tradition." In *Labyrinths*, translated by Donald A. Yates. New York: New Directions, 1962.

———. "On Dubbing." In *Borges in/and/on Film*, edited by Edgardo Cozarinsky; translated by Gloria Waldman and Ronald Christ. New York: Lumen, 1981.

Bourdieu, Pierre. *Les règles de l'art*. Paris: Seuil, 1998.

Brecht, Bertolt. "Brecht against Lukács." *Aesthetics and Politics*. London: Verso, 1980.

———. *On Film and Radio*. London: Methuen, 2000.

———. "The Radio as a Communication Apparatus." In *Radiotext(e)*, 15–17. New York: Autonomedia, 1993.

Brennan, Timothy. "Introduction to the English Edition." In *Music in Cuba*, by Alejo Carpentier, translated by Alan West-Durán, 1–58. Minneapolis: University of Minnesota Press, 2001.

Brinkley, Alan. *Franklin Delano Roosevelt*. Oxford: Oxford University Press, 2010.

———. *Voices of Protest: Huey Long, Father Coughlin and the Great Depression*. New York: Knopf, 1982.

Bronfman, Alejandra. "Weapons of the Geek: Romantic Narratives, Sonic Technologies, and Tinkerers in 1930s Santiago, Cuba." In *Media, Sound and Culture in Latin America and the Caribbean*, edited by Bronfman and Andrew Grant Wood. Pittsburgh: University of Pittsburgh Press, 2012.

Brown, Bill. *The Material Unconscious*. Cambridge: Harvard University Press, 1996.

———. *A Sense of Things*. Chicago: University of Chicago Press, 2003.

Brunette, Peter. "Two Wrights, One Wrong." In *The Modern American Novel and the Movies*, edited by Gerald Peary and Roger Shatzkin, 131–42. New York: Ungar, 1978.

Burks, Ruth Elizabeth. "Cinematic Censorship and the Film Adaptations of Richard Wright's Native Son." Ph.D. diss., University of California, Los Angeles, 1993.

Butts, J. J. "New World A-Coming: African American Documentary Intertexts of the Federal Writers' Project." *African American Review* 44, no. 4 (Winter 2011): 649–66.

———. "Writing Projects: New Deal Guidebooks, Community, and Housing Reform in New York City." *The Space Between* 2, no. 1 (2006): 113–38.

Caignet, Félix B. *A golpe de maracas: Poemas negros en papel mulato*. Havana, 1950.

Callow, Simon. *Orson Welles: Hello Americans*. New York: Viking, 2006.

Campbell, Timothy. *Wireless Writing in the Age of Marconi*. Minneapolis: University of Minnesota Press, 2006.

Campos, René A. "The Poetics of the Bolero in the Novels of Manuel Puig." *World Literature Today* 65, no. 4 (Autumn 1991): 637–42.

Cappetti, Carla. *Writing Chicago: Modernism, Ethnography, and the Novel*. New York: Columbia University Press, 1993.

Cárcamo-Huechante, Luis. "Indigenous Interference: Mapuche Use of Radio in Times of Acoustic Colonialism." *Latin American Research Review* 48 (2013): 50–68

Carpentier, Alejo. "El ocaso de la radio." *El Nacional* (Caracas), January 16, 1954.

———. "La radio y sus nuevas posibilidades." *Carteles*, December 17, 1933.

Reprinted in *Obras Completas*, 9:389–97. Mexico City: Siglo Veintiuno Editores, 1983.
Cassirer, Ernst. *The Logic of the Cultural Sciences*. New Haven: Yale University Press, 2000.
Castelli, Nora. "Entrevista con Manuel Puig: Una narrativa de lo melifluo." *Quimera*, no. 18 (April 1982).
Castro, Fidel. *Fidel en Radio Rebelde*. Havana: Gente Nueva, 1979.
———. *La historia me absolverá*. Lima, Peru: Ediciones Futuro, 1961.
Cather, Willa. *Not under Forty*. 1936. Lincoln: University of Nebraska Press, 1988.
Chanan, Michael. *Cuban Cinema*. Minneapolis: University of Minnesota Press, 2004.
Chandler, Raymond. *Later Novels and Other Writings*. New York: Library of America, 1995.
———. *Stories and Early Novels*. New York: Library of America, 1995.
Charters, Ann. Introduction to *Three Lives*, by Gertrude Stein. 1909. New York: Penguin, 1990.
Chion, Michel. *Audio-Vision*. New York: Columbia University Press, 1994.
Chumley, Lily Hope, and Nicholas Harkness, "Introduction: QUALIA." *Anthropological Theory* 13, no. 1/2 (2013): 3–11.
Chun, Wendy Hui Kyong, Anna Watkins Fisher, and Thomas Keenan, eds. *New Media, Old Media: A History and Theory Reader*. London: Routledge, 2015.
Clark, T. J. *Image of the People: Gustave Courbet and the 1848 Revolution*. London: Thames and Hudson, 1973.
Claxton, Robert. *From* Parsifal *to Perón*. Gainesville: University Press of Florida, 2007.
Cohen, Robert Carl. *Black Crusader: A Biography of Robert Franklin Williams*. Secaucus: Lyle Stuart, 1972.
Cohn, Deborah. *The Latin American Literary Boom and U.S. Nationalism during the Cold War*. Nashville: Vanderbilt University Press, 2012.
Collazos, Óscar, Julio Cortázar, and Mario Vargas Llosa. *Literatura en la revolución y revolución en la literatura*. Mexico City: Siglo Veintiuno, 1970.
Connor, Steven. *Dumbstruck: A Cultural History of Ventriloquism*. Oxford and New York: Oxford University Press, 2000.
———. "The Modern Auditory I." In *Rewriting the Self*, edited by Roy Porter, 203–23. London and New York: Routledge, 1997.
Copjec, Joan. "The Phenomenal Nonphenomenal: Private Space in *Film Noir*." In *Shades of Noir*, edited by Copjec. London: Verso, 1993.
Corradi, Juan E., Patricia Weiss Fargen, and Manuel Antonio Garretón, eds. *Fear at the Edge: State Terror and Resistance in Latin America*. Berkeley and Los Angeles: University of California Press, 1992.
Cortázar, Julio. "El noble arte." *La vuelta al día en ochenta mundos: Tomo II*. 1967. Buenos Aires: Siglo XXI, 2005.
Costa, Eduardo, and John Perreault. "Introduction to Tape Poems" (1969). In *Conceptualism and Other Fictions: The Collected Writings of Eduardo*

Costa, 1965–2015, edited by Patrick Greaney, 56–57. Los Angeles: Les Figues, 2016.

Craig, Douglas B. *Fireside Politics: Radio and Political Culture in the United States, 1920–1940*. Baltimore: Johns Hopkins University Press, 2000.

Cramer, Gisela. "The Word War at the River Plate: The Office of Inter-American Affairs and the Argentine Airwaves, 1940–1946." Edited by Cramer and Ursula Prutsch. In *¡Américas Unidas! Nelson A. Rockefeller's Office of Inter-American Affairs, 1940–1946*, 213–47. Frankfurt and Madrid: Vervuert, 2012.

Dallek, Robert. *Franklin Roosevelt and American Foreign Policy, 1932–1945*. New York: Oxford University Press, 1995.

Danto, Arthur. *The Transfiguration of the Commonplace*. Cambridge: Harvard University Press, 1981.

Debray, Régis. *Revolution in the Revolution? Armed Struggle and Political Struggle in Latin America*. Translated by Bobbye Ortiz. New York: MR Press, 1967.

Deharme, Paul. "Proposition for a Radiophonic Art." Translated by Anke Birkenmaier. *Modernism/modernity* 16, no. 2 (2009): 403–13.

Deleuze, Gilles. "He Stuttered." Translated by Constantin V. Boundas. In *Gilles Deleuze and the Theater of Philosophy*, edited by Boundas and Dorothea Olkowski. London: Routledge, 1994.

Deleuze, Gilles, and Félix Guatarri. *Kafka: Toward a Minor Literature*. Translated by Dana Polan. Minneapolis: University of Minnesota Press, 1986.

Denning, Michael. *The Cultural Front: The Laboring of American Culture in the Twentieth Century*. London: Verso, 1996.

———. "The Novelists' International." In *The Novel, Vol 1*, edited by Franco Moretti, 703–25. Princeton: Princeton University Press, 2006: 703–25.

Derrida, Jacques. *Archive Fever*. Translated by Eric Prenowitz. Chicago: University of Chicago Press, 1996.

———. *Specters of Marx*. Translated by Peggy Kamuf. New York: Routledge, 1994.

Doersken, Clifford. *America's Babel: Rogue Radio Broadcasters of the Jazz Age*. Philadelphia: University of Pennsylvania Press, 2005.

Dolar, Mladen. *A Voice and Nothing More*. Cambridge: MIT Press, 2006.

Donghi, Tulio Halperín. *The Peronist Revolution and Its Ambiguous Legacy*. London: University of London, Institute of Latin American Studies, 1998.

Dos Passos, John. *Number One*. New York: Houghton Mifflin, 1943.

———. "On the National Hookup: Spotlights and Microphones." In *In All Countries*, collected in *Travel Books and Other Writings 1916–1941*, 405–21. New York: Library of America, 2003.

———. *The Three Soldiers*. New York: Modern Library, 1932.

———. *U.S.A.* New York: Modern Library, 1938.

Douglas, Susan. *Listening In: Radio and the American Imagination*. Minneapolis: University of Minnesota Press, 2004.

Drabeck, Bernard A., and Helen E. Ellis, eds. *Archibald MacLeish: Reflections*. Amherst: University of Massachusetts Press, 1986.

Dupree, Mary Hellen. "From 'Dark Singing' to a Science of the Voice: Gustav Anton von Seckendorff and the Declamatory Concert Around 1800."

Deutsche Vierteljahrsschrift für Literaturwissenschaft und Geistesgeschichte 86, no. 3 (2012): 365–96.

Dyson, Frances. *Sounding New Media.* Berkeley: University of California Press, 2009.

Edison, Thomas Alva. "The Phonograph and Its Future." *North American Review* 126 (June 1878): 427–536.

Edwards, Brent Hayes. "The Literary Ellington." *Representations* 77 (Winter 2002): 1–29.

———. *The Practice of Diaspora: Literature, Translation, and the Rise of Black Internationalism.* Cambridge: Harvard University Press, 2003.

Edwards, Brian. "Fanon's *al-Jazáir*, or Algeria Translated." *Parallax* 8, no. 2 (2002): 99–115.

———. *Morocco Unbound.* Durham: Duke University Press, 2005.

Ehrick, Christine. *Radio and the Gendered Soundscape: Women and Broadcasting in Argentina and Uruguay, 1930–1950.* Cambridge: Cambridge University Press, 2015.

———. "'Savage Dissonance': Gender, Voice, and Women's Radio Speech in Argentina, 1930–1945." In *Sound in the Age of Mechanical Reproduction*, edited by David Suisman and Susan Strasser. Philadelphia: University of Pennsylvania Press, 2010.

Ellison, Ralph. *Invisible Man.* 1947. New York: Vintage, 1995.

Erlmann, Veit. *Reason and Resonance: A History of Modern Aurality.* New York: Zone, 2010.

Ezensberger, Hans Magnus. "Constituents of a Theory of Media." In *The Consciousness Industry.* New York: Seabury, 1974.

Fabre, Michel. *Richard Wright: Books and Writers.* Jackson: University Press of Mississippi, 1990.

———. *The Unfinished Quest of Richard Wright.* Translated by Isabel Barzun. 1973. Urbana and Chicago: University of Illinois Press, 1993.

Falicov, Tamara L. "Hollywood's Rogue Neighbor: The Argentine Film Industry during the Good Neighbor Policy, 1939–1945." *Americas* 63, no. 2 (October 2006): 245–60.

Fanon, Frantz. "Ici la voix de l'Algérie . . ." In *L'an V de la Révolution Algérienne*, 57–88. Paris: F. Maspero, 1962.

Farrell, James T. "*Partisan Review*, May 1938." In *Richard Wright: Critical Perspectives Past and Present*, edited by Henry Louis Gates Jr. and K. A. Appiah, 4–5. New York: Amistad, 1993.

Faulkner, William. *Absalom, Absalom!* New York: Vintage, 1991.

Federal Writers' Project (WPA). *New York City Guide.* New York: Random House, 1939.

Feitlowitz, Marguerite. *A Lexicon of Terror: Argentina and the Legacies of Torture.* New York: Oxford University Press, 1998.

Fernández, María. "Estri-dentistas: Taking the Teeth out of Futurism." In *At a Distance: Precursors to Art and Activism on the Internet*, edited by Annmarie Chandler and Norie Neumark, 342–71. Cambridge: MIT Press, 2005.

Ferrer, Ada. *Insurgent Cuba: Race, Nation, and Revolution, 1868–1898.* Chapel Hill: University of North Carolina Press, 1999.

Foucault, Michel. *L'archéologie du savoir.* Paris: Gallimard, 1969.

Fowler, Gene, and Bill Crawford. *Border Radio.* Austin: Texas Monthly Press, 1987.

Fox, Claire. *Making Art Panamerican: Cultural Policy and the Cold War.* Minneapolis: University of Minnesota Press, 2013.

Fraga, Enrique A. *La prohibición del lunfardo en la radiodifusión argentina, 1933–1953.* Buenos Aires: Lajouane Editorial, 2006.

Franco, Jean. "Trapped in Machismo." *New York Times*, September, 23, 1984.

Franqui, Carlos. *Cuba, la revolución ¿mito o realidad? Memorias de un fantasma socialista.* Mexico City: Océano de México, 2006.

———. *Family Portrait with Fidel: A Memoir.* Translated by Alfred MacAdam. New York: Random House, 1984.

Frederick, Howard H. *Cuban-American Radio Wars: Ideology in International Telecommunications.* Norwood, N.J.: Alex, 1986.

Freedgood, Elaine. *The Ideas in Things.* Chicago: University of Chicago Press, 2010.

Freeman, Judith. *The Long Embrace.* New York: Vintage, 2007.

Freud, Sigmund. *Beyond the Pleasure Principle.* Translated by James Strachey. New York: Norton, 1961.

Fridman, Daniel, and David Sheinin. "Wild Bulls, Discarded Foreigner, and Brash Champions: US Empire and the Cultural Constructions of Argentine Boxers." *Left History* 12, no. 1 (Spring/Summer 2007): 52–77.

Fuller, Matthew. *Media Ecologies: Materialist Energies in Art and Technoculture.* Cambridge: MIT Press, 2007.

Gallo, Ricardo. *La radio: Ese mundo tan sonoro.* Buenos Aires: Corregidor, 1991.

Gallo, Rubén. "John Dos Passos in Mexico." *Modernism / modernity* 14, no. 2 (April 2007): 329–45.

———. *Mexican Modernity: The Avant-Garde and the Technological Revolution.* Cambridge: MIT Press, 2010.

Gaylin, Ann. *Eavesdropping in the Novel from Austen to Proust.* Cambridge: Cambridge University Press, 2002.

Gelatt, Roland. *The Fabulous Phonograph: 1877–1977.* New York: Collier, 1977.

Gellman, Irwin F. *Roosevelt and Batista: Good Neighbor Policy in Cuba, 1933–1945.* Albuquerque: University of New Mexico Press, 1973.

Gilman, Richard. "Dos Passos and the Many Lives of 'U.S.A.'" *New York Times*, March 16, 1997.

Gilroy, John. *The Black Atlantic: Modernity and Double Consciousness.* Cambridge: Harvard University Press, 1993.

Giordano, Alberto. *Manuel Puig: La conversación infinita.* Buenos Aires: Beatriz Viterbo, 2001.

Gitelman, Lisa. *Scripts, Grooves, and Writing Machines: Representing Technology in the Edison Era.* Stanford: Stanford University Press, 1999.

Gitelman, Lisa, and Geoffrey Pingree, eds. *New Media: 1740–1915.* Cambridge: MIT Press, 2003.

Gladwell, Malcom. "Small Change." *New Yorker*, October 4, 2010.

Gleijeses, Piero. *Conflicting Missions: Havana, Washington, and Africa, 1959–1976*. Chapel Hill: University of North Carolina Press, 2002.

Goble, Mark. *Beautiful Circuits: Modernism and the Mediated Life*. New York: Columbia University Press, 2010.

Godden, Richard. "*Absalom, Absalom!* Haiti, and Labor History: Reading Unreadable Revolutions." In *William Faulkner's Absalom, Absalom!: A Casebook*, edited by Fred C. Hobson, 251–82. Oxford: Oxford University Press, 2003.

Goetz, Edward G. *New Deal Ruins: Race, Economic Justice, and Public Housing Policy*. Ithaca: Cornell University Press, 2013.

Goffman, Erving. *Forms of Talk*. Philadelphia: University of Pennsylvania Press, 1981.

Goldchluk, Graciela. "El archivo Puig: Los papeles privados." In Ñ, 10–11, July 24, 2010.

González, Reynaldo. *Llorar es un placer*. Havana: Letras Cubanas, 1988.

———. *El más humano de los autores*. Havana: Ediciones Unión, 2009.

González-Echevarría, Roberto. "*Biografía de un cimarrón* and the Novel of the Cuban Revolution." In *The Voice of the Masters: Writing and Authority in Modern Latin American Literature*. Austin: University of Texas Press, 1985.

Goyal, Yogita. "Theorizing Africa in Black Diaspora Studies: Caryl Phillips' *Crossing the River*." *Diaspora* 12, no. 1 (2003): 5–38.

Guevara, Ernesto Che. *La Guerra de guerrillas*. Havana: Editorial de Ciencias Sociales, 1985.

Haglund, David. *Latin America and the Transformation of U.S. Strategic Thought, 1936–1940*. Albuquerque: University of New Mexico Press, 1984.

Hallward, Peter. *Absolutely Postcolonial: Writing between the Singular and the Specific*. Manchester: Manchester University Press, 2001.

Hamberg, Jill. *Under Construction: Housing Policy in Revolutionary Cuba*. New York: Center for Cuban Studies, 1986.

Hammond, Cynthia. "The Interior of Modernism: Catherine Bauer and the American Housing Movement." In *Craft, Space, and Interior Design 1855–2005*, edited by Janice Helland and Sandra Alfoldy, 169–88. Burlington, Vt.: Ashgate, 2008.

Hanrahan, Brían. "Index and Diegesis in Weimar Broadcasting." Manuscript.

Harkness, Nicholas. *Songs of Seoul: An Ethnography of Voice and Voicing in Christian South Korea*. Berkeley: University of California Press, 2014.

Hayes, Joy. *Radio Nation: Communication, Popular Culture, and Nationalism in Mexico, 1920–1950*. Tucson: University of Arizona Press, 2000.

Hayles, N. Katherine. *How We Became Posthuman: Virtual Bodies in Cybernetics, Literature, and Informatics*. Chicago: University of Chicago Press, 1999.

Heel, Augilles. "Radio y TV: Panorámicas," *Revolución* (Havana), May 19, 1959, 12.

Henkin, David M. *The Postal Age: The Emergence of Modern Communications in Nineteenth-Century America*. Chicago: University of Chicago Press, 2007.

Higashida, Cheryl. “Aunt Sue’s Children: Re-Viewing the Gender(ed) Politics of Richard Wright’s Radicalism.” *American Literature* 75, no. 2 (2003): 395–425.

Hilmes, Michel. “NBC and the Network Idea: Defining the ‘American System.’” In *NBC: America’s Network*. Berkeley: University of California Press, 2007.

———. *Radio Voices: American Broadcasting, 1922–1952*. Minneapolis: University of Minnesota Press, 1997.

Hilmes, Michel, and Jason Loviglio, eds. *The Radio Reader*. New York: Routledge, 2002.

Hirschkind, Charles. *The Ethical Soundscape: Cassette Sermons and Islamic Counterpublics*. New York: Columbia University Press, 2006.

Hirschkop, Ken. *Mikhail Bakhtin: An Aesthetic for Democracy*. New York: Oxford University Press, 1999.

Ho, Fred, and Bill V. Mullen, eds. *Afro-Asia: Revolutionary Political and Cultural Connections between African Americans and Asian Americans*. Durham: Duke University Press, 2008.

Hollier, Denis. “The Death of Paper: A Radio Play.” *October* 78 (Autumn 1996): 3–20.

Hurston, Zora Neale. “Saturday Review of Literature, April 2, 1938.” In *Richard Wright: Critical Perspectives Past and Present,* edited by Henry Louis Gates Jr. and K. A. Appiah, 3–4. New York: Amistad, 1993.

Jagoda, Patrick. *Network Aesthetics*. Chicago: University of Chicago Press, 2016.

Jameson, Frederic. *The Antinomies of Realism*. New York: Verso, 2013.

———. “The Synoptic Chandler.” In *Shades of Noir*, edited by Joan Copjec, 33–56. London and New York: Verso, 1993.

Johnson, Barbara. “Melville’s Fist: The Execution of *Billy Budd*.” In *The Critical Difference*, 79–109. Baltimore: Johns Hopkins University Press, 1980.

———. “The Re(a)d and the Black.” In *Richard Wright: Critical Perspectives Past and Present*, edited by Henry Louis Gates Jr. and K. A. Appiah, 149–55. New York: Amistad, 1993.

Jones, Gavin. *Strange Talk: The Politics of Dialect Literature in Gilded Age America*. Berkeley: University of California Press, 1999.

Kahn, Douglas. *Noise, Water, Meat: A History of Sound in the Arts*. Cambridge: MIT Press, 1999.

Kahn, Douglas, and Kevin Whitehead, eds. *Wireless Imagination*. Cambridge: MIT Press, 1992.

Kaplan, Alice Yaeger. *Reproductions of Banality*. Minneapolis: University of Minnesota Press, 1986.

Kaplan, Amy. *The Anarchy of Empire in the Making of U.S. Culture*. Cambridge: Harvard University Press, 2002.

Karush, Matthew. *Culture of Class: Radio and Cinema in the Making of a Divided Argentina, 1920–1946*. Durham: Duke University Press, 2012.

Katz, Mark. *Capturing Sound: How Technology Has Changed Music*. Berkeley: University of California Press, 2007.

Keane, Damien. *Ireland and the Problem of Information: Irish Writing,*

Radio, Late Modernist Communication. Univeristy Park: Pennsylvania State University Press, 2014.

Kelly, Mary. *Post-Partum Document.* Berkeley: University of California Press, 1999.

Kittler, Friedrich. *Discourse Networks, 1800 /1900.* Stanford: Stanford University Press, 1990.

———. *Gramophone, Film, Typewriter.* Stanford: Stanford University Press, 1999.

Kiuchi, Toru, and Yoshinobu Hakutani. *Richard Wright: A Documented Chronology 1908–1960.* Jefferson, N.C.: McFarland, 2013.

Kolesnicov, Patricia. "Primeras joyas del archivo Puig: publican argumentos inéditos." *Clarín,* March, 8, 2004.

Kozak, Daniela. "Tesoros rescatados de las garras del tiempo." Ñ, December 11, 2012, www.revistaenie.clarin.com/escenarios/Tesoros-rescatados-garras-tiempo-Metropolis-hijo-otro-Sangre-negra_0_824917551.html.

Krebs, Edgardo C. "It's a Noir Thing." *filmcomment* 48, no. 5 (September/October 2012): 42.

———. "Native Son, Lost and Found." *filmcomment* 48, no. 5 (September and October 2012). http://filmcomment.com/article/native-son-film-richard-wright-lost-and-found.

Kuhnheim, Jill. *Beyond the Page: Poetry and Performance in Spanish America.* Tucson: University of Arizona Press, 2014.

Kutzinski, Vera. "Fearful Asymmetries: Langston Hughes, Nicolás Guillén, and 'Cuba Libre.'" *Diacritics* 34, no. 3/4 (Autumn–Winter 2004): 112–42.

———. *The Worlds of Langston Hughes: Modernism and Translation in the Americas.* Ithaca: Cornell University Press, 2012.

La règle de jeu. Directed by Jean Renoir. 1939. DVD. New York: Criterion, 2004.

LaBelle, Brandon. *Background Noise: Perspectives on Sound Art.* London: Continuum, 2006.

Lane, Jill. *Blackface Cuba, 1840–1895.* Philadelphia: University of Pennsylvania Press, 2005.

———. "Smoking *Habaneras,* or A Cuban Struggle with Racial Demons." *Social Text 104* 28, no. 3 (Fall 2010): 11–37.

Lavine, Amy. "From Slum Clearance to Economic Development: A Retrospective of Development Policies in New York State." *Albany Government Law Review* 212 (2011): 212–69.

Lawrence, Amy. *Echo and Narcissus: Women's Voices in Classical Hollywood Cinema.* Berkeley: University California Press, 1991.

Lejtman, Román. *Perón vuelve: Intrigas en el exilio y traiciones en el regreso.* Buenos Aires: Planeta, 2013.

Levine, Lawrence W., and Cornelia R. Levine. *The People and the President: America's Conversation with FDR.* Boston: Beacon, 2002.

Levine, Suzanne Jill. *Manuel Puig and the Spider Woman.* New York: Farrar, Straus and Giroux, 2000.

———. *The Subversive Scribe: Translating Latin American Fiction.* Minneapolis: Graywolf, 1991.

Levinson, Bret. *The Ends of Literature: The Latin American "Boom" in the Neoliberal Marketplace*. Palo Alto: Stanford University Press, 2002.

Link, Daniel. *Cómo se lee: y otras intervenciones críticas*. Buenos Aires: Grupo Editorial Norma, 2003.

———. "La obra como exigencia de vida." *Ñ* 7, no. 356 (July 24, 2010): 8–9.

Logie, Ilse, and Julia Romero. "Extranjeridad: Lengua y traducción en la obra de Manuel Puig." *Revista Iberoamericana* 74, no. 222 (January–March, 2008): 33–51.

Long, Marshall. "What Is So Special about Shoebox Halls? Envelopment, Envelopment, Envelopment." *Acoustics Today*, April 2009, 21–25.

López, Ana M., and Manuel Alvarado, eds. *Mediating Two Worlds: Cinematic Encounters in the Americas*. London: British Film Institute, 1993.

López, Antonio. *Unbecoming Blackness: The Diaspora Cultures of Afro-Cuban America*. New York: New York University Press, 2012.

Lott, Eric. *Love and Theft: Blackface Minstrelsy and the American Working Class*. Oxford: Oxford University Press, 1993.

Loviglio, Jason. *Radio's Intimate Public*. Minneapolis: University of Minnesota Press, 2005.

Lucey, Michael. *Never Say I*. Durham: Duke University Press, 2006.

———. "Simone de Beauvoir and Sexuality in the Third Person." *Representations* 109 (Winter 2010): 95–121.

Ludmer, Josefina. *El cuerpo del delito: Un manual*. Buenos Aires: Perfil, 1999. Translated by Glen S. Close as *The Corpus Delicti: A Manual of Argentine Fictions*. Pittsburgh: University of Pittsburgh Press, 2004.

Luis, William. *Culture and Customs of Cuba*. London: Greenwood, 2001.

Luis-Brown, David. *Waves of Decolonization: Discourses of Race and Hemispheric Citizenship in Cuba, Mexico, and the United States*. Durham: Duke University Press, 2008.

Luis López, Oscar. *Alejo Carpentier y la radio*. Havana: Letras Cubanas, 2003.

———. *La radio en Cuba*. Havana: Letras Cubanas, 1998.

Lukács, Georg. "From *Realism in Our Time:* Literature and the Class Struggle." In *Theory of the Novel: A Historical Approach,* edited by Michael McKeon, 759–64. Baltimore: Johns Hopkins University Press, 2000.

———. "Realism in the Balance." 1938. In *Aesthetics and Politics*. London: New Left Review Books, 1977.

———. *Realism in Our Time*. Translated by John and Necke Mander. London: Merlin, 1962.

Lye, Colleen. "The Afro-Asian Analogy." *PMLA* 123, no. 5 (October 2008): 1732–36.

———. *America's Asia: Racial Form and American Literature, 1893–1945*. Princeton: Princeton University Press, 2005.

———. "Racial Form." *Representations* 104 (Fall 2008): 92–101.

Mackey, Nathaniel. *Discrepant Engagement: Dissonance, Cross-Culturality, and Experimental Writing*. Cambridge: Cambridge University Press, 1993.

Maguire, Emily. "The Eusebia Cosme Show: Translating an Afro-Antillean

Identity." In *Writing the Afro-Hispanic: Essays on Africa and Africans in the Spanish Caribbean,* edited by Conrad James, 77–96. London: Adonis and Abbey, 2012. (Citations to article on academia.edu.)

Maine, Barry, ed. *John Dos Passos.* New York: Routledge, 1988.

Marcus, Sharon. *Apartment Stories: City and Home in Nineteenth-Century Paris and London.* Berkeley: University of California Press, 1999.

Márquez, Gabriel García. *Vivir para contarla.* Barcelona: Mondadari, 2002.

Martínez Víctores, Ricardo. *7RR: La historia de Radio Rebelde.* Havana: Editorial de Ciencias Sociales, 1978.

Marvin, Carolyn. *When Old Technologies Were New: Thinking About Electric Communication in the Late Nineteenth Century.* Oxford: Oxford University Press, 1990.

Matallana, Andrea. *"Locos por la radio": Una historia social de la radiofonía en la Argentina, 1923–1947.* Buenos Aires: Prometeo, 2006.

Masiello, Francine. *The Art of Transition: Latin American Culture and the Neoliberal Crisis.* Durham: Duke University Press, 2001.

———. "Rethinking Neocolonial Aesthetics: Literature, Politics, and Intellectual Community in Cuba's Revista de Avance." *Latin American Research Review* 28, no. 2 (1993): 3–31.

Maxwell, William J. *New Negro Old Left: African-American Writing between the Wars.* New York: Columbia University Press, 1999.

McClanahan, Annie. "Dead Pledges: Debt, Horror, and the Credit Crisis." *Post45 Peer Reviewed.* May 7, 2012.

McCracken, Ellen. "Manuel Puig's *Heartbreak Tango*: Women and Mass Culture." *Latin American Literary Review* 9, no. 18 (Spring 1981): 27–35.

———. "Metaplagiarism and the Critic's Role as Detective: Ricardo Piglia's Reinvention of Roberto Arlt." *PMLA* 106, no. 5 (October 1991): 1071–82.

McCullers, Carson. *The Heart Is a Lonely Hunter.* 1940. New York: Mariner, 2004.

McEnaney, Tom. "The Digital Revolution? History, Photography, and the Body of Fidel Castro." *La Habana Elegante*, no. 49 (Spring-Summer 2011).

———. "Realismo sonoro y fidelidad literaria: Las obras de cinta de Eduardo Costa." *Revista de estudios hispánicos* 50 (August 2016): 515–30.

McHale, Brian. "Talking U.S.A.: Interpreting Free Indirect Discourse in Dos Passos's *U.S.A.* Trilogy." *Degrés* 16 (1978).

McKay, Claude. *Banjo.* 1929. New York: Harcourt Brace, 1975.

McLuhan, Marshall. *Understanding Media.* 1964. Cambridge: MIT Press, 1994.

Melville, Herman. *Billy Budd and Other Stories.* New York: Viking Penguin, 1986.

Merrifield, Andy. "Crowd Politics, or Here Comes Everybuddy." *New Left Review* 71 (September/October 2011): 103–14.

Michaels, Walter Benn. "Romance and Real Estate." In *The Gold Standard and the Logic of Naturalism.* Berkeley: University of California Press, 1987.

Miller, Edward. *Emergency Broadcasting and 1930s American Radio.* Philadelphia: Temple University Press, 2003.

Miller, Eugene. "Richard Wright and Gertrude Stein." *Black American Literature Forum*. Indianapolis: Indiana State University, 1982.

———. *Voice of a Native Son*. Jackson: University Press of Mississippi, 1990.

Mills, Mara. "Deaf Jam: From Inscription to Reproduction to Information." *Social Text* 28, no. 1 (Spring 2008): 35–58.

Mindell, David. *Between Human and Machine: Feedback, Control and Computing before Cybernetics*. Baltimore: Johns Hopkins University Press, 2002.

Misemer, Sarah M. *Moving Forward, Looking Back: Trains, Literature, and the Arts in the River Plate*. Lewisburg: Bucknell University Press, 2010.

Mitchell, W. J. T. "Addressing Media." *MediaTrope* 1 (2008): 1–18.

Monegal, Emir Rodríguez. "Los sueños de Evita: A propósito de la última novela de Manuel Puig." *Plural* 22 (July 1973): 34–36.

Monegal, Emir Rodríguez, and Juan José Sebreli, *Comediantes y mártires: Ensayo contra los mitos*. Barcelona: Random House Mondadori, 2008.

Moore, Robin. *Nationalizing Blackness: Afrocubanismo and the Artistic Revolution in Havana, 1920–1940*. Pittsburgh: University of Pittsburgh Press, 1997.

Moreno, María. "Doble casetera." *Página 12*, October 25, 2010.

Moretti, Franco. *Modern Epic*. London: Verso, 1995.

Mormino, Gary R., and George E. Pozzetta. "The Reader Lights the Candle: Cuban and Florida Cigar Workers' Oral Tradition." *Labor's Heritage 5* (Spring 2003): 4–28.

Moten, Fred. *In the Break: The Aesthetics of the Black Radical Tradition*. Minneapolis: University of Minnesota Press, 2003.

Mowitt, John. "Breaking up Fanon's Voice." In *Frantz Fanon: Critical Perspectives*, edited by Anthony C. Allesandrini. New York: Routledge, 1999.

Muldoon, Paul. *Meeting the British*. New York: Faber and Faber, 1987.

Negroes with Guns: Rob Williams and Black Power. Independent Lens. May 3, 2011.

Ngai, Sianne. *Ugly Feelings*. Cambridge: Harvard University Press, 2005.

North, Michael. *The Dialect of Modernism: Race, Language, and Twentieth-Century Culture*. Oxford: Oxford University Press, 1994.

Ortíz, Alicia Dujovne. *Eva Perón: A Biography*. New York: St. Martin's, 1996.

Panofsky, Erwin. *Perspective as Symbolic Form* and *Gothic Architecture and Scholasticism*. Cambridge: MIT Press, 1996.

Patel, Kiran Klaus. *The New Deal: A Global History*. Princeton: Princeton University Press, 2016.

Pauls, Alan. *Manuel Puig, La traición de Rita Hayworth*. Buenos Aires: Librería Hachette, 1986.

Pérez Jr., Louis A. *Cuba under the Platt Amendment 1902–1934*. Pittsburgh: University of Pittsburgh Press, 1986.

———. *To Die in Cuba: Suicide and Society*. Chapel Hill: University of North Carolina Press, 2005.

Peters, John Durham. *Speaking into the Air*. Chicago: University of Chicago Press, 1999.

Phu, Thy. "Bigger at the Movies: Sangre Negra and the Cinematic Projection of Native Son." *Black Camera* 2, no. 1 (Winter 2010): 36–57.

Pickard, Victor. *America's Battle for Media Democracy: The Triumph of Corporate Libertarianism and the Future of Media Reform*. Cambridge: Cambridge University Press, 2014.

———. "Before Net Neutrality: The Surprising 1940s Battle for Radio Freedom." *Atlantic*, January 29, 2015.

Picker, John. *Victorian Soundscapes*. Oxford: Oxford University Press, 2003.

Piglia, Ricardo. *La argentina en pedazos*. Buenos Aires: Ediciones de Urraca, 1993.

———. *La ciudad ausente*. Buenos Aires: Seix Barral, 1992. Translated by Sergio Waisman as *The Absent City*. Durham: Duke University Press, 2000.

———. "Luba." *Nombre Falso*. Buenos Aires: Siglo Veintiuno, 1975.

———. "Writing and Translation." In *Voice-Overs: Translation and Latin American Literature*, edited by Daniel Balderston and Marcy E. Schwartz, 64–67. Albany: State University of New York Press, 2002.

Pizer, Donald. *Toward a Modernist Style: Dos Passos*. London: Bloomsbury, 2013.

Plotz, John. *Portable Property: Victorian Culture on the Move*. Princeton: Princeton University Press, 2008.

Portelli, Alessandro. "*Everybody's Healing Novel*: *Native Son and Its Contemporary Critical Context*." *Mississippi Quarterly: The Journal of Southern Cultures* 50, no. 2 (Spring 1997): 255–65.

Puig, Manuel. *Boquitas pintadas*. Barcelona: Seix Barral, 1972.

———. *Mi querida familia*. Vol. 1, *Cartas Europeas 1956–1962*. Buenos Aires: Editorial Entropía, 2005.

———. *Eternal Curse on the Reader of These Pages*. New York: Vintage, 1982.

———. Interview by Kathleen Wheaton. "The Art of Fiction No. 114." *Paris Review*, no. 113 (Winter II 1989).

Puig, Manuel, and Jorgelina Corbatta. "Interview." *Review of Contemporary Fiction* 11, no. 3 (Fall 1991).

Radford, Gail. *Modern Housing for America: Policy Struggles in the New Deal Era*. Chicago: University of Chicago Press, 1996.

Rancière, Jacques. *The Ignorant Schoolmaster: Five Lessons in Intellectual Emancipation*. Translated by Kristen Ross. Stanford: Stanford University Press, 1991.

———. *Politics of Literature*. 2006. Translated by Julie Rose. Malden, Mass.: Polity, 2011.

Reilly, John M. "Giving Bigger a Voice: The Politics of Narrative in *Native Son*." In *New Essays on "Native Son,"* edited by Keneth Kinnamon, 35–62. Cambridge: Cambridge University Press, 1990.

Rheingold, Howard. *Smart Mobs: The Next Social Revolution*. Cambridge: Perseus, 2003.

Rivero, Yeidy M. *Tuning Out Blackness: Race and Nation in the History of Puerto Rican Television*. Durham: Duke University Press, 2005.

Romero, Julia. "Las traiciones de Manuel Puig: Otra lectura de *La traición de*

Rita Hayworth." In *Homenaje a Manuel Puig,* edited by José Amícola, 15–20. Buenos Aires: Universidad Nacional de la Plata, 1994.

Rosen, Fred. *Empire and Dissent: The United States and Latin America.* Durham: Duke University Press, 2008.

Rozitchner, León, ed. *Vías Argentinas: (Ensayos sobre el ferrocarril).* Buenos Aires: Milena Caserola, 2010.

Ruttenburg, Nancy. *Democratic Personality.* Stanford: Stanford University Press, 1998.

Sadlier, Darlene. *Americans All: Good Neighbor Cultural Diplomacy in World War II.* Austin: University of Texas Press, 2012.

Salwen, Michael B. "The Origins of CMQ: Pre-Castro Cuba's Leading Radio Network." *Historical Journal of Film, Radio and Television* 13, no. 3 (1993): 315–32.

———. *Radio and Television in Cuba: The Pre-Castro Era.* Ames: Iowa State University Press, 1994.

Sánchez, Luis Rafael. *La guaracha del Macho Camacho.* 1976. Madrid: Cátedra, 2005.

Sarduy, Severo. *The Ant Killers.* Translated by Philip Barnard. In *For Voice,* 111–36. Pittsburgh: Latin American Literary Review Press, 1985.

———. *Cartas a mi hermana en La Habana.* Miami: Severo Sarduy Cultural Foundation, 2013.

———. *Los matadores de hormigas.* In *Obra Completa*, vol. 2, edited by Gustavo Guerrero and François Wahl, 1077–93. Madrid: Galaxia Gutenburg, Círcluo de Lectores, 1999.

———. *Para la voz.* Madrid: Fundamentos, 1978.

———. "Para una biografía pulverizada en el número de *Quimera.*" In *Obra Completa*, vol. 1, edited by Gustavo Guerrero and François Wahl. Madrid: Galaxia Gutenburg, Círculo de Lectores, 1999.

———. "Soy una Juana de Arco electrónica, actual." In *Obra Completa*, vol. 2, edited by Gustavo Guerrero and François Wahl, 30–31. Madrid: Galaxia Gutenburg, Círcluo de Lectores, 1999.

———. "Notas a las Notas a las Notas . . . A Propósito de Manuel Puig." *Revista Iberoamericana* 37, nos. 76–77 (July-December, 1971): 555–67. Translated by Suzanne Jill Levine as "Note to the Notes to the Notes . . . A Propos of Manuel Puig." *World Literature Today* 65, no. 4 (Autumn 1991): 625–30.

Sarlo, Beatriz. "El brillo, la parodia, Hollywood y la modestia." *Página 12*, July 29, 1990.

———. *La imaginación técnica.* Buenos Aires: Ediciones Nueva Visión, 1992.

———. "Oralidad y lenguas extranjeras." In *Oralidad y Argentinidad*, edited by Walter Bruno Berg and Markus Klaus Schäffauer. Tübingen: Gunter Narr, 1997.

———. *Tiempo Pasado: Cultura de la memoria y giro subjetivo.* Buenos Aires: Siglo Veintiuno, 2005.

Sartre, Jean-Paul. "Collectives." "A Propos de Dos Passos et de '1919.'" *Critiques littéraires (Situations, I).* Paris: Gallimard, 1947.

———. *Critique of Dialectical Reason.* Vol. 1. New York: Verso, 2006.

———. "Prière d'insérer." *Le sursis*. 1945. Paris: Gallimard, 1972.

Schoultz, Lars. *Beneath the United States: A History of U.S. Policy towards Latin America*. Cambridge: Harvard University Press, 1998.

Sconce, Jeffrey. *Haunted Media: Electronic Presence from Telegraphy to Television*. Durham: Duke University Press, 2000.

Scott, Ellen. "Blacker Than Noir: The Making and Unmaking of Richard Wright's 'Ugly' *Native Son* (1951)." *Adaptation* 6, no. 1 (October 2012): 93–119.

Scott, Joan. "The Evidence of Experience." *Critical Inquiry* 17, no. 4 (Summer 1991): 773–91.

Sebreli, Juan José. *Escritos sobre escritos, ciudades bajo ciudades*. Buenos Aires: Editorial Sudamericana, 1997.

Sexton, Jay. *The Monroe Doctrine: Empire and Nation in Nineteenth-Century America*. New York: Hill and Wang, 2011.

Shirky, Clay. *Here Comes Everybody: The Power of Organizing without Organizations*. New York: Penguin, 2008.

Silverman, Kaja. *The Acoustic Mirror*. Bloomington: University of Indiana Press, 1988.

Silverstein, Michael. "Relative Motivation in Denotational and Indexical Sound Symbolism of Wasco-Wishram Chinookan." In *Sound Symbolism*, edited by Leanne Hinton, Johanna Nichols, and John J. Ohala, 40–53. Cambridge: Cambridge University Press, 2003.

Simon, Roger D. "Housing: Past and Present." *Journal of Urban History* 26 (2000): 687–91.

Sirvén, Pablo. *El rey de la TV: Goar Mestre y la historia de la televisión*. Buenos Aires: Clarín Aguilar, 1996.

Sitkoff, Harvard. *A New Deal for Blacks*. New York: Oxford University Press, 1978.

Smith, Zadie. "*Their Eyes Were Watching God*: What Does Soulful Mean?" In *Changing My Mind*, by Smith, 3–13. New York: Penguin, 2009.

Sohn, Stephen Hong. "Introduction: Alien/Asian: Imagining the Racialized Future." *MELUS* 33, no. 4 (Winter 2008): 5–22.

Soley, Lawrence. *Free Radio: Electronic Civil Disobedience*. Boulder, Colo.: Westview, 1999.

Soley, Lawrence, and John S. Nichols. *Clandestine Radio Broadcasting: A Study of Revolutionary and Counterrevolutionary Electronic Communication*. New York: Praeger, 1987.

Sollors, Werner. "Modernization as Adultery: Richard Wright, Zora Neale Hurston and American Culture of the 1930s and 1940s." *Hebrew University Studies in Literature and the Arts* 18 (1990): 109–55.

Spellman, A. B. "Billie Holiday: 'Lady in Satin.'" Interview by Murray Horowitz. *National Public Radio*. August 1, 2001.

Speranza, Graciela. *Manuel Puig: Después del fin de la literatura*. Buenos Aires: Grupo Editorial Norma, 2000.

———. "Ricardo Piglia o el arte del desvío." In *Fuera de campo*. Buenos Aires: Anagrama, 2006.

Spitzer, Leo. "Linguistic Perspectivism in the Quijote." In *Linguistics and Literary History*. Princeton: Princeton University Press, 1948.

Spivak, Gayatri. "Three Feminist Readings: McCullers, Drabble, Habermas." *Union Seminary Quarterly Review* 35, no. 1 (Fall–Winter 1979–80): 15–34.

Stadler, Gustavus. "My Wife': The Tape Recorder and Warhol's Queer Ways of Listening." *Criticism* 56, no. 3 (Summer 2014): 425–56.

Sterne, Jonathan. *The Audible Past: The Cultural Origins of Sound Reproduction.* Durham: Duke University Press, 2003.

———. "Nature Builds No Telephones." In *MP3: Meaning of a Format*, 61–91. Durham: Duke University Press, 2012.

Stewart, Susan. *Crimes of Writing: Problems in the Containment of Representation.* New York: Oxford University Press, 1991.

Stoever, Jennifer Lynn. *The Sonic Color Line: Race and the Cultural Politics of Listening.* New York: New York University Press, 2016.

Stoever-Ackerman, Jennifer. "Splicing the Sonic Color-Line: Tony Schwartz Remixes Postwar *Nueva York*." *Social Text* 28, no. 1 (Spring 2010): 59–85.

Szalay, Michael. *New Deal Modernism.* Durham: Duke University Press, 2000.

Taffet, J. F. *Foreign Aid as Foreign Policy: The Alliance for Progress in Latin America.* New York: Routledge, 2007.

Tamen, Miguel. *Friends of Interpretable Objects.* Cambridge: Harvard University Press, 2001.

Tanner, Laura E. "Uncovering the Magical Disguise of Language: The Narrative Presence in Richard Wright's *Native Son*." In *Richard Wright: Critical Perspectives Past and Present*, edited by Henry Louis Gates Jr. and K. A. Appiah, 132–48. New York: Amistad, 1993.

Thacker, Eugene, and Andrew Galloway. *The Exploit: A Theory of Networks.* Minneapolis: University of Minnesota Press, 2007.

Thomas, Hugh. *Cuba: The Pursuit of Freedom.* New York: Harper and Row, 1971.

Thompson, Currie. "From the Margins to the Margins: The Representation of Blacks in Classic Argentine Cinema." *Post Script—Essays in Film and the Humanities* 29, no. 1 (Fall 2009): 2–13.

Thompson, Emily. *The Soundscape of Modernity.* Cambridge: MIT Press, 2002.

Tippins, Sherill. *February House.* New York: Mariner, 2006.

Townsend, Sarah J. "Radio/Puppets; or, The Institutionalization of a (Media) Revolution and the Afterlife of a Mexican Avant-Garde." *Cultural Critique* 91 (Fall 2015): 32–71.

Turner, Frederick C., and José Enrique Miguens, eds. *Juan Perón and the Reshaping of Argentina.* Pittsburgh: University of Pittsburgh Press, 1983.

Tyson, Timothy. *Radio Free Dixie.* Chapel Hill: University of North Carolina Press, 1999.

Ulanovsky, Carlos, ed. *Días de radio (1920–1959).* 1995. Buenos Aires: Emecé, 2004.

Ulin, Julieann Veronica. "The Astonishing Humanity: Domestic Discourses in the Friendship and Fiction of Richard Wright and Carson McCullers."

In *Richard Wright: New Readings for the 21st Century*, edited by Alice Mikal Craven and William Dow. London: Palgrave Macmillan, 2011.

———. "Talking to Bessie: Richard Wright's Domestic Servants." *American Literature* 85, no. 1 (March 2013): 151–76.

Vale, Lawrence. *From the Puritans to the Projects: Public Housing and Public Neighbors*. Cambridge: Harvard University Press, 2000.

———. *Purging the Poorest: Public Housing and the Design Politics of Twice-Cleared Communities*. Chicago: University of Chicago Press, 2013.

Vanpée, Janie. "Rousseau's *Emile ou de l'education*: A Resistance to Reading." *Yale French Studies* 77 (1990).

Venegas, Cristina. *Digital Dilemmas: The State, the Individual, and Digital Media in Cuba*. New Brunswick, N.J.: Rutgers University Press, 2010.

Verma, Neil. *Theater of the Mind: Imagination, Aesthetics and American Radio Drama*. Chicago: University of Chicago Press, 2012.

Wagner, Bryan. *Disturbing the Peace: Black Culture and the Police Power after Slavery*. Cambridge: Harvard University Press, 2009.

Walker, Paul. "Charles Abrams vs. Robert Moses: Contested Rhetorics of Urban Housing." *Rhetoric Review* 31, no. 3 (2012): 289–308.

Ward, Brian. *Radio and the Struggle for Civil Rights in the South*. Gainesville: University Press of Florida, 2004.

Washburn, Philo C. *Broadcasting Propaganda*. London: Praeger, 1992.

Webb, Constance. *Richard Wright*. New York: Putnam, 1968.

Weheliye, Alexander G. *Phonographies: Grooves in Sonic Afro-Modernity*. Durham: Duke University Press, 2005.

———. "'Feenin': Posthuman Voices in Contemporary Black Popular Music." *Social Text* 20, no. 2 (Summer, 2002): 21–47.

Weiss, Allen. *Breathless: Sound Recording, Disembodiment, and the Transformation of Lyrical Nostalgia*. Middleton, Conn: Wesleyan University Press, 2002.

———, ed. *Experimental Sound and Radio*. Cambridge: MIT Press, 2001.

Welles, Orson. "Pan-Americanism." *Hello Americans*. Program 11. Radio. CBS. https://archive.org/details/otr_helloamericanswithorsonwelles

Wershler-Henry, Darren Sean. *The Iron Whim: A Fragmented History of Typewriting*. Toronto: McClelland and Stewart, 2005.

Williams, Gareth. *The Other Side of the Popular: Neoliberalism and Subalternity in Latin America*. Raleigh: Duke University Press, 2002.

Williams, Linda. *Playing the Race Card: Melodramas of Black and White from Uncle Tom to O. J. Simpson*. Princeton: Princeton University Press, 2002.

Williams, Robert F. *Crusader* 3, no. 8 (April 1962): 1–8.

———. *Crusader* 3, no. 9 (May 1962): 1–8.

———. *Negroes with Guns*. New York: Marzani and Munsell, 1962.

———. Papers. Bentley Historical Library, University of Michigan.

Willihnganz, Jonah. "The Voice of America in Richard Wright's *Lawd Today!*" In *Broadcasting Modernism*, edited by Debrah Rae Cohen, Michael Coyle, and Jane A. Lewty, 124–41. Gainesville: University Press of Florida, 2009.

Wilson, Sarah. "Gertrude Stein and the Radio." *Modernism/modernity* 11, no. 2 (2004) 261–78.

Winkler, Jonathan Reed. *Nexus: Strategic Communications and American Security in World War I.* Cambridge: Harvard University Press, 2008.

Woloch, Alex. *The One vs. The Many: Minor Characters and the Space of the Protagonist in the Novel.* Princeton: Princeton University Press, 2003.

Wood, Jamie. "Madness at the Whitney: Resistance to Genre in *Dempsey and Firpo*." *Dandelion: Post-Graduate Arts Journal and Research Network* 1, no. 1 (2010).

Wright, Richard. "Between Laughter and Tears." *New Masses* 5 (October 1937): 22–23.

———. "Blueprint for Negro Writing" (1937). In *Within the Circle: An Anthology of African American Literary Criticism from the Harlem Renaissance to the Present*, edited by Angelyn Mitchell. Durham: Duke University Press, 1994.

———. "*The Heart Is a Lonely Hunter*: Inner Landscape." *New Republic* 103 (August 5, 1940): 195.

———. "Huddie Ledbetter, Famous Negro Folk Artist, Sings the Songs of Scottsboro and His People." *Daily Worker*, August 12, 1937.

———. *Lawd Today!* 1938. New York: Avon, 1963.

———."Long Black Song." In *Uncle Tom's Children.* 1938. New York: Harper and Row, 1965.

———. *Native Son.* 1940. New York: HarperPerennial, 1993.

———. Papers. Beinecke Rare Book and Manuscript Library, Yale University.

———. *Uncle Tom's Children.* 1938. New York: Harper and Row, 1965.

———. "WPA Slashes Are Bar to Harlem Homes." *Daily Worker*, July 8, 1937.

Yahav, Amit. "Sonorous Duration: *Tristram Shandy* and the Temporality of Novels." *PMLA* 128, no. 4 (2013): 872–87.

Yampolsky, Mikhail. "Voice Devoured: Artaud and Borges on Dubbing." Translated by Larry P. Joseph. *October* 64 (Spring 1993): 57–77.

Index

FLASHPOINTS

1. *On Pain of Speech: Fantasies of the First Order and the Literary Rant,* Dina Al-Kassim
2. *Moses and Multiculturalism,* Barbara Johnson, with a foreword by Barbara Rietveld
3. *The Cosmic Time of Empire: Modern Britain and World Literature,* Adam Barrows
4. *Poetry in Pieces: César Vallejo and Lyric Modernity,* Michelle Clayton
5. *Disarming Words: Empire and the Seductions of Translation in Egypt,* Shaden M. Tageldin
6. *Wings for Our Courage: Gender, Erudition, and Republican Thought,* Stephanie H. Jed
7. *The Cultural Return,* Susan Hegeman
8. *English Heart, Hindi Heartland: The Political Life of Literature in India,* Rashmi Sadana
9. *The Cylinder: Kinematics of the Nineteenth Century,* Helmut Müller-Sievers
10. *Polymorphous Domesticities: Pets, Bodies, and Desire in Four Modern Writers,* Juliana Schiesari
11. *Flesh and Fish Blood: Postcolonialism, Translation, and the Vernacular,* S. Shankar
12. *The Fear of French Negroes: Transcolonial Collaboration in the Revolutionary Americas,* Sara E. Johnson
13. *Figurative Inquisitions: Conversion, Torture, and Truth in the Luso-Hispanic Atlantic,* Erin Graff Zivin
14. *Cosmopolitan Desires: Global Modernity and World Literature in Latin America,* Mariano Siskind
15. *Fiction Beyond Secularism,* Justin Neuman
16. *Periodizing Jameson: Dialectics, the University, and the Desire for Narrative,* Phillip E. Wegner
17. *The Practical Past,* Hayden White
18. *The Powers of the False: Reading, Writing, Thinking Beyond Truth and Fiction,* Doro Wiese
19. *The Object of the Atlantic: Concrete Aesthetics in Cuba, Brazil, and Spain, 1868–1968,* Rachel Price
20. *Epic and Exile: Novels of the German Popular Front, 1933–1945,* Hunter Bivens
21. *An Innocent Abroad: Lectures in China,* J. Hillis Miller
22. *Form and Instability: East Europe, Literature, Postimperial Difference,* Anita Starosta
23. *Bombay Modern: Arun Kolatkar and Bilingual Literary Culture,* Anjali Nerlekar
24. *Intimate Relations: Social Reform and the Late Nineteenth-Century South Asian Novel,* Krupa Shandilya
25. *Media Laboratories: Late Modernist Authorship in South America,* Sarah Ann Wells
26. *Acoustic Properties: Radio, Narrative, and the New Neighborhood of the Americas,* Tom McEnaney